T0360520

THE FUTURE OF
HIGH ENERGY PHYSICS
Some Aspects

THE FUTURE OF
HIGH ENERGY PHYSICS
Some Aspects

Editors

L R Flores Castillo
Chinese University of Hong Kong

K Prokofiev
Institute for Advanced Study
Hong Kong University of Science and Technology

World Scientific

NEW JERSEY · LONDON · SINGAPORE · BEIJING · SHANGHAI · HONG KONG · TAIPEI · CHENNAI · TOKYO

Published by

World Scientific Publishing Co. Pte. Ltd.

5 Toh Tuck Link, Singapore 596224

USA office: 27 Warren Street, Suite 401-402, Hackensack, NJ 07601

UK office: 57 Shelton Street, Covent Garden, London WC2H 9HE

Library of Congress Cataloging-in-Publication Data
Names: Flores Castillo, Luis Roberto, editor. | Prokofiev, K., editor.
Title: The future of high energy physics : some aspects / editors,
    L.R. Flores Castillo (The Chinese University of Hong Kong, Hong Kong),
    K. Prokofiev (The Hong Kong University of Science and Technology, Hong Kong).
Description: Singapore ; Hackensack, NJ : World Scientific, [2017]
Identifiers: LCCN 2017001759| ISBN 9789813209916 (hardcover) | ISBN 9813209917 (hardcover)
Subjects: LCSH: Particles (Nuclear physics) | Higgs bosons. | Colliders (Nuclear physics)
Classification: LCC QC793.2 .F88 2017 | DDC 539.7/9--dc23
LC record available at https://lccn.loc.gov/2017001759

British Library Cataloguing-in-Publication Data
A catalogue record for this book is available from the British Library.

Printed in Singapore

Preface

In January 2016, for the second time, the Jockey Club Institute for Advanced Study (JCIAS) of the Hong Kong University of Science and Technology hosted its annual program on High Energy Physics.[a] The aim of this three-week long event is to bring together key players of the Theory, Experiment and Accelerator physics communities to discuss the future high-energy physics projects. An emphasis is given to the design parameters and the physics potential of accelerator facilities currently under consideration. These include the Future Circular Collider (FCC), proposed to be built in Europe near the CERN accelerator complex, the International Linear Collider (ILC) which would be built in Japan, and the Circular Electron Positron Collider, proposed to be built in China. The two circular colliders, FCC and CEPC, share an important design challenge, that is, after the first few years of operation of colliding electron and positron beams, they will be given necessary upgrades to produce proton–proton collisions instead. In the case of the FCC, these two modes of operation would be denoted as FCC-ee (for electron–positron) and FCC-pp (for proton–proton) colliders, respectively. For the CEPC, the upgraded collider would be called the SPPC: the Super Proton–Proton Collider. This ulterior transformation will vastly extend the physics reach of these facilities, but it entails design constraints that need to be considered even at the very early stage of conceptual design.

The program was attended by over a hundred of participants, who were encouraged to maintain an intense dialogue in an informal and enticing environment provided by the JCIAS. The program was divided into three main tracks: Accelerator Physics, Theory, and Experimental and Detector. Similarly, this volume is divided into the three aspects, each of which comprise some of the contributions shared during the program. The section on Accelerator Physics includes the study of beam-beam effects, different alternatives for the final design of the CEPC, preliminary studies of the high-field magnets required for the construction of the proton–proton machines, and the study of optimization methods used in

---

[a]The program was successfully launched in January 2015 and the articles were published in August 2015.[1–6]

v

accelerator design.[7-14] The Theory section includes general visions of the physics to be probed at these new facilities, as well as examples of the expected reach in the search for a number of specific models.[15-25] These include the nonunitarity of the leptonic mixing matrix, supersymmetric spectra from Grand Unification Theories. The resonant mono-Higgs production measurements of the Higgs boson potential, and Triple Gauge Coupling measurements are also discussed. The Experimental section touches on detector design and timing.[26-32] It includes contributions about the importance of high-precision hadronic calorimetry, design of future calorimeters, and fast timing in collider detectors. Overviews of the detector and experimental issues to be considered in future are also included in this section.

We thank all participants, contributors, and organizing committee members for the success of the program. We hope that the contributions in this volume serve as a good reference for the community involved, or interested, in the future developments in the field of high-energy physics.

L. R. Flores Castillo

The Chinese University of Hong Kong, Hong Kong

castillo@phy.cuhk.edu.hk

K. Prokofiev

The Hong Kong University of Science and Technology, Hong Kong

kprok@ust.hk

Editors

References

1. Y. Tu and S.-H. Henry Tye, *Int. J. Mod. Phys. A* **30**, 1502004 (2015), doi: 10.1142/S0217751X15020042.
2. V. D. Shiltsev, *Int. J. Mod. Phys. A* **30**, 1544001 (2015), doi: 10.1142/S0217751X15440017.
3. I. Hinchliffe, A. Kotwal, M. L. Mangano, C. Quigg and L.-T. Wang, *Int. J. Mod. Phys. A* **30**, 1544002 (2015), doi: 10.1142/S0217751X15440029.
4. R. Talman, *Int. J. Mod. Phys. A* **30**, 1544003 (2015), doi: 10.1142/S0217751X15440030.
5. S. Antusch and O. Fischer, *Int. J. Mod. Phys. A* **30**, 1544004 (2015), doi: 10.1142/S0217751X15440042.
6. J. Hajer, A. Ismail, F. Kling, Y.-Y. Li, T. Liu and S. Su, *Int. J. Mod. Phys. A* **30**, 1544005 (2015), doi: 10.1142/S0217751X15440054.
7. Y. Cai, in this volume.
8. H. Geng, in this volume.
9. K. Ohmi, in this volume.
10. Y. Zhang, in this volume.
11. D. Wang *et al.*, in this volume.
12. F. Su *et al.*, in this volume.
13. C. Wang, K. Zhang and Q. Xu, in this volume.
14. Y. Li and L. Yang, in this volume.
15. C. Quigg, in this volume.

16. J. Ellis, in this volume.
17. M. Reece, in this volume.
18. S.-F. Ge, H.-J. He and R.-Q. Xiao, in this volume.
19. Z. Liu, in this volume.
20. S. Antusch and O. Fischer, in this volume.
21. S. Antusch, E. Cazzato and O. Fischer, in this volume.
22. L. Bian, J. Shu and Y. Zhang, in this volume.
23. N. Chen, in this volume.
24. Q.-S. Yan, in this volume.
25. S. Antusch and C. Sluka, in this volume.
26. J. Hauptman, in this volume.
27. S. V. Chekanov and M. Demarteau, in this volume.
28. C. G. Tully, in this volume.
29. J. Hauptman, in this volume.
30. RD52 Collab. (M. Cascella, S. Franchino and S. Lee), in this volume.
31. R. Talman and J. Hauptman, in this volume.
32. CEPC Working Group (H. Yang), in this volume.

CONTENTS

Theory

Future Colliders Symposium in Hong Kong: Scientific Overview

Chris Quigg

Theoretical Physics Department, Fermi National Accelerator Laboratory,
P.O. Box 500, Batavia, Illinois 60510, USA
quigg@fnal.gov

Opening Lecture at the Hong Kong University of Science and Technology Jockey Club Institute for Advanced Study Program on High Energy Physics Conference, January 18–21, 2016.

Keywords: Higgs boson; hadron colliders; electron–positron colliders.

1. Introductory Remarks

I am grateful for the chance to open this conference on future colliders, especially because we will have a rich and stimulating program of talks on accelerator science and technology, on experimental results, plans, and detector concepts, and on the implications of current theoretical understanding for the next decades of research.

1.1. *The most important question*

One of my goals in this talk is to offer many questions, so I would like to begin with a big one that applies to everything we do — to theory, experiment, and accelerators alike — and to which we should give our scrupulous attention:

How are we prisoners of conventional thinking?

One aspect of this question pertains to the way we address well-identified problems. Refining an approach we have taken before may not be the optimal response to known challenges. Just over the past few days I have been encouraged to hear mind-expanding ways of thinking about potential remedies for the high synchrotron-radiation-induced heat load in future proton–proton colliders, or of integrating final-focus beam elements with detectors in future electron–proton colliders. A second aspect has to do with the specific questions we are asking. Are we asking the right questions, or are we missing something essential? Have we framed our questions in the right way, or are we merely rehearsing conventional formulations, without reexamining our premises and preconceptions?

1.2. *Accelerator milestones* ...

Since we have come to Hong Kong to discuss future colliders, it is worth taking a moment to review the inventions, insights, and technologies that make this discussion possible. I heartily commend to your attention the volume by Sessler and Wilson[1] for an authoritative, approachable, and considerably more complete survey.

The idea of cyclic acceleration embodied in Lawrence's cyclotron — more generally, that repeated applications of achievable gradients could accelerate charged particles to extremely high energies — underlies both the circular and linear colliders we are contemplating.

A practical limitation of the cyclotron was the need to evacuate an entire cylindrical volume to accommodate the accelerating particle as it spiraled out from an initial small radius to a final large radius. By raising the confining magnetic field in synchrony with the increasing momentum, one could contain the particles in a beam pipe of fixed radius, dispensing with the hole in the doughnut. Coupled with the notion of phase stability — that particles lagging or leading the nominal phase (have less or more than the nominal energy) are accelerated more or less than the particles at nominal energy, the varying magnetic field leads to the idea of a synchrotron, the basis for all circular colliders.

Early proton synchrotrons, such as the Berkeley Bevatron, still required apertures that were, by today's standards, gigantic. That changed dramatically with the invention of alternate-gradient (strong) focusing.[a] This advance put accelerator builders on the path to dense, well-controlled beams, making possible vacuum chambers only a few centimeters across. The subsequent development of active optics, including the breakthrough of stochastic cooling,[3] led to the intense beams required for high-luminosity colliders.

We take for granted the elementary fact that the c.m. energy of a beam of momentum p incident on a fixed target of mass M is $\sqrt{s_{\text{ft}}} \approx \sqrt{2Mp}$, whereas the c.m. energy of beams in head-on collision is $\sqrt{s_{\text{cb}}} \approx 2p$. Rolf Wideröe filed a patent application based on this observation in 1943.[4] The concept of colliding beams was first realized at Frascati, Novosibirsk, and Stanford in the early 1960s.[5,6] Of these, the Princeton–Stanford Colliding Beams Experiment (CBX), carried out by O'Neill, Barber, Gittelman, and Richter, entered my student consciousness through a story in the New York *Times* reporting the first electron–electron collisions at a c.m. energy of 600 MeV.[7] If that seems puny, consider that to achieve the same result in a fixed-target setting would require an electron beam of about 350 GeV, which we have still not attained! The *Times* reported that "two electrons come close enough for a collision only once every 15 or 20 minutes." Soon thereafter, I read that the

---

[a]The classic description of how using quadrupole lenses to squeeze the beams sequentially in the horizontal and vertical planes leads to net focusing is in Ref. 2.

scientists did not know what would happen when they made the high-energy electrons collide head-on. "What," I thought, "could be more exciting than not knowing the answer?" So began my fascination with high-energy colliders. Let us remember, when we seek to motivate new colliders, the power of *We do not know* . . . .

To implement the Big Idea of particle colliders, we have required efficient radio-frequency accelerating cavities and superb vacuum technology, superconducting magnets and materials, and cryogenic technology. We have gone beyond the readily available stable beam particles, electrons and protons, using to excellent effect positrons and antiprotons as well. Perhaps we will see dedicated $\gamma\gamma$ colliders, muon storage rings as neutrino sources, and even $\mu^+\mu^-$ colliders. Novel acceleration methods may someday take us more efficiently to energies and luminosities of interest.

1.3. *Our science holds many opportunities*

About a decade ago, I was asked to present the issues before us to a panel charting the course for particle physics as part of the *Physics 2010* decadal survey in the United States.[8] To illustrate the richness, diversity, and intellectual depth of our field, the liveliness of our conversations with nearby disciplines, and the timeliness of our aspirations, I composed the list of goals shown in Fig. 1.

In a decade or two, we can hope to . . .

Understand electroweak symmetry breaking	*Detect neutrinos from the universe*
Observe the Higgs boson	Learn how to quantize gravity
Measure neutrino masses and mixings	*Learn why empty space is nearly weightless*
Establish Majorana neutrinos ($\beta\beta_{0\nu}$)	Test the inflation hypothesis
Thoroughly explore CP violation in B decays	*Understand discrete symmetry violation*
Exploit rare decays (K, D, . . .)	Resolve the hierarchy problem
Observe n EDM, pursue e EDM	*Discover new gauge forces*
Use top as a tool	Directly detect dark-matter particles
Observe new phases of matter	*Explore extra spatial dimensions*
Understand hadron structure quantitatively	Understand origin of large-scale structure
Uncover QCD's full implications	*Observe gravitational radiation*
Observe proton decay	Solve the strong CP problem
Understand the baryon excess	*Learn whether supersymmetry is TeV-scale*
Catalogue matter and energy of universe	Seek TeV-scale dynamical symmetry breaking
Measure dark-energy equation of state	*Search for new strong dynamics*
Search for new macroscopic forces	Explain the highest-energy cosmic rays
Determine the (grand) unifying symmetry	*Formulate the problem of identity*

. . . learn the right questions to ask   . . . and rewrite the textbooks!

Fig. 1. A to-do (wish) list for particle physics and neighboring fields, circa 2005.

Beyond the significance of individual entries, what is striking is the scale diversity and variety of experimental techniques, and the range of energies and distance scales involved.

I hope you will agree that it is an impressive list of opportunities, including many for the LHC, and that it was plausible to anticipate very significant achievements over a 20 year time horizon. Indeed, if we look back over the decade past, we and our scientific neighbors can claim a lot of progress. (I invite you to make your own report card!) Happily, and as expected, there is still much to accomplish. Please think about how you would update or improve the list, and how we can best advance the science.

2. Discovery of the Higgs Boson in LHC Run 1

We entered the LHC era having established two new laws of Nature, quantum chromodynamics and the electroweak theory. We had identified six flavors of quarks (u, d, s, c, b, t) and six flavors of leptons $(e, \mu, \tau$ and three neutrinos) as spin-$\frac{1}{2}$ fermions that we may idealize, provisionally, as pointlike particles. Interactions are derived from $SU(3)_c \otimes SU(2)_L \otimes U(1)_Y$ gauge symmetry, which reflects the curious fact that — in our experience — charged-current weak interactions apply only to the left-handed quarks and leptons. We do not know whether that reflects a fundamental asymmetry in the laws of Nature, or arises because right-handed charged-current interactions are so feeble that they have eluded detection.

The $SU(3)_c$ color symmetry that generates the strong interaction is unbroken, but the electroweak symmetry must be hidden because the weak interactions are short-range and standard Dirac masses for the quarks and leptons would conflict with the gauge symmetry. The surviving symmetry is the phase symmetry that generates electromagnetism: $SU(2)_L \otimes U(1)_Y \to U(1)_{EM}$. An essential task for the LHC has been to illuminate the nature of the previously unknown agent that hides electroweak symmetry. We have imagined a number of possibilities, including (i) A force of a new character, based on interactions of an elementary scalar; (ii) A new gauge force, perhaps acting on hitherto undiscovered constituents; (iii) A residual force that emerges from strong dynamics among electroweak gauge bosons; (iv) An echo of extra spacetime dimensions. The default option has been the first, an example of spontaneous symmetry breaking[b] analogous to the Ginzburg–Landau[10] phenomenology of the superconducting phase transition and the Meissner effect.

2.1. *The importance of the* 1 *TeV scale*

The footprint of spontaneous symmetry breaking in the electroweak theory is the massive scalar particle known as the Higgs boson. While the electroweak theory does not predict the Higgs-boson mass, a thought experiment yields a conditional upper bound, or tipping point, for M_H.[11] It is informative to consider scattering

---

[b]See Ref. 9 for a narrative of the historical development and references to the original literature.

of longitudinal gauge bosons and Higgs bosons at high energies. The two-body reactions involving $W_L^+ W_L^-$, $Z_L Z_L$, HH, HZ_L satisfy s-wave unitarity, provided that $M_H \leq \left(8\pi\sqrt{2}/3G_F\right)^{1/2} \approx 1$ TeV. If the bound is respected, perturbation theory is reliable (except near resonance poles), and a Higgs boson is to be found below 1 TeV in mass. If not, weak interactions among $W_L^\pm$, Z_L, H become strong on 1 TeV scale. One way or the other, *new phenomena are to be found around 1 TeV*. This analysis shows us that the role of the "Brout–Englert–Higgs mechanism" in the electroweak theory is not only to break $SU(2)_L \otimes U(1)_Y \to U(1)_{EM}$ and to generate masses for the electroweak gauge bosons and the fermions, but also — through the action of the Higgs boson — to regulate gauge boson interactions at high energies.

In the years leading up to experiments at the LHC, the analysis of precise measurements of electroweak observables, within the standard electroweak theory, pointed to a light Higgs boson, with a mass no greater than about 200 GeV.[c]

We have not (yet) found an argument — based either on theoretical consistency or on the analysis of observations within a particular framework — that points to a specific scale beyond the 1 TeV scale.

2.2. *Searches at the Large Hadron Collider*

Let us quickly review what experiments at the LHC have revealed so far about the Higgs boson. The LHC makes possible searches in many channels of production (gluon fusion $gg \to H$, associated production $q\bar{q}' \to H(W, Z)$, vector-boson fusion, and the $Ht\bar{t}$ reaction) and decay ($\gamma\gamma, WW^*, ZZ^*, b\bar{b}, \tau^+\tau^-, \ldots$). Since the discovery of $H(125) \to (\gamma\gamma, \ell^+\ell^-\ell^+\ell^-)$ was announced by the ATLAS[14] and CMS[15] Collaborations in 2012, the evidence has developed as it would for a standard-model Higgs boson.[16,17]

In addition to the $\gamma\gamma$ and ZZ discovery modes, the W^+W^- mode[18,19] is established, and the spin-parity assignment $J^P = 0^+$ is overwhelmingly favored.[20,21] A combined measurement of the Higgs-boson mass yields $M_H = 125.09 \pm 0.24$ GeV.[22] A grand average of the combined ATLAS and CMS measurements of the Higgs-boson signal yield (i.e. production times branching fraction) is 1.09 ± 0.11 times the standard-model expectation.[23] Within the uncertainties, individual modes are in line with the standard-model predictions.

If $H(125)$ is to be unambiguously identified as the standard-model Higgs boson of our textbooks, what remains to be demonstrated? We need to investigate, through precise measurements of the HWW and HZZ couplings, whether it fully accounts for electroweak symmetry breaking. We must extend the indications[23] that $H(125)$ couples to fermions, test whether the $Hf\bar{f}$ couplings are proportional to the fermion masses, and indeed whether the interaction of fermions with the Higgs field accounts entirely for their masses. The predicted branching fractions are collected in

---

[c]See, for example, Refs. 12 and 13.

Table 1.   Branching fractions $\mathcal{B}$ for a 125 GeV standard-model Higgs boson (from Ref. 24).

Mode	$b\bar{b}$	WW	gg	$\tau^+\tau^-$	$c\bar{c}$	ZZ	$\gamma\gamma$	$Z\gamma$	$\mu^+\mu^-$
$\mathcal{B}$	0.577	0.215	0.0857	0.0632	0.0291	0.0264	0.00228	0.00154	0.00022

Table 1. It is noteworthy, and completely expected at the current level of sensitivity, that we have only observed Higgs couplings to fermions of the third generation — top from the production rate attributed to gluon fusion, direct observations of decays into $b\bar{b}$ and $\tau^+\tau^-$. It is essential to learn whether the same mechanism is implicated in the masses of the lighter fermions. Detection of $H \to \mu^+\mu^-$ is foreseen at the LHC. The observation of the decay into charm pairs looks highly challenging in the LHC environment, but merits very close consideration.

Another significant test is the total width of $H(125)$, predicted to be $\Gamma(H(125)) = 4.07$ MeV for a standard-model Higgs boson. This is well below the experimental resolution for a direct determination at the LHC, but by applying the clever insight that — within a framework that resembles the standard model — measurements of the off-shell coupling strength in the WW and ZZ channels at invariant masses above M_H constrain the Higgs-boson width, the LHC experiments restrict $\Gamma(H(125))$ to be less than a few tens of MeV.[d]

We will continue to search for admixtures of spin-parity states other than the dominant $J^P = 0^+$, and to test that all production modes are as expected.

Much exploration remains as well. Does $H(125)$ have partners? Does it decay to new particles, perhaps serving as a portal to unseen sectors? Are there any signs of compositeness, of new strong dynamics? Finally, we can contemplate the implications of a 125 GeV Higgs boson.

2.3. *Why does discovering the agent of electroweak symmetry breaking matter?*

An instructive way to respond to this question is to imagine a world without a symmetry-breaking (Higgs) mechanism at the electroweak scale. A full analysis of that *Gedanken* world is rather involved,[29] but here are the main points, restricted for simplicity to one generation of quarks and leptons. The electron and quarks would have no mass. QCD would confine quarks into nucleons and other hadrons, and the nucleon mass — to which the up- and down-quark masses contribute only small amounts in the real world — would be little changed.[e] In the Lagrangian, the massless quarks exhibit an $SU(2)_L \otimes SU(2)_R$ chiral symmetry that is spontaneously broken, near the confinement scale, to $SU(2)$ isospin symmetry. The resultant

---

[d]For early theoretical analyses, see Refs. 25 and 26. First experimental determinations are presented in Refs. 27 and 28.
[e]Whether the proton or neutron would be the lighter — hence stable — nucleon is too close a call for us to settle.

linkage of left-handed and right-handed quarks gives rise to the "constituent-quark" masses, and hides the electroweak symmetry because the left-handed and right-handed quarks transform differently under $SU(2)_L \otimes U(1)_Y$. The electroweak gauge bosons $W^\pm$ and Z acquire tiny masses, about 2500 times smaller than those we observe in the real world. The scale is set, not by the vacuum expectation value v of the (absent) Higgs field, but by the pion decay constant f_π.

Now suppose that protonuclei — say, alpha particles — are created in the early universe and survive to late times (whatever that might mean). A massless electron means that the Bohr radius of an atom would be infinite, so it is not possible to identify an electron as belonging to a specific atom. In other words, "atoms" lose integrity. If an electron cannot be assigned to a particular nucleus, the notion of valence bonding evaporates. No atoms means no chemistry, no stable composite structures like liquids, solids, … no template for life!

Returning to our world, it is important that we not get ahead of the evidence. We have good indications that $H(125)$ couples approximately as expected to top and bottom quarks and to the tau lepton. We anticipate that $H \to \mu^+\mu^-$ can be established at the High-Luminosity LHC, if not before. Measuring the coupling of $H(125)$ to charm seems highly challenging at the LHC; to achieve that, we need either new insights or a Higgs factory. Demonstrating $H \to e^+e^-$, with its predicted branching fraction $\approx 5 \times 10^{-9}$, is beyond challenging, but to my mind showing that spontaneous symmetry breaking gives mass to the electron would merit a Nobel Prize in Chemistry!

3. Looking Ahead to Future Colliders

3.1. *A Higgs factory?*

The discovery of $H(125)$ motivates consideration of an e^+e^- Higgs factory (or a stage of a linear collider), and comparison with what LHC will do, and when experiments will happen. An excellent starting point is Ref. 30. The initiatives under active discussion include the International Linear Collider in Japan,[31] the Circular Electron–Positron Collider in China,[32] and the FCC-ee Design Study centered at CERN.[33] The performance of a Higgs factory is addressed in the following talk by Matt Reece, so my comments will be brief.

There is little question that if a Higgs factory allowing the detailed study of the associated-production reaction $e^+e^- \to HZ$ at $\sqrt{s} \approx 240$ GeV were available today, it would be a superb complement to ongoing experiments at the LHC, and would attract many users. That is not the case, and so we need to assess what a purpose-built machine can do when its experimental program begins. At any moment, the telling measurements depend on what is already know. For example, will $H(125)$ continue to match the textbook description, or will it begin to show nonstandard properties? How would the discovery of another "Higgs-like object" change the picture? And what would direct evidence for or against new degrees of freedom mean for our goals for a new machine?

It is also worthwhile to examine the benefits and (opportunity) cost of parameter variations for a projected collider. What would be the value of extending a Higgs factory to the top threshold? How well could we hope to determine the top mass, the strong coupling α_s, and the top Yukawa coupling to the Higgs boson? How much running time would be required? If we imagine running on the Z-peak, what is the goal — Giga-Z or Tera-Z? What are the implications of high-luminosity running at the Z for the machine design? How much running time would be required to significantly improve our knowledge of electroweak observables, or to exploit the copious source of boosted b-hadrons? What would it take to significantly improve the precision of M_W by mapping the excitation curve?

Many of these Higgs-factory enhancements are easy to dream about, but may be costly to deliver and take much time to exploit!

3.2. *A Very Large Hadron Collider: Generalities*

We are looking beyond CERN's Large Hadron Collider to a ring with circumference two to four times that of the LHC. Superconducting dipoles with field strength between 15 and 20 teslas would support a proton–proton collider with c.m. energy $\sqrt{s} \approx 50$ to 100 TeV.[f] The goal of a "100 TeV" hadron collider has been set for a machine study, but it important to keep in mind that a feasible, scientifically desirable pp collider might have a different energy. At this point, we should undertake physics studies over a range of energies, bearing in mind that different combinations of energy and luminosity can, to some degree, yield comparable discovery potential.[g] The work we do for "100 TeV" can enhance what we achieve with LHC. It is important to consider search and measurement examples that will stretch detector capabilities, to examine the role of special-purpose detectors, including concepts that have been set aside in the past.[37] There is great value in developing tools that enable others to extend the work.

I think it premature to enunciate the scientific case for the "100 TeV" hadron collider, but the right time to explore possibilities. We can cite plenty of reasons that such a machine could be highly exciting and scientifically rewarding,[h] but it will be many years before we will be able to make a credible technical proposal. It is overwhelmingly likely that we will learn a great deal in the intervening time (not only from the LHC), and I would not bet against discoveries that alter our conception of the Great Questions in some dramatic way. What we learn from the LHC (and elsewhere) might point to an energy landmark for the next great machine. I recall that for nearly two decades, the central pillar of the case put forward for a linear collider has been that it would unravel the rich spectrum of light superpartners. That case has vanished identically.

---

[f]Such dipoles could enable a 33 TeV pp collider in the LHC tunnel. See Ref. 34 for a reality check.
[g]In Ref. 35, back at the beginning of time, we explored the reach of $p^{\pm}p$ colliders at $\sqrt{s} = 2, 10, 20, 40, 70,$ and 100 TeV. I am pleased to note that the LHC Higgs Cross Section Working Group, Ref. 36, is providing cross sections at $\sqrt{s} = 14, 33, 40, 60, 80,$ and 100 TeV.
[h]See Ref. 38 for a good start.

In a world with multiple, widely separated, physical scales, the electroweak scale and the light Higgs-boson mass present a puzzle: Why are M_W and M_H so much smaller than the unification scale or the Planck scale, presuming those to be physically significant? In quantum field theory, distant scales tend to be linked through quantum corrections, and so large hierarchies seem to ask for "natural" explanations[i] more satisfying than "just-so stories." We have not yet found any direct evidence on the 1 TeV scale for new dynamics or a new symmetry that could explain the many orders of magnitude between the electroweak scale and the others. (Supersymmetry, in particular, is hiding very effectively.) Experiment has not established a pattern of serious quantitative failures of electroweak theory, nor have we uncovered any clear sign of the flavor-changing neutral currents that occur generically in "new-physics" extensions to the standard model. Searches for forbidden or suppressed processes that might reveal something about flavor-changing neutral currents are consequently of great interest, as is the ongoing campaign to make ever-more-precise tests of the electroweak theory.

Opinions about how to respond to a possible hierarchy problem have evolved over many years. We originally sought once-and-done remedies, such as supersymmetry or technicolor, that invoked new physics on the TeV scale to exorcise the problem once and for all. Maybe that is not the right approach. Should we instead favor a stepwise approach with a sequence of effective theories? Might we have misunderstood the hierarchy problem, and so need to reframe it? Perhaps it is time to ask whether the unreasonable effectiveness of the standard model[40] (to borrow a turn of phrase from Eugene Wigner[41]) is itself a deep clue to what lies beyond.

All this is to argue that we should continue to examine our notion of the hierarchy problem. It is, after all, a problem for our feelings about how nature should work, not a contradiction that we arrive at from first principles. Ken Wilson, one of the founders of naturalness, continued to think about the issues. For a revealing counterpoint to appeals to authority, see Sec. 5 of his historical survey, Ref. 42, in the passage beginning with "The final blunder ... ."

How we conceive of the hierarchy problem will help determine how theorists invest their intellectual capital. But I am skeptical of the assertion that — by some arbitrary measure — the 100 TeV machine will test naturalness at the 10^{-4} level, rather than 1% at the LHC, and that those two orders of magnitude will somehow settle the matter. As a justification for a new collider, it is unpersuasive.

While the standard model gives an excellent account of a wealth of experimental information, it has nothing to say about a number of important questions, including the nature of dark matter, the origin of the matter excess in the universe, the riddle of dark energy, and the pattern of fermion masses and mixing angles. And although the LHC — still in its exploratory phase — has not yet presented us with new physics on the TeV scale, we may have some hints.[43]

---

[i]See Ref. 39 for a perceptive review of the naturalness principle.

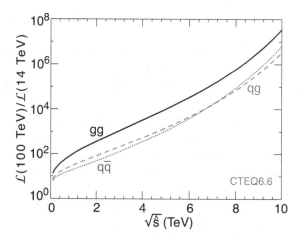

Fig. 2. Parton luminosity ratios (from Ref. 50) at $\sqrt{s} = 100$ and 14 TeV as a function of parton–parton subenergy $\sqrt{\hat{s}}$, evaluated using the CTEQ6.6 parton distributions[51] with $Q^2 = \hat{s}$.

Both CMS[44] and ATLAS[45] report indications of excesses in diboson invariant mass distributions in the neighborhood of 2 TeV in their event samples at $\sqrt{s} = 8$ TeV. The most recent data of the LHCb Experiment[46] display continuing tensions in the quark-mixing matrix element V_{ub} measured by different techniques. The ratio of branching fractions $\mathcal{B}(B^+ \to K^+\mu^+\mu^-)/\mathcal{B}(B^+ \to K^+e^+e^-)$ in the interval $1\ \mathrm{GeV}^2 \leq q^2 \leq 6\ \mathrm{GeV}^2$ is determined by LHCb as $0.745^{+0.090}_{-0.074}\,(\text{stat}) \pm 0.036\,(\text{syst})$, which differs by 2.6σ from the lepton-universality expectation.[47] Both ATLAS[48] and CMS[49] have shown provocative indications of a diphoton resonance near 750 GeV in their 2015 run at $\sqrt{s} = 13$ TeV. If real, this will be a sensational discovery on its own, and will almost certainly indicate other new phenomena to follow.

3.3. *A Very Large Hadron Collider: Some specifics*

What new opportunities will a "100 TeV" pp collider offer? Figure 2 shows the ratios of parton luminosities for collisions of gg, $q\bar{q}$, and qg (gluons g and light quarks q) in pp collisions at $\sqrt{s} = 100$ and 14 TeV. At what will be modest parton subenergies at a 100 TeV collider, $\sqrt{\hat{s}} \lesssim 1$ TeV, the parton luminosities increase by an order of magnitude or more. This advantage could, in principle, be overcome by increasing the 14 TeV pp luminosity by 1–2 orders of magnitude beyond the High-Luminosity LHC, but that is a somewhat daunting prospect. At higher values of $\sqrt{\hat{s}}$, there is a decisive advantage to increasing $\sqrt{s}$.

An instructive example at modest scales is the increase in Higgs-boson production cross sections shown in Table 2.[j] Beyond giving us the means to learn more about $H(125)$ and other particle that come into view at the LHC, a 100 TeV–class

---

[j]Michelangelo Mangano shows how we can expect to refine our knowledge of Higgs-boson properties in his talk at this symposium.

Table 2. Ratio $\mathcal{R}_{100:14} \equiv \sigma(\sqrt{s} = 100 \text{ TeV})/\sigma(\sqrt{s} = 14 \text{ TeV})$ for various Higgs-production reactions, according to Ref. 36.

Process	$gg \to H$	$q\bar{q} \to WH$	$q\bar{q} \to ZH$	$qq \to qqH$	$t\bar{t}H$	$b\bar{b}H$	$gg \to HH$
$\mathcal{R}_{100:14}$	14.7	9.7	12.5	18.6	61	15	42

collider will enhance the discovery reach at low masses, making accessible rare processes and phenomena characterized by low detection efficiencies and challenging backgrounds.

Consider as well particles in the upper reaches of the HL-LHC discovery range, for example a gauge boson of mass around parton subenergy $\sqrt{\hat{s}} = 6$ TeV produced singly in the $q\bar{q}$ channel, or pair production of ≈ 3 TeV particles in the gg channel, for which the parton luminosities increase by factors of 10^4 and 10^5, respectively. If we contemplate an order-of-magnitude increase in the integrated pp luminosity, this implies event samples up to a million times larger.

At still higher energy scales, the 100 TeV collider enters unexplored terrain, where we may find new particles and new phenomena.[k] In addition to all the usual suspects of the LHC era — supersymmetry, strong dynamics, extra dimensions, and all the rest — we might have access to $(B + L)$-violating phenomena. Tye and Wong, for example, have argued that a 9.3 TeV sphaleron produced in collisions of left-handed light quarks would give rise to final states containing multiple same-sign leptons and multiple b quarks.[52]

New phenomena may arise with relatively large cross sections, should hitherto unknown collective effects emerge as increasing energies create unusual conditions in proton–proton collisions. I have in mind novel event structures, perhaps reflecting the partonic structure of the protons, or evidence for a new component of particle production such as thermalization or hydrodynamical behavior.

I show in Fig. 3 two examples of how the discovery reach increases as the pp energy is raised beyond $\sqrt{s} = 14$ TeV. The left panel depicts the cross section times branching fraction at next-to-leading order for a sequential standard-model W'-boson decaying into electron + antineutrino — an artificial benchmark, but one that is straightforward to state and adapt to other cases.[l] If the discovery limit at the 14 TeV HL-LHC is taken to be 7 TeV, then (at constant branching fraction and pp luminosity) the 100 TeV limit would be approximately 30 TeV. With an order of magnitude increase in integrated luminosity, the discovery limit approaches 40 TeV.

The right panel of Fig. 3 shows the dijet invariant mass distribution evaluated at next-to-leading order. A 5 TeV reach in dijet mass at the HL-LHC grows to 20 TeV at $\sqrt{s} = 100$ TeV, with fixed pp luminosity, while 10 TeV at the HL-LHC increases to over 50 TeV at the 100 TeV collider, opening much space for discovery.

---

[k]Matthew McCullough exhibited a selection of these in his talk at the Symposium.
[l]This stylized W' has standard-model couplings to fermions, but no decays into gauge bosons.

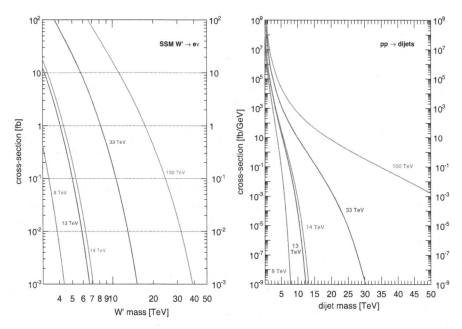

Fig. 3. Left panel: Cross section for the production and decay of a sequential standard-model $W' \to e\nu$ boson calculated at next-to-leading order with MCFM,[53] at c.m. energies $\sqrt{s} = 8, 13, 14, 33, 100$ TeV. Right panel: Cross section $d\sigma/d\mathcal{M}$ for the production of dijets with invariant mass $\mathcal{M}$ calculated at next-to-leading order with MCFM,[53] at c.m. energies $\sqrt{s} = 8, 13, 14, 33, 100$ TeV.

3.4. *Provisional luminosity recommendations*

At the 2015 Hong Kong workshop, we examined various arguments for the luminosity that would be required for a productive 100 TeV hadron collider. Our assessment,[50] which should be revisited during the ongoing studies, was this:

> "The goal of an integrated luminosity in the range of 10–20 ab^{-1} per experiment, corresponding to an ultimate instantaneous luminosity[54] approaching 2×10^{35} cm^{-2}s^{-1} seems well-matched to our current perspective on extending the discovery reach for new phenomena at high mass scales, high-statistics studies of possible new physics to be discovered at (HL-)LHC, and incisive studies of the Higgs boson's properties. Specific measurements may set more aggressive luminosity goals, but we have not found generic arguments to justify them. The needs of precision physics arising from new physics scenarios to be discovered at the HL-LHC, to be suggested by anomalies observed during the e^+e^- phase of a future circular collider, or to be discovered at 100 TeV, may well drive the need for even higher statistics. Such requirements will need to be established on a case-by-case basis, and no general scaling law gives a robust extrapolation from 14 TeV. Further work on *ad hoc* scenarios, particularly for low-mass phenomena and elusive signatures, is therefore desirable."

3.5. *Hadron Colliders and Unified Theories*

The neutrality of matter — with its implication that proton and electron charges exactly balance — is a powerful encouragement for a unified theory of the strong, weak, and electromagnetic interactions. An attractive possibility is a simple unifying gauge group $\mathcal{G}$ that contains the $SU(3)_c \otimes SU(2)_L \otimes U(1)_Y$ bits that we have discovered in our relatively low-energy experiments. Taking into account the degrees of freedom we know, the appropriately normalized $SU(3)_c$, $SU(2)_L$, and $U(1)_Y$ coupling constants evolve toward a common value at very high energies. Coupling-constant unification is more promising in supersymmetric $SU(5)$ than in the original $SU(5)$ theory, provided that the change in evolution due to a full spectrum of super-partners occurs near 1 TeV.[55]

Plotted as a function of $\ln Q$, $1/\alpha_s$ evolves with slope $7/2\pi$ if the standard model is embedded in $SU(5)$, but the slope changes to $3/2\pi$ above the energy at which a full spectrum of superpartners is active. Could experiments at the LHC, or a future collider, test the hypothesis of supersymmetric unification by measuring the strong coupling constant (or the weak mixing parameter $\sin^2 \theta_W$) as a function of scale? ATLAS[56] and CMS[57] have already made what I would characterize as exploratory measurements of α_s that extend to scales above 1 TeV by determining the ratio of three-jet to two-jet rates in pp collisions at $\sqrt{s} = 7$ TeV.

Seeing, or not seeing, a change of slope would be powerful evidence for or against the existence of a new set of colored particles that would complement ongoing searches for specific new-particle signatures. Considerable thought will be required to determine the most promising classes of measurements. I suspect that the study of Z^0 + jets will be fruitful. A continuing conversation between theory and experiment will be needed to isolate $\alpha_s(Q)$ measured at a high scale.

4. Issues for the Future (Starting Now!)

Let us conclude with a short list of questions we would like to answer:

(1) *There is a Higgs boson!* Might there be several?
(2) Does the Higgs boson regulate WW scattering at high energies?
(3) Is the Higgs boson elementary or composite? How does it interact with itself? What triggers electroweak symmetry breaking?
(4) Does the Higgs boson give mass to fermions, or only to the weak bosons? What sets the masses and mixings of the quarks and leptons? (How) is fermion mass related to the electroweak scale?
(5) Will new flavor symmetries give insights into fermion masses and mixings?
(6) What stabilizes the Higgs-boson mass below 1 TeV?
(7) Do the different charged-current behaviors of left-handed and right-handed fermions reflect a fundamental asymmetry in Nature's laws?
(8) What will be the next symmetry that we recognize? Are there additional heavy gauge bosons? Is nature supersymmetric? Is the electroweak theory contained in a unified theory of the strong, weak, and electromagnetic interactions?

(9) Are all flavor-changing interactions governed by the standard-model Yukawa couplings? Does "minimal flavor violation" hold? If so, why? At what scale?

(10) Are there additional sequential quark and lepton generations? Or new exotic (vector-like) fermions?

(11) What resolves the strong CP problem?

(12) What are the dark matters? Is there any flavor structure?

(13) Is electroweak symmetry breaking an emergent phenomenon connected with strong dynamics? How would that alter our conception of unified theories of the strong, weak, and electromagnetic interactions?

(14) Is electroweak symmetry breaking related to gravity through extra spacetime dimensions?

(15) What resolves the vacuum energy problem?

(16) (When we understand the origin of electroweak symmetry breaking,) what lessons does electroweak symmetry breaking hold for unified theories? ... for inflation? ... for dark energy?

(17) What explains the baryon asymmetry of the universe? Are there new (charged-current) CP-violating phases?

(18) Are there new flavor-preserving phases? What would observation, or more stringent limits, on electric-dipole moments imply for theories beyond the standard model?

(19) (How) are quark-flavor dynamics and lepton-flavor dynamics related (beyond the gauge interactions)?

(20) At what scale are neutrino masses set? Do they speak to the TeV scale, the unification scale, the Planck scale, or ... ?

(21) Could our Laws of Nature be environmentally determined?

And finally, the question that looms over all the others,
How are we prisoners of conventional thinking?

Acknowledgments

Fermilab is operated by Fermi Research Alliance, LLC under Contract No. De-AC02-07CH11359 with the United States Department of Energy. I thank the conference organizers for the kind invitation to speak. I am grateful to Henry Tye and members of the Jockey Club Institute for Advanced Study for their generous hospitality, to the participants for their contributions to a stimulating environment, and to Prudence Wong for her gracious practical assistance. I thank John Campbell for providing Fig. 3, and for helpful discussions.

References

1. A. Sessler and E. Wilson, *Engines of Discovery: A Century of Particle Accelerators* (World Scientific, Singapore, 2007).

2. E. D. Courant and H. S. Snyder, *Ann. Phys.* **3**, 1 (1958) [*Ann. Phys.* **281**, 360 (2000)], http://j.mp/1U6Y3ON.

3. S. van der Meer, Nobel Lecture: Stochastic cooling and the accumulation of antiprotons, http://j.mp/1VHCbYF.

4. P. Waloschek (ed.) and R. Wideröe, The infancy of particle accelerators: Life and work of Rolf Wideröe, DESY-94-039. The Colliding Beams patent, *Anordnung zur Herbeiführung Kernreaktionen*, submitted 6 September 1943 and issued 11 May 1953, can be seen as pp. 95 and 96 of http://j.mp/1Q0EkPg.

5. B. Richter, The rise of colliding beams, in *The Rise of the Standard Model: Particle Physics in the 1960s and 1970s*, eds. L. H. Hoddeson, L. Brown, M. Riordan and M. Dresden (Cambridge University Press, Cambridge, 1997), pp. 261–284.

6. V. Shiltsev, The first colliders: AdA, VEP-1 and Princeton-Stanford, in *Challenges and Goals for Accelerators in the XXI Century*, eds. O. Brüning and S. Myers (World Scientific, Singapore, 2016), arXiv:1307.3116 [physics.hist-ph].

7. Atom Smasher Test Shows Way to Save on Energy, *New York Times*, 13 March 1965, p. 9.

8. EPP 2010: Elementary Particle Physics in the 21st Century (National Research Council, Washington, 2006), http://sites.nationalacademies.org/BPA/BPA_048230.

9. C. Quigg, *Annu. Rev. Nucl. Part. Sci.* **65**, 25 (2015), arXiv:1503.01756 [hep-ph].

10. V. L. Ginzburg and L. D. Landau, *Zh. Eksp. Teor. Fiz.* **20**, 1064 (1950) [English translation: *Men of Physics: Landau*, Vol. II, ed. D. ter Haar (Pergamon, New York, 1965)].

11. B. W. Lee, C. Quigg and H. B. Thacker, *Phys. Rev. D* **16**, 1519 (1977).

12. ALEPH and CDF and D0 and DELPHI and L3 and OPAL and SLD and LEP Electroweak Working Group and Tevatron Electroweak Working Group and SLD Electroweak Working Group and Heavy Flavour Group Collabs., Precision electroweak measurements and constraints on the Standard Model, arXiv:0811.4682 [hep-ex], See also http://lepewwg.web.cern.ch and http://sanc.jinr.ru/users/zfitter/.

13. H. Flächer, M. Goebel, J. Haller, A. Hoecker, K. Mönig and J. Stelzer, *Eur. Phys. J. C* **60**, 543 (2009) [Erratum: *ibid.* **71**, 1718 (2011)], arXiv:0811.0009 [hep-ph], see also http://project-gfitter.web.cern.ch.

14. ATLAS Collab. (G. Aad *et al.*), *Phys. Lett. B* **716**, 1 (2012), arXiv:1207.7214 [hep-ex].

15. CMS Collab. (S. Chatrchyan *et al.*), *Phys. Lett. B* **716**, 30 (2012), arXiv:1207.7235 [hep-ex].

16. ATLAS Higgs-boson publications, http://j.mp/1PO6qgD and other public results, http://j.mp/1A1kzA5.

17. CMS Higgs-boson papers, http://j.mp/1PO6I6V.

18. CMS Collab. (S. Chatrchyan *et al.*), *J. High Energy Phys.* **1401**, 096 (2014), arXiv:1312.1129 [hep-ex].

19. ATLAS Collab. (G. Aad *et al.*), *Phys. Rev. D* **92**, 012006 (2015), arXiv:1412.2641 [hep-ex].

20. CMS Collab. (V. Khachatryan *et al.*), *Phys. Rev. D* **92**, 012004 (2015), arXiv:1411.3441 [hep-ex].

21. ATLAS Collab. (G. Aad *et al.*), *Eur. Phys. J. C* **75**, 476 (2015), arXiv:1506.05669 [hep-ex].

22. ATLAS and CMS Collabs. (G. Aad *et al.*), *Phys. Rev. Lett.* **114**, 191803 (2015), arXiv:1503.07589 [hep-ex].

23. ATLAS and CMS Collabs., Measurements of the Higgs boson production and decay rates and constraints on its couplings from a combined ATLAS and CMS analysis of the LHC *pp* collision data at $\sqrt{s} = 7$ and 8 TeV, ATLAS-CONF-2015-044, 15 September 2015, http://cds.cern.ch/record/2052552.

24. S. Dittmaier *et al.*, Handbook of LHC Higgs cross sections: 2. Differential distributions, arXiv:1201.3084 [hep-ph]; SM Higgs branching ratios and partial-decay widths (2012 update), http://j.mp/1mirOhP.

25. N. Kauer and G. Passarino, *J. High Energy Phys.* **1208**, 116 (2012), arXiv:1206.4803 [hep-ph].

26. F. Caola and K. Melnikov, *Phys. Rev. D* **88**, 054024 (2013), arXiv:1307.4935 [hep-ph].

27. CMS Collab. (V. Khachatryan *et al.*), *Phys. Lett. B* **736**, 64 (2014), arXiv:1405.3455 [hep-ex].

28. ATLAS Collab. (G. Aad *et al.*), *Eur. Phys. J. C* **75**, 335 (2015), arXiv:1503.01060 [hep-ex].

29. C. Quigg and R. Shrock, *Phys. Rev. D* **79**, 096002 (2009), arXiv:0901.3958 [hep-ph].

30. S. Dawson *et al.*, Higgs Working Group Report of the Snowmass 2013 Community Planning Study, arXiv:1310.8361 [hep-ex].

31. The International Linear Collider, https://www.linearcollider.org/ILC.

32. The Circular Electron–Positron Collider, http://cepc.ihep.ac.cn.

33. The FCC-ee Design Study, http://tlep.web.cern.ch/.

34. S. Gourlay, High field magnets for *pp* colliders, talk at the *Hong Kong University of Science and Technology Jockey Club Institute for Advanced Study Program on The Future of High Energy Physics*, 5–30 January 2015, http://j.mp/1WBWiYR.

35. E. Eichten, I. Hinchliffe, K. D. Lane and C. Quigg, *Rev. Mod. Phys.* **56**, 579 (1984) [Erratum: *ibid.* **58**, 1065 (1986)].

36. LHC Higgs Cross Section Working Group, Higgs cross sections for HL-LHC and HE-LHC, http://j.mp/1ZXfNM6.

37. For example, FELIX: Proposal for a Forward ELastic and Inelastic EXperiment at the LHC, http://felix.web.cern.ch.

38. N. Arkani-Hamed, T. Han, M. Mangano and L. T. Wang, Physics opportunities of a 100 TeV proton–proton collider, arXiv:1511.06495 [hep-ph].

39. M. Dine, *Annu. Rev. Nucl. Part. Sci.* **65**, 43 (2015), arXiv:1501.01035 [hep-ph].

40. C. Quigg, *Annu. Rev. Nucl. Part. Sci.* **59**, 505 (2009), arXiv:0905.3187 [hep-ph]; *Contemp. Phys.* **57**, 177 (2106), arXiv:1507.02977 [hep-ph].

41. E. P. Wigner, *Commun. Pure Appl. Math.* **13**, 1 (1960), http://j.mp/1PngVae.

42. K. G. Wilson, *Nucl. Phys. B* (*Proc. Suppl.*) **140**, 3 (2005), arXiv:hep-lat/0412043.

43. P. Koppenburg, CP violation and CKM physics (including LHCb news from run 2), LHCb-PROC-2015-029, https://cds.cern.ch/record/2039659.

44. CMS Collab. (V. Khachatryan *et al.*), *J. High Energy Phys.* **1408**, 174 (2014), arXiv:1405.3447 [hep-ex].

45. ATLAS Collab. (G. Aad *et al.*), *J. High Energy Phys.* **1512**, 055 (2015), arXiv:1506.00962 [hep-ex].

46. LHCb Collab. (R. Aaij *et al.*), *Nature Phys.* **11**, 743 (2015), arXiv:1504.01568 [hep-ex].

47. LHCb Collab. (R. Aaij *et al.*), *Phys. Rev. Lett.* **113**, 151601 (2014), arXiv:1406.6482 [hep-ex].

48. ATLAS Collab., Search for resonances decaying to photon pairs in 3.2 fb^{-1} of *pp* collisions at $\sqrt{s} = 13$ TeV with the ATLAS detector, ATLAS-CONF-2015-081, http://cds.cern.ch/record/2114853.

49. CMS Collab., Search for new physics in high mass diphoton events in proton–proton collisions at 13 TeV, CMS-PAS-EXO-15-004, http://cds.cern.ch/record/2114808.

50. I. Hinchliffe, A. Kotwal, M. L. Mangano, C. Quigg and L. T. Wang, *Int. J. Mod. Phys. A* **30**, 1544002 (2015), arXiv:1504.06108 [hep-ph].

51. P. M. Nadolsky, H. L. Lai, Q. H. Cao, J. Huston, J. Pumplin, D. Stump, W. K. Tung and C.-P. Yuan, *Phys. Rev. D* **78**, 013004 (2008), arXiv:0802.0007 [hep-ph].

52. S.-H. H. Tye and S. S. C. Wong, *Phys. Rev. D* **92**, 045005 (2015), arXiv:1505.03690 [hep-th]; See also J. Ellis and K. Sakurai, *J. High Energy Phys.* **1604**, 086 (2016), arXiv:1601.03654 [hep-ph].

53. J. M. Campbell, R. K. Ellis and C. Williams, *J. High Energy Phys.* **1107**, 018 (2011), arXiv:1105.0020 [hep-ph], see also http://mcfm.fnal.gov.

54. M. Benedikt, FCC study overview and status, talk at *FCC Week 2015*, Washington D.C., 23–29 March 2015, http://j.mp/1Upsuhp.

55. S. Raby, M. Ratz and K. Schmidt-Hoberg, *Phys. Lett. B* **687**, 342 (2010), arXiv:0911.4249 [hep-ph].

56. ATLAS Collab., Measurement of multi-jet cross-section ratios and determination of the strong coupling constant in proton-proton collisions at $\sqrt{s} = 7$ TeV with the ATLAS detector, ATLAS-CONF-2013-041, http://cds.cern.ch/record/1543225.

57. CMS Collab. (V. Khachatryan *et al.*), *Eur. Phys. J. C* **75**, 288 (2015), arXiv:1410.6765 [hep-ex].

Prospects for Future Collider Physics*

John Ellis

Theoretical Particle Physics and Cosmology Group, Department of Physics,
King's College London, Strand, London WC2R 2LS, UK
Theoretical Physics Department, CERN, CH 1211 Geneva 23, Switzerland
John.Ellis@cern.ch

One item on the agenda of future colliders is certain to be the Higgs boson. *What is it trying to tell us?* The primary objective of any future collider must surely be to identify physics beyond the Standard Model, and supersymmetry is one of the most studied options. *Is supersymmetry waiting for us and, if so, can LHC Run 2 find it?* The big surprise from the initial 13 TeV LHC data has been the appearance of a possible signal for a new boson X with a mass $\simeq 750$ GeV. *What are the prospects for future colliders if the $X(750)$ exists?* One of the most intriguing possibilities in electroweak physics would be the discovery of nonperturbative phenomena. *What are the prospects for observing sphalerons at the LHC or a future collider?*

Keywords: Higgs boson; beyond the Standard Model; supersymmetry; LHC; future colliders.

1. The Higgs Boson

We already know the mass of the Higgs boson with an accuracy $\sim 0.2\%$:

$$m_H = 125.09 \pm 0.21 \pm 0.11 \text{ GeV} , \tag{1}$$

where the first (dominant) uncertainty is statistical and the second is systematic.[1] We can expect that the LHC experiments will reduce the overall uncertainty to below 100 MeV, setting a hot pace for future collider experiments to follow. Precise knowledge of the mass of the Higgs boson will be important for precision tests of

*Contribution to the Hong Kong UST IAS Programme and Conference on High-Energy Physics, based largely on personal research with various collaborators. KCL-PH-TH-2016-16, LCTS-2016-12, CERN-TH-2016-075.

the Standard Model — some Higgs decay rates depend on it quite sensitively — but is also crucial for understanding the stability of the electroweak vacuum,[2] as discussed later.

One of the most basic questions about the Higgs boson is whether it is elementary or composite. In the former case, the large sizes of loop corrections pose the problem of the naturalness (fine-tuning) of the electroweak scale. The solution to this problem that I personally prefer is to postulate an effective cutoff around a TeV due to supersymmetry, though the absence of supersymmetric particles at the LHC (so far) is putting a dent in some people's confidence in this solution. The alternative idea that the Higgs boson is composite has some historical precedents in its side, namely the composite mesons of QCD and the Cooper pairs of superconductivity. Early versions of this idea tended to fail electroweak precision tests and predicted a relatively heavy Higgs-like scalar boson. However, more recent versions interpret the Higgs boson as a relatively light pseudo-Nambu–Goldstone boson and interest in composite models may be resurrected if the existence of the $X(750)$ boson is confirmed. For the moment, though, there is as little evidence for composite models as for supersymmetry.

Under these circumstances, a favored approach is to assume that the Higgs boson and all other known particles are described by Standard Model fields, and parametrize the possible effects of new physics beyond the Standard Model via higher-dimensional combinations of them, e.g. at dimension six:[3]

$$\mathcal{L}_{\text{eff}} = \sum_n \frac{c_n}{\Lambda^2} \mathcal{O}_n \,, \tag{2}$$

where $\Lambda \gg m_Z$, m_W, m_H is the mass scale of new physics and the coefficients c_n help characterize it. They are to be constrained by experiment, e.g. by precision electroweak measurements, Higgs data and measurements of triple-gauge couplings (TGCs).

The left panel of Fig. 1 shows results from one analysis of dimension-six coefficients,[4] expressed in terms of constraints on the $\Lambda' \equiv \Lambda/\sqrt{c}$ currently provided by these measurements. The lowest (black) error bars are from a global fit in which all relevant operators are included, whereas the top (green) error bars are from fits with the operators switched on individually, and intermediate (blue and red) error bars show the effects of Higgs and TGC measurements, respectively. We see that the current constraints imply that the $\Lambda' \gtrsim 0.5$ TeV, in general.

There have been various studies of the sensitivities of future e^+e^- colliders within this framework. The right panel of Fig. 1 displays results from one such analysis,[5] showing the prospective sensitivities of measurements at FCC-ee, whose design foresees much greater luminosities at low energies than the ILC. The upper (green) error bars are for individual operators, whereas the lower (red) error bars are for a global including all operators. We see that the prospective FCC-ee constraints would yield sensitivity to $\Lambda' \gg$ TeV, in general.

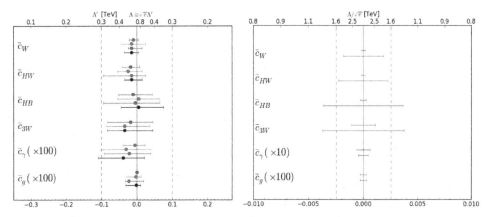

Fig. 1. (Color online) Left panel: The 95% CL constraints obtained for single-coefficient fits (green bars), and the marginalized 95% ranges for the LHC signal-strength data combined with the kinematic distributions for associated $H + V$ production measured by ATLAS and D0 (blue bars), combined with the LHC TGC data (red lines), and the global combination with both the associated production and TGC data (black bars).[4] Right panel: Summary of the 95% CL limits on dimension-six operator coefficients affecting Higgs and TGC observables at FCC-ee.[5] The individual (marginalized) limits are shown in green (red).

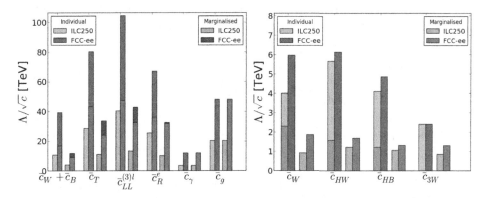

Fig. 2. (Color online) Summary of the reaches for the dimension-six operator coefficients with TeV scale sensitivity, when switched on individually (green) and when marginalized (red), from projected precision measurements at the ILC250 (lighter shades) and FCC-ee (darker shades). The left plot shows the operators that are most strongly constrained by EWPTs and Higgs physics, where the different shades of dark green and dark red represent the effects of EWPT theoretical uncertainties at FCC-ee. The right plot shows constraints from Higgs physics and TGCs, and the different shades of light green demonstrate the improved sensitivity when TGCs are added at ILC250. Plots from Ref. 5.

Comparisons between the prospective ILC and FCC-ee constraints are shown in Fig. 2, with the (green) bars on the left representing individual constraints and the right (red) bars marginalized constraints from a global fit.[5] The left panel compares the constraints on a set of dimension-six operators from Higgs and precision electroweak measurements, with the different darker shadings showing the impact of the theoretical uncertainties in the latter. We see that FCC-ee has prospective sensitivities in the tens of TeV. The right panel compares the prospective constraints from Higgs and TGC measurements. We see here that FCC-ee could reach into the multi-TeV range, as could the ILC when TGC measurements are included (lighter shading).

2. Supersymmetry

Although the LHC has not yet found any signs of supersymmetric particles, I would argue that Run 1 of the LHC has actually provided three additional indirect arguments for supersymmetry. (i) In the Standard Model, the measurements of m_H (1) and m_t indicate *prima facie* that the electroweak vacuum is un/metastable, and supersymmetry would stabilize it. (ii) Simple supersymmetric models predicted successfully the Higgs mass, saying that it should be < 130 GeV.[6] Moreover, (iii) simple supersymmetric models predicted successfully that the couplings should be within few percent of their values in the Standard Model.[7] These arguments are in addition to the traditional arguments for supersymmetry based on the naturalness of the electroweak scale, GUTs, string theory, dark matter, etc.

Let us review the vacuum stability argument. In the Standard Model the Higgs quartic self-coupling λ is renormalized by itself, but the dominant renormalization is by loops of top quarks, which drive $\lambda < 0$ at some scale Λ:[2]

$$\log \frac{\Lambda}{\text{GeV}} = 11.3 + 1.0 \left(\frac{M_h}{\text{GeV}} - 125.66 \right) - 1.2 \left(\frac{m_t}{\text{GeV}} - 173.10 \right)$$

$$+ 0.4 \left(\frac{\alpha_3(M_Z) - 0.1184}{0.0007} \right). \tag{3}$$

The current experimental values of the Higgs mass (1), the official world average top quark mass $m_t = 173.34 \pm 0.27 \pm 0.71$ GeV (Ref. 8) and the QCD coupling $\alpha_3(M_Z) = 0.1177 \pm 0.0013$ (Ref. 9) indicate that the Higgs self-coupling λ turns negative at $\ln(\Lambda/\text{GeV}) = 10.0 \pm 1.0$ within the Standard Model. This turndown implies that our present electroweak vacuum is in principle unstable, though its lifetime may be much larger than the age of the Universe. However, even in this case there is a problem, since most of the initially hot Universe would not have cooled down into our electroweak vacuum.[10]

This problem would be completely avoided in a supersymmetric extension of the Standard Model, where the effective potential is guaranteed to be positive semidefinite. Indeed, one can argue that vacuum stability may require something very like supersymmetry.[11] Unfortunately, there are many possible supersymmetric extensions of the Standard Model and no signs in superspace, and we do not know which superdirection Nature may have taken.

What do the data tell us? In the absence of any clues, we use the available electroweak, flavor, Higgs, LHC and cosmological dark matter constraints in global fits to constrain the parameters of specific supersymmetric models.

The simplest possibility is to consider models with universal soft supersymmetry breaking at some input GUT scale. The scenario in which universality is assumed for the soft supersymmetry-breaking gaugino masses and those of all the scalar partners of Standard Model particles and the Higgs multiplets is called the constrained minimal supersymmetric extension of the Standard Model (CMSSM),[12] and models in which this assumption is relaxed for the Higgs multiplets are called nonuniversal Higgs models (NUHM1,2).[13] These models are under quite strong pressure from the LHC, with p values ~ 0.1.[14,15] On the other hand, a model in which the soft supersymmetry-breaking masses are treated as phenomenological inputs at the electroweak scale (the pMSSM) is less strongly constrained by LHC data. For example, assuming limited universality motivated by upper limits on flavor-changing neutral interactions (the pMSSM10), one finds a higher value of $p \sim 0.3$.[16]

Specifically, assuming that the cosmological dark matter is provided by the lightest neutralino, in the CMSSM the dark matter density constraint provides an upper limit on the supposedly universal fermion mass $m_{1/2}$ for fixed scalar mass m_0, whereas at low values of m_0 and $m_{1/2}$ there is tension between the LHC searches for missing-energy events and the anomalous magnetic moment of the muon, $g_\mu - 2$. The $\sim 3\sigma$ discrepancy between the experimental measurement and the Standard Model calculation could be resolved via low-mass supersymmetry, but this cannot be achieved within the CMSSM and related models. Figure 3 displays the $(m_0, m_{1/2})$ plane in the CMSSM, with the region favored at the 68% CL bounded by the red contour and that allowed at the 95% CL bounded by the blue contour, and the best-fit point indicated by a green star. The region in which coannihilation with

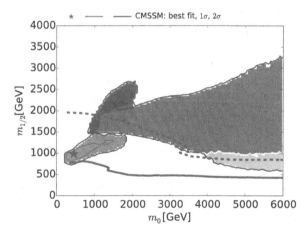

Fig. 3.   (Color online) The $(m_0, m_{1/2})$ planes in the CMSSM. Regions in which different mechanisms bring the dark matter density into the allowed range are shaded as described in the paper.[17] The red and blue contours represent the 68% and 95% CL contours, with the green star indicating the best-fit point. The solid purple contour shows the current LHC 95% exclusion from $\not{E}_T$ searches, and the dashed purple contour shows the prospective 5σ discovery reach for $\not{E}_T$ searches at the LHC with 3000/fb at 14 TeV, which corresponds approximately to the 95% CL exclusion sensitivity with 300/fb at 14 TeV.

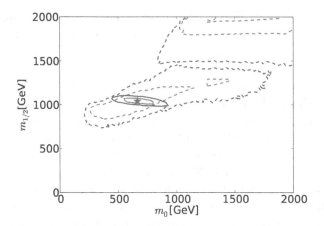

Fig. 4.   (Color online) The 68% and 95% CL regions in the $(m_0, m_{1/2})$ planes (solid red and blue lines) obtained by combining prospective cross-section, $\not{E}_T$ and jet measurements with 3000/fb of luminosity at the LHC at a center-of-mass energy of 14 TeV with the current global fit (here shown as dashed lines). Plot from Ref. 18.

the lighter $\tilde{\tau}$ slepton brings the dark matter density into the range allowed by cosmology is shaded pink, that where rapid annihilation via direct-channel heavy Higgs bosons is shaded dark blue, that where both mechanisms are important is shaded purple, and the stop coannihilation region is shaded lighter blue.[17] We see that the current LHC constraint is important at low $(m_0, m_{1/2})$, and that the estimated future LHC sensitivity covers all the $\tilde{\tau}$ coannihilation region and part of the rapid heavy Higgs annihilation and stop coannihilation regions.

Let us be optimistic, and assume that Nature is described by the current best-fit point in the CMSSM, namely the green star inside the stau coannihilation region in Fig. 3. In this case it would be possible not only to discover supersymmetry in future runs of the LHC, but also to measure some of its parameters quite accurately, as seen in Fig. 4.[18] The prediction of the supersymmetric mass scale and such a detailed confrontation between direct and indirect constraints on supersymmetry would provide tests of the underlying theory akin to those of the Standard Model provided by direct and indirect constraints on the masses of the top quark and the Higgs boson.

The left panel of Fig. 5 shows the $(m_0, m_{1/2})$ plane in the NUHM1, using the same coloring scheme as in Fig. 3. We see again that the LHC should be able to explore an interesting area of the NUHM1 parameter space,[14] and the same is true of the NUHM2 parameter space (not shown).[15] What would be a key distinctive signature of supersymmetry in the CMSSM and the NUHM1,2? Much of the parameter spaces of these models accessible to the LHC lies in the stau coannihilation region, where the mass difference between the lighter stau $\tilde{\tau}_1$ and the lightest neutralino $\tilde{\chi}_1^0$ is quite small. In such a case, the lifetime of the next-to-lightest supersymmetric particle (NLSP), the $\tilde{\tau}_1$, may be quite long, as seen in Fig. 6.[17] Thus,

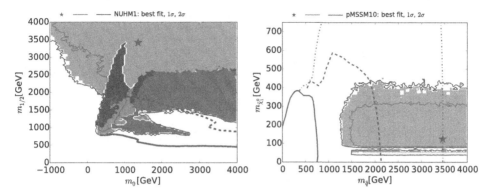

Fig. 5. (Color online) Left panel: The $(m_0, m_{1/2})$ plane in the NUHM1. Right panel: The $(m_{\tilde{q}}, m_{\tilde{\chi}_1^0})$ plane in the pMSSM10. In the green regions the dark matter density is brought into the allowed range by chargino coannihilation and in the pink and yellow strips on the right panel by rapid annihilation via the h and Z poles: the other colors on the left panel have the same significances as in Fig. 3. Plots from Ref. 17.

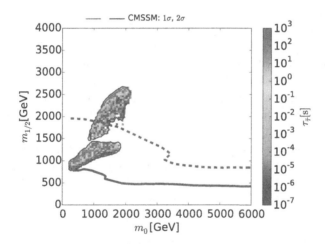

Fig. 6. (Color online) The $(m_0, m_{1/2})$ plane in the CMSSM, showing the regions where the lowest-χ^2 points within the 95% CL region that have $10^3 \mathrm{s} > \tau_{\tilde{\tau}_1} > 10^{-7} \mathrm{s}$: the lifetimes of the $\tilde{\tau}_1$ at these points are color-coded, as indicated in the legends.[17] Also shown in these panels as solid purple contours are the current LHC 95% exclusions from $\not{E}_T$ searches in the $\tilde{\tau}_1$ coannihilation regions, and as dashed purple contours the prospective 5-σ discovery reaches for $\not{E}_T$ searches at the LHC with 3000/fb at 14 TeV, corresponding approximately to the 95% CL exclusion sensitivity with 300/fb at 14 TeV. The sensitivities of LHC searches for metastable $\tilde{\tau}_1$'s in the $\tilde{\tau}_1$ coannihilation region are expected to be similar.[19]

possible signatures could include long-live charged particles that decay outside the detector or with a separated decay vertex within it.[19]

The situation is rather different within the pMSSM10, whose $(m_{\tilde{q}}, m_{\tilde{\chi}_1^0})$ plane is shown on the right panel of Fig. 5.[17] We see again that future runs of the LHC have a fair chance of discovering supersymmetry also in this scenario (the dashed

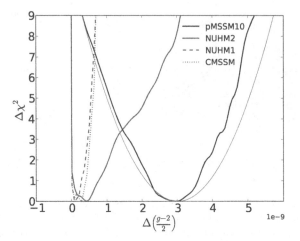

Fig. 7. The χ^2 likelihood functions for the anomalous magnetic moment of the muon, $g_\mu - 2$, in the CMSSM, NUHM1, NUHM2 and pMSSM10, taking account of LHC Run 1 and other constraints, as described in Ref. 16.

line is for $m_{\tilde{q}} \ll m_{\tilde{g}}$ and the dash–dotted line for $m_{\tilde{g}} = 4.5$ TeV), but we do not expect a long-lived charged particle signature. However, the pMSSM10 can resolve the tension between LHC searches and the measurement of $g_\mu - 2$.[16] Figure 7 shows that, whereas the CMSSM and related models (blue curves) predict values of $g_\mu - 2$ that are very similar to those in the Standard Model, the pMSSM10 (black curve)[16] can accommodate the experimental value (red curve) without falling foul of the LHC constraints.

The left panel of Fig. 8 displays the dependences on the gluino mass, $m_{\tilde{g}}$, of the χ^2 functions from global fits to the CMSSM and related models (blue curves) and the pMSSM10 (black curve). We see that the LHC data, in particular, set 95% CL constraints $m_{\tilde{g}} \gtrsim 1.5$ TeV in the CMSSM and related models, which may be relaxed

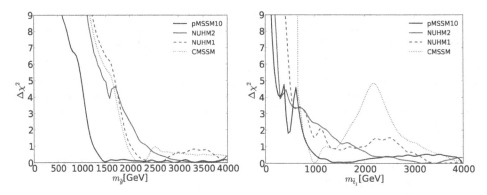

Fig. 8. The χ^2 likelihood functions for the gluino mass (left panel) and the lighter stop squark (right panel) in the CMSSM, NUHM1, NUHM2 and pMSSM10, taking account of LHC Run 1 and other constraints, as described in Ref. 16.

Table 1. Summary of the detectability of supersymmetry in the CMSSM, NUHM1, NUHM2 and pMSSM10 models at the LHC in searches for $\not{E}_T$ events, long-lived charged particles (LL) and heavy A/H Higgs bosons, and in direct DM search experiments, according to the dominant mechanism for bringing the DM density into the cosmological range.[17] The symbols $\checkmark$, $(\checkmark)$ and $\times$ indicate good prospects, interesting possibilities and poorer prospects, respectively. The symbol $-$ indicates that a DM mechanism is not important for the corresponding model.

DM mechanism	Exp't	Models			
		CMSSM	NUHM1	NUHM2	pMSSM10
$\tilde{\tau}_1$	LHC	$\checkmark$ $\not{E}_T$, $\checkmark$ LL	$(\checkmark$ $\not{E}_T$, $\checkmark$ LL$)$	$(\checkmark$ $\not{E}_T$, $\checkmark$ LL$)$	$(\checkmark$ $\not{E}_T)$, $\times$ LL
coann.	DM	$(\checkmark)$	$(\checkmark)$	$\times$	$\times$
$\tilde{\chi}_1^\pm$	LHC	$-$	$\times$	$\times$	$(\checkmark$ $\not{E}_T)$
coann.	DM	$-$	$\checkmark$	$\checkmark$	$(\checkmark)$
$\tilde{t}_1$	LHC	$-$	$-$	$\checkmark$ $\not{E}_T$	$-$
coann.	DM	$-$	$-$	$\times$	$-$
A/H	LHC	$\checkmark$ A/H	$(\checkmark$ $A/H)$	$(\checkmark$ $A/H)$	$-$
funnel	DM	$\checkmark$	$\checkmark$	$(\checkmark)$	$-$
Focus	LHC	$(\checkmark$ $\not{E}_T)$	$-$	$-$	$-$
point	DM	$\checkmark$	$-$	$-$	$-$
h, Z	LHC	$-$	$-$	$-$	$(\checkmark$ $\not{E}_T)$
funnels	DM	$-$	$-$	$-$	$(\checkmark)$

to $m_{\tilde{g}} \gtrsim 1.0$ TeV in the pMSSM10. The good news is that future runs of the LHC should have sensitivity to $m_{\tilde{g}} \lesssim 3$ TeV, so there are significant chances that the LHC may discover supersymmetry within these scenarios, though no guarantees. The right panel of Fig. 8 displays the dependences of the global χ^2 functions on the lighter stop squark mass, $m_{\tilde{t}_1}$, in the same line styles as on the left panel. In this case, we see that a "natural" light stop with $m_{\tilde{t}_1} \sim 400$ GeV is allowed in the pMSSM10 at the $\Delta\chi^2 \simeq 2$ level. This region may be accessible to future LHC searches for compressed sparticle spectra.

Table 1 summarizes the prospects for discovering supersymmetry in the CMSSM, NUHM1, 2 and pMSSM10 either at the LHC and/or in direct dark matter search experiments, organized according to the dominant mechanism for bring in the dark matter density into the range allowed by cosmology.[17] A hyphen (-) indicates that the corresponding mechanism is not important in the given supersymmetric model. We are encouraged to see that in every box without a hyphen there are prospects for discovering supersymmetry at the LHC and/or in a planned direct dark matter search experiment. No wonder we are excited about the prospects for Run 2 of the LHC! If supersymmetry does escape us at the LHC, a 100 TeV collider would have great capabilities for discovering heavy squarks and/or gluinos.[18]

3. Who Ordered That?

This is the famous quip by Rabi about the muon. The same might be said about the $\gamma\gamma$ "bump" with an invariant mass $\simeq 750$ GeV reported by the ATLAS[20] and CMS experiments[21] in a preliminary analysis of their 13 TeV data in December 2015. Both experiments now also report insignificant hints in their 8 TeV data. At the time of writing, the data shown by ATLAS at the Moriond conference in early March 2016 exhibit a 3.9σ enhancement,[22] whereas the CMS data display a 3.4σ enhancement.[23] A naive combination of the p-values of the two peaks corresponds to a 4.99σ signal, whose significance is reduced by the "look-elsewhere effect" to 3.89σ. This is insufficient to claim a discovery, but according to CERN Director-General Fabiola Gianotti, we "are allowed to be slightly excited."[24]

If interpreted as a new particle X, the reported signal would correspond to $\sigma(pp \to X) \times \text{BR}(X \to \gamma\gamma) \sim$ few fb. Needless to say, any such $X(750)$ would itself definitely constitute physics beyond the Standard Model, though what role it may play in resolving any of the oft-touted outstanding problems of the Standard Model is most unclear. Even more exciting than the existence of $X(750)$ itself is the prospect that it would be merely the tip of an iceberg of new physics, a harbinger of a whole new layer of matter.[25]

Let us be conservative, and assume that the $X(750)$ has spin zero.[26] In this case, its $\gamma\gamma$ decays would presumably be mediated by anomalous triangle diagrams of massive charged particles. Fermions may be the most plausible candidates, as scalar loops generally have smaller numerical values, and postulating new charged vector bosons is disfavored by Occam's razor. The form factors for loop diagrams are suppressed for light fermions in the loops with masses $\ll m_X/2$, such as the top quark, and are maximized for fermions with masses $\sim m_X/2$. However, a heavy conventional fourth generation is strongly excluded by other constraints, and would require nonperturbative Yukawa couplings. The most likely possibility seems to be one or more vector-like fermions, which may have masses larger than the electroweak scale. If some of these are colored, they could also mediate X production via gluon–gluon fusion, which would accommodate the energy dependence of the signal more easily than light $\bar{q}q$ collisions.

The minimal model is (1) a single vector-like charge-2/3 quark (a single bottom-like quark would make a contribution to the $\gamma\gamma$ decay rate that is smaller by a factor 4). Alternatively, one could postulate (2) an SU(2) doublet of vector-like quarks, or (3) a doublet and two singlet vector-like quarks. Finally, one may go the whole hog, and postulate (4) a full vector-like generation, including leptons as well as quarks.[26]

Figure 9 shows the $XF\bar{F}$ couplings λ (assumed for simplicity to be universal) that would be required to explain the possible $X(750)$ signal in these different models, as functions of the vector-like fermion mass (also assumed for simplicity to be universal), under the assumption that $X \to gg$ is the dominant decay mode.[26] In each panel, the black line corresponds to $\sigma(pp \to X) \times \text{BR}(X \to \gamma\gamma) = 6$ fb, and the colored band corresponds to ± 1 fb around this central value.[26] If $\lambda/4\pi > 1/2$ the

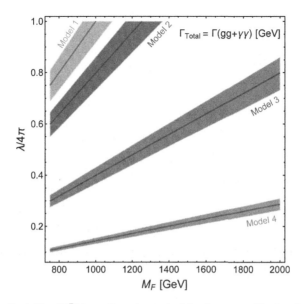

Fig. 9. (Color online) The $X\bar{F}F$ couplings λ required in the vector-like fermion models (1)–(4) described in the text to yield $\sigma(pp \to X) \times \mathrm{BR}(X \to \gamma\gamma) = 6 \pm 1$ fb (solid black lines and colored bands), assuming that diboson decays dominate. Plot from Ref. 27, adapted from Ref. 26.

Table 2. Ratios of X decay rates for the various models introduced in the text, assuming $\alpha_s(m_X) \simeq 0.092$. The upper limits on $\frac{\mathrm{BR}(X \to VV)}{\mathrm{BR}(X \to \gamma\gamma)}$ are obtained from LHC 8 TeV data, as described in Ref. 26.

Model	$\frac{\mathrm{BR}(X \to gg)}{\mathrm{BR}(X \to \gamma\gamma)}$	$\frac{\mathrm{BR}(X \to Z\gamma)}{\mathrm{BR}(X \to \gamma\gamma)}$	$\frac{\mathrm{BR}(X \to ZZ)}{\mathrm{BR}(X \to \gamma\gamma)}$	$\frac{\mathrm{BR}(X \to W^\pm W^\mp)}{\mathrm{BR}(X \to \gamma\gamma)}$
(1)	180	1.2	0.090	0
(2)	460	10	9.1	61
(3)	460	1.1	2.8	15
(4)	180	0.46	2.1	11
Current limit	$\sim 2 \times 10^4$	7	13	30

coupling λ is nonperturbative, whereas it is perturbative for smaller values. We see, therefore, that models (1) and (2) may well require a nonperturbative treatment, whereas models (3) and (4) could well be perturbative.[a] In the case of model (4), which includes neutral vector-like leptons, these could constitute the dark matter if the common mass $\lesssim 1500$ GeV.

In each of the models studied, it is possible to calculate the ratios of the decay rates of $X \to gg$, $Z\gamma$, W^+W^-, ZZ and $\gamma\gamma$ via the triangular loop diagrams, with the results shown in Table 2. Also shown in this table are the upper limits on

[a]On the other hand, all the models would have to be nonperturbative if $\Gamma_X \simeq 45$ GeV, as slightly (dis)favored by ATLAS (CMS) data.

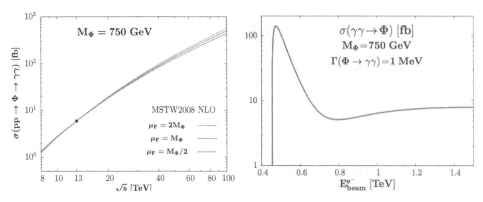

Fig. 10. Left panel: Increase of the production in pp collisions at different center-of-mass energies of a singlet boson Φ with mass 750 GeV produced by gluon–gluon collisions and decaying into $\gamma\gamma$, assuming that two-boson decays are dominant and normalized to the possible LHC signal at 13 TeV. Right panel: Cross-section for its production in $\gamma\gamma$ collisions at an e^+e^- collider as a function of the electron beam energy. Plots from Ref. 27.

these ratios inferred from LHC 8 TeV data, as discussed in Ref. 26. We see that model (2) is formally in conflict with the upper limits on $X \to Z\gamma$ and W^+W^-, though it may be premature to conclude that the model is excluded. The good news is that the models are potentially accessible to experimental searches in other diboson channels. As discussed below, there are also interesting possibilities to look for heavy fermions at the LHC and future colliders.[27] All in all, there is both experimental and theoretical work for a generation if the X particle exists, and we should know the answer to this question in 2016.

The left panel of Fig. 10 shows how rapidly $\sigma(pp \to X) \times \mathrm{BR}(X \to \gamma\gamma)$ would grow with the pp center-of-mass energy, assuming production via gluon–gluon fusion.[27] At 100 TeV the cross-section would increase by two orders of magnitude, with PDF and higher-order QCD uncertainties that are $\sim 30\%$. The right panel of Fig. 10 displays, as a function of the e^- beam energy, the cross-section for $\gamma\gamma \to X$ production at an e^+e^- collider that is optimized for $\gamma\gamma$ collisions. Needless to say, an e^+e^- collider with $E^{e^-}_{\mathrm{Beam}} < 375$ GeV would not be able to produce the $X(750)$, and we see that $E^{e^-}_{\mathrm{beam}} \simeq 500$ GeV would be preferred.

Figure 11 displays the cross-sections for the production of vector-like fermions in pp collisions as functions of the center-of-mass energy.[27] The left panel shows the cross-sections for vector-like quark production at different collider center-of-mass energies as functions of the quark mass, and the right panel shows the cross-sections for producing different types of vector-like leptons (doublets L, charged and neutral leptons E, N and associated N, L pairs) as functions of the center-of-mass energy for a mass of 0.4 TeV. As we see in Table 3, the LHC sensitivity for vector-like quarks in the models (1)–(4) introduced previously should extend to ~ 2 TeV and for vector-like leptons in model (4) to 0.7 TeV, and the corresponding sensitivities of a 100 TeV collider would be to ~ 13 and ~ 5 TeV, respectively. The LHC should

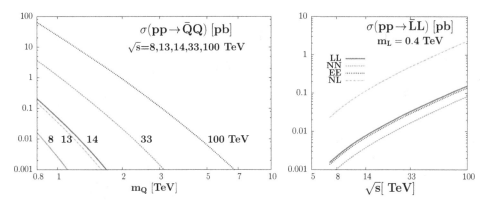

Fig. 11. Left panel: Cross-sections for vector-like quark pair-production in pp collisions at different center-of-mass energies. Right panel: Cross-sections for the pair-production of vector-like leptons with masses 0.4 TeV in pp collisions as functions of the center-of-mass energy. Plots from Ref. 27.

Table 3. Prospective sensitivities to vector-like quarks (left) and leptons (right) [particle masses indicated in TeV] for various pp collider scenarios.

Model	Vector-like quark mass sensitivity				Vector-like lepton mass sensitivity			
	100 fb^{-1} 13 TeV	300 fb^{-1} 14 TeV	300 fb^{-1} 33 TeV	20 ab^{-1} 100 TeV	100 fb^{-1} 13 TeV	300 fb^{-1} 14 TeV	300 fb^{-1} 33 TeV	20 ab^{-1} 100 TeV
(1)	1.4	1.7	3.1	11.7				—
(2)	1.5	1.8	3.4	12.7				—
(3)	1.6	2.0	3.7	13.7				—
(4)	1.6	2.0	3.7	13.7	0.56	0.73	1.7	5.3

be able to explore the possible range of vector-like quark masses in plausible models of $X(750)$ production and decay, and a 100 TeV collider would be able to explore their dynamics in some detail, e.g. probing how they mix with the Standard Model quarks.[27]

As an alternative to the minimal singlet scenario for the $X(750)$ enhancement, one may also consider a two-Higgs-doublet scenario,[27] in which it could be interpreted as a superposed pair of heavy Higgs bosons H, A. In many such models, such as supersymmetry, these bosons are nearly degenerate. For example, if the Higgs potential is the same as in the minimal supersymmetric extension of the Standard Model, one finds $M_H - M_A \simeq 15$ GeV if $\tan\beta = 1$. This choice is motivated by consideration of the dominant $H/A \to \bar{t}t$ decays, which yield $\Gamma_{H,A} = 32, 35$ GeV in this case.[b] In this model, one can also calculate the ratio

---

[b]Since $H/A \to \bar{t}t$ decays dominate over the decays into boson pairs considered in the previous singlet scenario, the loop diagrams responsible for H/A production and decay must be enhanced compared to that scenario, e.g. by postulating relatively light charged leptons with masses $\sim m_{H/A}/2$, or additional (multiply-?)charged particles.

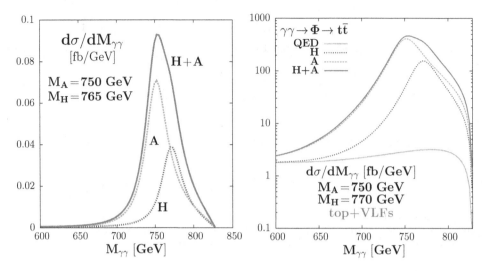

Fig. 12.   Left panel: The $\Phi = H$, A line-shape for $M_A = 750$ GeV, $M_H = 765$ GeV, $\Gamma_H = 32$ GeV and $\Gamma_A = 35$ GeV for $\tan\beta = 1$. Right panel: The invariant mass distribution $d\sigma/dM_{\gamma\gamma}$ for the process $\gamma\gamma \to t\bar{t}$ in the $\gamma\gamma$ mode of a linear e^+e^- collider, with H/A parameters as on the left panel. Plots from Ref. 27.

$\sigma(pp \to A) \times \mathrm{BR}(A \to \gamma\gamma)/\sigma(pp \to H) \times \mathrm{BR}(H \to \gamma\gamma) \simeq 2$. The combined H/A signal in pp collisions is therefore an "asymmetric Breit–Wigner" as shown on the left panel of Fig. 12, with a full width at half maximum of ~ 45 GeV, corresponding to the signal width favored (slightly) by ATLAS. The right panel of Fig. 12 shows the corresponding line shape in $\gamma\gamma$ collisions at an e^+e^- collider with center-of-mass energy 1 TeV. In addition to single H/A production, there is rich bosonic phenomenology in associated $\bar{t}tH/A$ production and pair production, as discussed in Ref. 27.

Before we get too excited, though, we should remember the wise words of Laplace: "*Plus un fait est extraordinaire, plus il a besoin d'être appuyé de fortes preuves,*" i.e. "*The more extraordinary a claim, the stronger the proof required to support it.*" The Higgs boson was (to some extent) expected, and the possible range of its mass was quite restricted before its discovery. In contrast, the $X(750)$ is totally unexpected. For this reason, we certainly should wait and see how the hint develops with increased luminosity before getting much more than "slightly excited."[24]

4. Search for Sphalerons

Let me now turn to a topic even more speculative than the existence of the $X(750)$, namely the search for sphalerons.[28] These are nonperturbative configurations in the electroweak sector of the Standard Model that would mediate processes that change the SU(2) Chern–Simons number: $\Delta n \neq 0$, and thereby violate baryon and lepton numbers, with significance for generating the cosmological baryon asymmetry.[29] It used to be thought that sphaleron-induced transitions would be very suppressed at accessible energies, but this conventional wisdom has recently been challenged by Tye and Wong (TW).[30] They argue that, since the effective Chern–Simons potential is periodic, one should use Bloch wave functions $\Psi(Q)$ to calculate the transition rate:

$$\left(-\frac{1}{2m}\frac{\partial^2}{\partial Q^2} + V(Q)\right)\Psi(Q) = E\Psi(Q),\qquad(4)$$

$$V(Q) \simeq 4.75\ (1.31\sin^2(Qm_W) + 0.60\sin^4(Qm_W))\ \mathrm{TeV},\qquad(5)$$

where Q is related to the Chern–Simons number by $Q \equiv \mu/m_W : n\pi = \mu - \sin(2\mu)/2$. The Bloch wave function approach of TW yields a rate similar a tunneling calculation for transitions at quark–quark collision energies E below the sphaleron energy $E_{\mathrm{Sph}} \simeq 9$ TeV, and an enhanced rate at higher energies that we parametrize as:[31]

$$\sigma(\Delta n = \pm 1) = \frac{1}{m_W^2}\sum_{ab}\int dE\,\frac{d\mathcal{L}_{ab}}{dE}\,p\exp\left(c\frac{4\pi}{\alpha_W}S(E)\right),\qquad(6)$$

where p is an unknown factor, $S(E) = 0$ for $E > E_{\mathrm{Sph}}$, and the results are largely independent of c over a plausible range.

The left panel of Fig. 13 shows how the sphaleron transition rate would grow, according to (6), for $E_{\mathrm{Sph}} = 9 \pm 1$ TeV.[31] We see that the cross-section grows significantly at the LHC between 13 and 14 TeV, and by a factor $\sim 10^6$ between 13 and 100 TeV. It should be remembered that the normalization factor p is unknown, and that it might depend on the transition energy E. However, as seen on the right panel of Fig. 13, most of the transitions take place for $E \sim E_{\mathrm{Sph}}$, so this energy dependence may not be important.

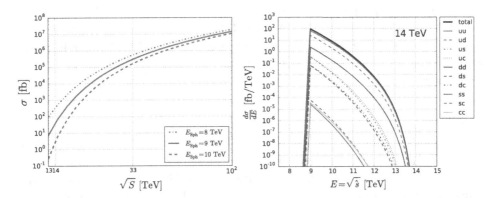

Fig. 13.   Left panel: The energy dependence of the total cross-section for sphaleron transitions for the nominal choices $E_{\mathrm{Sph}} = 9$ TeV, $c = 2$ and $p = 1$ in (6) (solid curve), and for $E_{\mathrm{Sph}} = 8$ and 10 TeV (dot-dashed and dashed lines, respectively). Right panel: Contributions to the cross-section for sphaleron transitions from the collisions of different flavors of quarks, for $E_{\mathrm{CM}} = 14$ TeV, $E_{\mathrm{Sph}} = 9$ TeV and $p = 1$ in (6). Plots from Ref. 31.

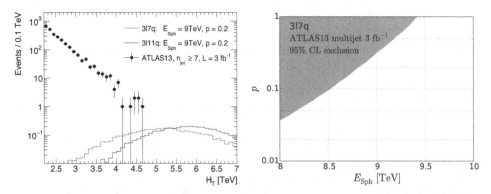

Fig. 14.   (Color online) Left panel: Comparison of the numbers of events with $n_{\text{jet}} \geq 3$ measured by ATLAS in $\sim 3/\text{fb}$ of data at 13 TeV in bins of H_T, compared with simulations for $E_{\text{Sph}} = 9$ TeV of $\Delta n = -1$ sphaleron transitions to final states with three antileptons and seven antiquarks (red histogram) and $\Delta n = +1$ transitions to final states with three leptons and 11 quarks (blue histogram). Right panel: The exclusion in the (E_{Sph}, p) plane for $\Delta n = -1$ transitions obtained by recasting the ATLAS 2015 search for microscopic black holes using $\sim 3/\text{fb}$ of data at 13 TeV. Plots from Ref. 31.

We have simulated the final states in sphaleron-induced transitions, and found that they are quite similar to the simulated final states for microscopic black hole decay. Accordingly, we have recast an ATLAS search for microscopic black holes in 13 TeV collisions with 3/fb of luminosity,[32] and used it to constrain the normalization factor p. The left panel of Fig. 14 compares the H_T distribution in the final states of sphaleron transitions with $\Delta n = \mp 1$ (labeled 3l7q and 3l11q, respectively) with the results of the ATLAS black hole search. We see that there are no events at large H_T where the sphaleron signal would peak, and set the upper limit on p shown on the right panel of Fig. 14. The ATLAS data already set the upper limit $p \lesssim 0.3$ for $\Delta n = -1$ transitions and the stronger constraint $p \lesssim 0.2$ for $\Delta n = +1$ transitions if $E_{\text{Sph}} = 9$ TeV. With 3000/fb of data at 14 TeV, the LHC would be sensitive to $p \sim 10^{-4}$, and a 100 TeV collider with 20/ab would be sensitive to $p \sim 10^{-11}$ for $E_{\text{Sph}} = 9$ TeV. The suggestion of TW[30] certainly needs close scrutiny, and the outcome could open exciting prospects for future pp collider experiments.[33]

5. Summary

In my opinion, rumors of the death of supersymmetry are greatly exaggerated: it is still the most interesting framework for TeV-scale physics, and still provides the best candidate for cold dark matter. As discussed in this talk, simple models with universal soft supersymmetry breaking such as the CMSSM are under pressure, with p-values around 0.1, but this is not enough to reject them. More general models such as the pMSSM quite healthy, with p-values around 0.3, and there are good prospects for discovering sparticles during LHC Run 2 and/or in direct dark matter detection experiments.

More speculatively, particle physics will enter a brave new world if the $X(750)$ signal is confirmed, with exciting prospects for future pp collider experiments in particular. Let us keep our fingers crossed and await the verdict of ATLAS and CMS during 2016.

Finally, it may be time to think again about sphalerons and the possibility that they could have detectable effects at the LHC and future colliders.

Acknowledgments

The author's research was supported partly by the London Center for Terauniverse Studies (LCTS), using funding from the European Research Council via the Advanced Investigator Grant 26732, and partly by the STFC Grant ST/L000326/1. He thanks Henry Tye for hospitality at the Hong Kong UST IAS, and Kirill Prokofiev and Luis Flores Castillo for the invitation to give this talk.

References

1. ATLAS and CMS Collabs. (G. Aad *et al.*), *Phys. Rev. Lett.* **114**, 191803 (2015), doi:10.1103/PhysRevLett.114.191803, arXiv:1503.07589 [hep-ex].
2. D. Buttazzo, G. Degrassi, P. P. Giardino, G. F. Giudice, F. Sala, A. Salvio and A. Strumia, *J. High Energy Phys.* **1312**, 089 (2013), doi:10.1007/JHEP12(2013)089, arXiv:1307.3536 [hep-ph].
3. W. Buchmuller and D. Wyler, *Nucl. Phys. B* **268**, 621 (1986), doi:10.1016/0550-3213(86)90262-2.
4. J. Ellis, V. Sanz and T. You, *J. High Energy Phys.* **1503**, 157 (2015), doi:10.1007/JHEP03(2015)157, arXiv:1410.7703 [hep-ph].
5. J. Ellis and T. You, *J. High Energy Phys.* **1603**, 089 (2016), doi:10.1007/JHEP03(2016)089, arXiv:1510.04561 [hep-ph].
6. J. R. Ellis, G. Ridolfi and F. Zwirner, *Phys. Lett. B* **257**, 83 (1991), doi:10.1016/0370-2693(91)90863-L; H. E. Haber and R. Hempfling, *Phys. Rev. Lett.* **66**, 1815 (1991), doi:10.1103/PhysRevLett.66.1815; Y. Okada, M. Yamaguchi and T. Yanagida, *Prog. Theor. Phys.* **85**, 1 (1991), doi:10.1143/PTP.85.1.
7. J. R. Ellis, S. Heinemeyer, K. A. Olive and G. Weiglein, *J. High Energy Phys.* **0301**, 006 (2003), doi:10.1088/1126-6708/2003/01/006, arXiv:hep-ph/0211206.
8. ATLAS, CDF, CMS and D0 Collabs., arXiv:1403.4427 [hep-ex].
9. D. d'Enterria and P. Z. Skands (eds.), arXiv:1512.05194 [hep-ph].
10. M. Fairbairn and R. Hogan, *Phys. Rev. Lett.* **112**, 201801 (2014), doi:10.1103/PhysRevLett.112.201801, arXiv:1403.6786 [hep-ph]; A. Hook, J. Kearney, B. Shakya and K. M. Zurek, *J. High Energy Phys.* **1501**, 061 (2015), doi:10.1007/JHEP01(2015)061, arXiv:1404.5953 [hep-ph].
11. J. R. Ellis and D. Ross, *Phys. Lett. B* **506**, 331 (2001), doi:10.1016/S0370-2693(01)00156-3, arXiv:hep-ph/0012067.
12. M. Drees and M. M. Nojiri, *Phys. Rev. D* **47**, 376 (1993), doi:10.1103/PhysRevD.47.376, arXiv:hep-ph/9207234; G. L. Kane, C. F. Kolda, L. Roszkowski and J. D. Wells, *Phys. Rev. D* **49**, 6173 (1994), doi:10.1103/PhysRevD.49.6173, arXiv:hep-ph/9312272.
13. H. Baer, A. Mustafayev, S. Profumo, A. Belyaev and X. Tata, *Phys. Rev. D* **71**, 095008 (2005), doi:10.1103/PhysRevD.71.095008, arXiv:hep-ph/0412059; H. Baer, A. Mustafayev, S. Profumo, A. Belyaev and X. Tata, *J. High Energy Phys.* **0507**, 065 (2005), doi:10.1088/1126-6708/2005/07/065, arXiv:hep-ph/0504001.
14. O. Buchmueller *et al.*, *Eur. Phys. J. C* **74**, 2922 (2014), doi:10.1140/epjc/s10052-014-2922-3, arXiv:1312.5250 [hep-ph].

15. O. Buchmueller *et al.*, *Eur. Phys. J. C* **74**, 3212 (2014), doi:10.1140/epjc/s10052-014-3212-9, arXiv:1408.4060 [hep-ph].

16. K. J. de Vries *et al.*, *Eur. Phys. J. C* **75**, 422 (2015), doi:10.1140/epjc/s10052-015-3599-y, arXiv:1504.03260 [hep-ph].

17. E. A. Bagnaschi *et al.*, *Eur. Phys. J. C* **75**, 500 (2015), doi:10.1140/epjc/s10052-015-3718-9, arXiv:1508.01173 [hep-ph].

18. O. Buchmueller, M. Citron, J. Ellis, S. Guha, J. Marrouche, K. A. Olive, K. de Vries and J. Zheng, *Eur. Phys. J. C* **75**, 469 (2015), doi:10.1140/epjc/s10052-015-3675-3, arXiv:1505.04702 [hep-ph].

19. M. Citron, J. Ellis, F. Luo, J. Marrouche, K. A. Olive and K. J. de Vries, *Phys. Rev. D* **87**, 036012 (2013), doi:10.1103/PhysRevD.87.036012, arXiv:1212.2886 [hep-ph]; N. Desai, J. Ellis, F. Luo and J. Marrouche, *Phys. Rev. D* **90**, 055031 (2014), doi:10.1103/PhysRevD.90.055031, arXiv:1404.5061 [hep-ph].

20. ATLAS Collab., http://cds.cern.ch/record/2114853/files/ATLAS-CONF-2015-081.pdf.

21. CMS Collab., https://cds.cern.ch/record/2114808/files/EXO-15-004-pas.pdf.

22. ATLAS Collab., http://cds.cern.ch/record/2141568/files/ATLAS-CONF-2016-018.pdf.

23. CMS Collab., https://cds.cern.ch/record/2139899/files/EXO-16-018-pas.pdf.

24. F. Gianotti, talk to CERN staff, 18 January 2016.

25. See http://inspirehep.net/record/1410174/citations and http://inspirehep.net/record/1409807/citations.

26. J. Ellis, S. A. R. Ellis, J. Quevillon, V. Sanz and T. You, doi:10.1007/JHEP03(2016)176, arXiv:1512.05327 [hep-ph].

27. A. Djouadi, J. Ellis, R. Godbole and J. Quevillon, arXiv:1601.03696 [hep-ph].

28. N. S. Manton, *Phys. Rev. D* **28**, 2019 (1983), doi:10.1103/PhysRevD.28.2019; F. R. Klinkhamer and N. S. Manton, *Phys. Rev. D* **30**, 2212 (1984), doi:10.1103/PhysRevD.30.2212.

29. M. Fukugita and T. Yanagida, *Phys. Lett. B* **174**, 45 (1986), doi:10.1016/0370-2693(86)91126-3.

30. S.-H. H. Tye and S. S. C. Wong, *Phys. Rev. D* **92**, 045005 (2015), doi:10.1103/PhysRevD.92.045005, arXiv:1505.03690 [hep-th].

31. J. Ellis and K. Sakurai, arXiv:1601.03654 [hep-ph].

32. ATLAS Collab. (G. Aad *et al.*), *J. High Energy Phys.* **1603**, 026 (2016), doi:10.1007/JHEP03(2016)026, arXiv:1512.02586 [hep-ex].

33. J. Ellis, K. Sakurai and M. Spannowsky, The IceCube experiment is also interesting for sphaleron searches, arXiv:1603.06573 [hep-ph].

Physics at a Higgs Factory

Matthew Reece

Department of Physics, Harvard University,
17 Oxford St., Cambridge, MA 02138, USA
mreece@physics.harvard.edu

I give an overview of the physics potential at possible future e^+e^- colliders, including the ILC, FCC-ee, and CEPC. The goal is to explain some of the measurements that can be done in the context of electroweak precision tests and Higgs couplings, to compare some of the options under consideration, and to put the measurements in context by summarizing their implications for some new physics scenarios. This is a writeup of a plenary talk at the Hong Kong University of Science and Technology Jockey Club Institute for Advanced Study Program on High Energy Physics Conference, 18–21 January 2016. Some previously unpublished electroweak precision results for FCC-ee and CEPC are included.

Keywords: Future colliders; Higgs boson; electroweak precision tests.

1. Introduction

One of the most exciting developments in high-energy physics in recent years is the design and planning of multiple large-scale future experiments. These include both electron–positron and proton–proton colliders. A major goal of the electron–positron machines is to precisely measure the couplings of the Higgs boson. For this reason they are often referred to as "Higgs factories," and all of the machines being planned will run at energies near 240 GeV where the $e^+e^- \to Zh$ cross-section is largest. This note grew out of a talk that I was asked to give at the Hong Kong IAS on the topic "Physics at a Higgs Factory." One of the major physics questions, especially for the CEPC collider whose planning is still at an early stage, is the extent to which the collider should be solely focused on the Higgs. How important is Z-pole physics? How important are measurements on the $t\bar{t}$ or W^+W^- threshold? In this paper, I will explain some of the physics that I think is most useful for making informed decisions about such questions. I also want to give some context. Rather than just asking about how accurately measurements can be done, I will ask: what will these measurements tell us about what could lie beyond the Standard Model?

The talk was based in part on previous studies in collaboration with JiJi Fan and Lian-Tao Wang on electroweak precision observables at future colliders,[1-3] as well as work by others that I will cite below. I have also added some new material in light of discussions at the Hong Kong meeting and other recent workshops.

2. Electroweak Precision

2.1. *Projected reach in S and T*

I will focus my discussion on a few dimension-six operators that involve the Higgs boson and generally receive important contributions in natural theories. For instance, in a supersymmetric theory, the stops that cancel the leading quadratic divergence also run in the loop to produce these operators. Namely,

$$S \text{ parameter:} \quad S\left(\frac{\alpha}{4s_W c_W v^2}\right) h^\dagger \sigma^i h W^i_{\mu\nu} B^{\mu\nu}, \tag{1}$$

$$T \text{ parameter:} \quad -T\left(\frac{2\alpha}{v^2}\right) |h^\dagger D_\mu h|^2, \tag{2}$$

$$\text{Higgs decays:} \quad c_{hgg} h^\dagger h G^a_{\mu\nu} G^{a\mu\nu} + c_{h\gamma\gamma} h^\dagger h F_{\mu\nu} F^{\mu\nu}. \tag{3}$$

Of course, there are many other dimension-six operators and it is interesting to constrain them all.[4] Flavor-violating operators, for example, can be leading probes of new physics. But this set of operators is very common in any new physics coupling to the Higgs boson. Other familiar operators in the electroweak sector tend to be subdominant: the TGC operator W^3 and the W-parameter $(DW)^2$, for instance, have tiny coefficients when SU(2) multiplets are integrated out at one loop, while the U-parameter is dimension 8.

The electroweak precision fit depends on a number of experimental inputs. Among the most important ones to improve are the W mass, the effective weak mixing angle $\sin^2 \theta_{\text{eff}}$ (measured through quantities like left–right asymmetries), the top mass, and the Z mass and width. A first question we can ask is: given our *current* knowledge of these observables, which ones are the bottlenecks? In other words, which quantities are the most important ones to improve our knowledge of *first* if we want a better electroweak fit? The plots in Fig. 1 answer this question for the T and S parameters. The most effective way to improve the bound on the T parameter is to obtain a better measurement of the W boson mass. This can be done in Higgs factories operating at 240 GeV, where W pairs can be produced in abundance. It does not require Z pole physics. On the other hand, improving the bound on the S parameter demands better measurements of $\sin^2 \theta_{\text{eff}}$. For this, high luminosity on and around the Z pole is the preferred strategy. This is one motivation for operating future e^+e^- colliders near the Z pole: repeating much of the LEP physics program with higher precision can significantly improve our knowledge of the S parameter. (That said, I don't know of studies investigating how well we

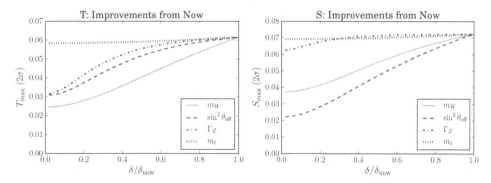

Fig. 1. Effects on the current T and S parameter constraints of reducing uncertainties on individual quantities m_W, $\sin^2\theta_{\text{eff}}$, Γ_Z, and m_t. In this plot δ includes both experimental and theoretical uncertainties. In the T plot, we have set $S = 0$ (performing a one-parameter fit), and vice versa.

could extract $\sin^2\theta_{\text{eff}}$ from observables at 240 GeV.) Figure 1 also illustrates that improvements saturate at some point. If we measure the W boson mass a factor of 5 or so better than we now know it — that is, to a precision of about 3 MeV, which CEPC, for instance, would accomplish — it is no longer the bottleneck in our knowledge of the T parameter, and we might then want to improve measurements of other quantities like Γ_Z and m_t as secondary priorities.

The leading observables that matter for probing S and T will be measured with significantly higher precision than we have now at *any* of the e^+e^- colliders under discussion. The ILC will measure[5] the W mass to 5 MeV and $\sin^2\theta_{\text{eff}}$ to 1.3×10^{-5}; CEPC will measure[3] the W mass to 3 MeV and $\sin^2\theta_{\text{eff}}$ to 2.3×10^{-5}; and the FCC-ee will (according to more conservative estimates[6]) measure the W mass to 1.2 MeV and $\sin^2\theta_{\text{eff}}$ to 0.3×10^{-5}. When the experimental uncertainties become particularly small, theory uncertainties matter a great deal as well, and we have included estimates of the remaining theory uncertainty after 3-loop calculations are performed.[7–10] (Such calculations will be a crucial task for theorists to complete in the coming years.)

The projected fits in the (S, T) plane for the various future experiments are shown in Fig. 2. The ILC and CEPC are projected to do comparably well, although they have slightly different strengths. FCC-ee is more ambitious in terms of its projected mass resolution and luminosity, with correspondingly better projected fits.

2.2. *Experimental choices*

Now that we have seen the estimates of how well the different colliders can do, it is useful to step back and ask which features of the colliders lead to these results, and which observables are the most important ones to optimize. We have seen that the first step is to produce more precise measurements of m_W and $\sin^2\theta_{\text{eff}}$. What next? To help answer this question, in Fig. 3 we have plotted the change in the S and T parameter constraints if we begin with the CEPC baseline design and then improve

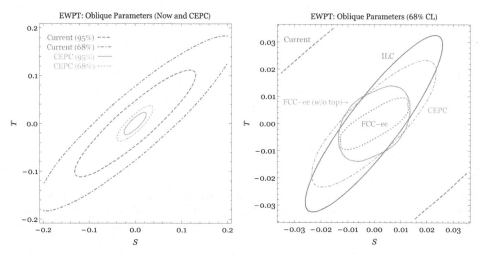

Fig. 2.   Precision that will be achieved for the S and T parameters at future colliders. At left: comparison of the current electroweak precision fit (artificially recentered at $S = T = 0$) with the expectations for CEPC, using projections from the pre-CDR.[3] At right: 68% contours for current data, CEPC, ILC, FCC-ee without running at the top threshold, and FCC-ee with running at the top threshold. (The latter two fits were referred to as "TLEP-W" and "TLEP-t" in our previous work.[1])

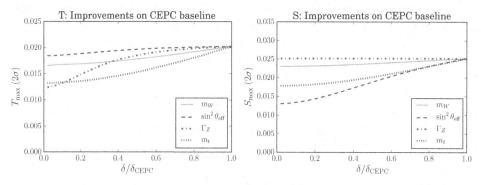

Fig. 3.   Like Fig. 1, except relative to the baseline CEPC measurements (as tabulated in Table 4.5 of the pre-CDR[3]) rather than to current measurements.

one measurement at a time. We see that once the CEPC baseline is achieved — which includes a measurement of m_W to 3 MeV accuracy — the most efficient way to improve the T parameter fit is to obtain a better measurement of the top quark mass. To improve the S parameter fit, we should either improve $\sin^2 \theta_{\text{eff}}$ (which, as discussed in the CEPC pre-CDR, is likely possible for at least a factor of 2 beyond the baseline design) or the top quark mass. Further improvements in m_W are of limited use and Γ_Z can be useful only with dramatic improvements (at which point, because it depends on a different linear combination of S and T than other observables, it can yield a better bound).

The CEPC collider design does not, at this time, include a plan to run at the top threshold. FCC-ee does, with corresponding improvements in precision for the T parameter. Top mass measurements at e^+e^- colliders can be much more precise than at the LHC, because the threshold can determine the top mass and width in the 1S scheme which is less subject to theoretical uncertainties than kinematic measurements of the mass.[11] One argument in favor of a linear collider is that it is easier to go to the top threshold and do precision measurements there — and to go to even higher energies and measure couplings of the top quark to heavy gauge bosons and the Higgs boson.

On the other hand, refining the Z mass and width measurements can also be a route to higher accuracy in the electroweak fits once the initial bottlenecks are overcome. Here the circular colliders have an advantage: they have a useful energy calibration based on resonant spin depolarization.[12] The idea is that polarized beams will precess in the magnetic field, so applying an orthogonal field at the right frequency can depolarize the electrons. Similar to NMR, this technique allows a very precise energy calibration.

Both CEPC and FCC-ee can take advantage of this precise energy calibration to measure m_Z and Γ_Z — and possibly other quantities like m_W — more accurately than the ILC. The FCC-ee electroweak fits we have presented in the past[1] largely come from projections in the Snowmass Electroweak Working Group studies[6] that are typically more conservative than those in the TLEP "First Look" report.[13] For instance, resonant spin depolarization might allow the W mass to be measured to 500 keV accuracy or better. (This could be a major motivation for accumulating luminosity on the WW threshold at $\sqrt{s} \approx 160$ GeV.) On the other hand, theoretical uncertainties might ultimately limit the utility of high experimental precision. To shed some light on this issue, we have plotted in Fig. 4 the (S, T) ellipses that would be achieved with purely statistical and systematic uncertainties — which are exquisitely precise, especially for the T-parameter — and the effect of also including theoretical systematics. The inputs to these fits are tabulated in Table 1. Depending on one's point of view, this is either an indication that aggressive estimates of the experimentally achievable precision are overly optimistic or that theorists will have to work harder to overcome significant obstacles to higher-precision calculations.

The FCC-ee project has continued to investigate the possibility of more high precision measurements. The experimental issues involved in resonant depolarization for energy calibration at FCC-ee were recently studied systematically.[16] Other continuing studies at FCC-ee include the measurement of top quark couplings[17] and of the value of α at the Z mass scale,[18] avoiding the need to directly understand hadronic contributions to the running. Such studies, beyond the preliminary estimates, are important to assess whether bottlenecks can be overcome to achieve higher precision results like the inner ellipses in Fig. 4.

Table 1. FCC-ee measurement uncertainties. These are drawn from a number of references.[5,6,9,10,13–15] Further explanation of the "TLEP-t" column, which is largely based on the Snowmass Electroweak Working Group report and used for the inner FCC-ee contour in Fig. 2, may be found in our earlier work.[1] The "stat." and "syst." columns draw more heavily from the TLEP "First Look" report,[13] and are used in Fig. 4.

	"TLEP-t"	Stat.	Stat. + Syst.	Theory
$\alpha_s(M_Z^2)$	$\pm 1.0 \times 10^{-4}$	$\pm 1.0 \times 10^{-4}$	—	—
$\Delta\alpha_{\text{had}}^{(5)}(M_Z^2)$	$\pm 4.7 \times 10^{-5}$	$\pm 4.7 \times 10^{-5}$	—	—
m_Z	± 100 keV	± 5 keV	± 100 keV	—
m_t	$(\pm 0.02_{\text{exp}} \pm 0.1_{\text{th}})$ GeV	± 10 MeV	± 14 MeV	± 100 MeV
m_h	± 0.1 GeV	± 0.1 GeV	—	—
m_W	$(\pm 1.2_{\text{exp}} \pm 1_{\text{th}})$ MeV	± 300 keV	± 500 keV	± 1 MeV
$\sin^2\theta_{\text{eff}}^{\ell}$	$(\pm 0.3_{\text{exp}} \pm 1.5_{\text{th}}) \times 10^{-5}$	$\pm 1.0 \times 10^{-6}$	$\pm 1.0 \times 10^{-6}$	$\pm 1.5 \times 10^{-5}$
Γ_Z	$(\pm 1_{\text{exp}} \pm 0.8_{\text{th}}) \times 10^{-4}$ GeV	± 8 keV	± 100 keV	± 80 keV

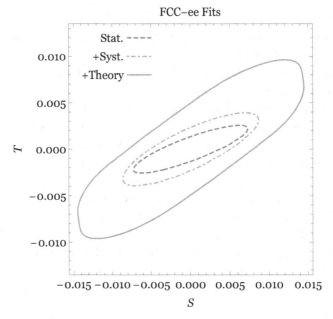

Fig. 4. (Color online) 68% CL contours in the $S - T$ plane for various estimates of the capabilities of FCC-ee. The dashed blue and dot-dashed orange contours are based on purely statistical uncertainties and statistical plus systematic uncertainties from the TLEP "First Look" report.[13] The green contour also adds in theoretical uncertainties on m_t, m_W, $\sin^2\theta_{\text{eff}}$, and Γ_Z as in our earlier work[1] and is almost indistinguishable from the inner FCC-ee curve shown in Fig. 2.

2.3. *Summary of the (S, T) fits*

Of course, we want to have the best measurements possible of many different quantities. But as a reasonable set of baselines that we should ask for from future experiments, we suggest:

- Measure m_W to better than 5 MeV. The current uncertainty is 15 MeV. All designs being discussed meet this standard.
- Measure $\sin^2 \theta_W$ to better than 2×10^{-5}. The current uncertainty is 16×10^{-5}. Again, all designs being discussed can deliver this.
- Measure m_Z and Γ_Z to 500 keV precision (currently 2 MeV). The future circular colliders would deliver this accuracy, but the ILC would not.
- Measure m_t to 100 MeV precision (currently somewhere around 0.8 GeV, with difficult-to-quantify theoretical uncertainties). The ILC and FCC-ee promise this accuracy, but CEPC does not.
- Have precise enough theory to make use of these results. This requires at least 3-loop calculations.

Each of the possible future electron–positron colliders delivers at least a substantial subset of this wish list (and some do far better for some quantities!), and would provide order-of-magnitude improvements on our current knowledge of electroweak precision. The circular and linear colliders would have interesting complementarity if both are constructed.

As one illustration of how to put the S and T parameter numbers in context, we show in Fig. 5 the translation of these bounds into constraints on stop soft masses. The electroweak fit — mostly the T-parameter — translates into a constraint

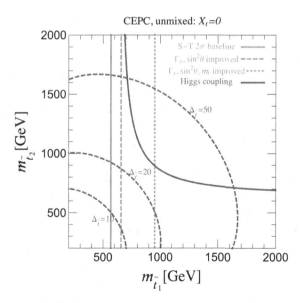

Fig. 5. (Color online) Probing stop masses with measurements at CEPC. The orange curves are the reach for the CEPC baseline electroweak precision fit and the extensions if Γ_Z and m_t measurements are improved as well. The purple curve comes from measurements of Higgs couplings to photons and gluons. The blue dashed lines indicate fine-tuning levels.

on the left-handed stop soft mass of around 700 GeV. This is comparable to already existing direct search bounds from the LHC, but more robust. Direct search bounds can be evaded if the neutralinos are heavy or if the stop decays in unusual ways. Future electron–positron colliders are nicely complementary, probing particles like the stop directly through their coupling to the Higgs boson and definitively closing loopholes. The blue dashed lines in Fig. 5 show that these measurements would constrain supersymmetry to be tuned down to at least the few percent level. There is an interesting "blind spot" for stops with a left–right mixing $X_t \approx \sqrt{m_{\tilde{t}_{\text{heavy}}}^2 - m_{\tilde{t}_{\text{light}}}^2}$, for which the lighter stop mass eigenstate decouples from the Higgs boson, but in this limit it also plays no role for naturalness so the basic conclusion is unchanged.[2,19] Furthermore, other precision observables like $b \to s\gamma$ limit the most natural parts of the blind spot parameter space.

3. Higgs Boson Measurements

3.1. *The central Higgs factory physics: Coupling measurements*

Of course, a major part of the precision physics program at a Higgs factory is measuring the Higgs couplings! This offers exciting opportunities to probe new physics, in addition to testing a portion of the Standard Model that even after the high-luminosity LHC we will still have only coarse information about. Unlike the case of electroweak precision, *all* of the e^+e^- colliders under consideration will pursue a very similar physics program at 240 GeV (so there are fewer contrasts to draw between different choices). For that reason I will focus less on comparing the options and more on explaining how they constrain various new physics scenarios.

The prospects for Higgs coupling measurements at future e^+e^- colliders have been extensively studied.[1,3,20–27] A summary of the precision that will be attained is shown in Fig. 6, which is extracted from the CEPC pre-CDR and offers a comparison of measurements at CEPC and at the ILC. The figure also shows that the high-luminosity LHC can complement a future lepton collider, for instance by obtaining a precise measurement of the ratio $\Gamma(h \to \gamma\gamma)/\Gamma(h \to ZZ^*)$. The summary is that all of the largest Higgs couplings will be measured to roughly percent-level accuracy. Because Higgs factories dominantly rely on the Higgsstrahlung process $e^+e^- \to Z^* \to Zh$, the coupling to the Z boson will be especially well-measured, particularly at circular colliders with their very large projected luminosity. Even the small coupling to photons will be measured to within a few percent accuracy (folding in knowledge from LHC), and the tiny coupling to muons to within about 10%. Any of the future e^+e^- colliders under consideration will take us from the LHC's fuzzy, out-of-focus picture of a "Standard Model-like" Higgs to a truly precise knowledge that the Higgs is (or, more excitingly, is not!) as predicted.

The coupling of the Z to the Higgs will be very precisely measured through the Higgsstrahlung cross-section itself. This single measurement is already a powerful

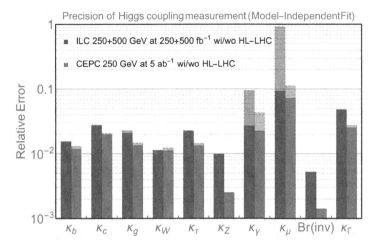

Fig. 6. Uncertainties on Higgs couplings from a 10-parameter fit at CEPC and ILC, lifted directly from Fig. 3.20 of the CEPC pre-CDR.[3] The bars with dashed edges include a combination with the high-luminosity LHC.

$$h \dashv\!\!\!\!\bullet\!\!\!\!\vdash h \;\Rightarrow\; \partial_\mu(h^\dagger h)\partial^\mu(h^\dagger h)$$

Fig. 7. Any new physics coupling to the Higgs boson can enter in a loop — here represented by an abstract blob — and produce a wave function renormalization effect through the dimension-six operator shown. This affects the rate for any Higgs-related process, including the Higgsstrahlung process that can be exquisitely measured.[28,29]

probe of new physics.[19,28–30] As shown in Fig. 7, any new physics coupling to the Higgs will lead to a wave function renormalization effect that will affect the overall rate of Higgs production. In theories like Twin Higgs models,[31] Folded Supersymmetry,[32] or other realizations of "neutral naturalness," the new physics coupling to the Higgs can be difficult to see directly. It is encouraging that even in the most pessimistic such scenario — new singlet scalars coupling to the Higgs portal — the Higgsstrahlung precision achievable at CEPC or FCC-ee can probe masses up to several hundred GeV and provide a very general test of fine-tuning.[29] This is a useful complement to more model-specific tests, which are often more precise when they are available. For instance, in the case of Folded Supersymmetry, the folded stops have the same electroweak quantum numbers as normal stops, and the strongest probe of these particular models at a future e^+e^- collider is likely to be the T-parameter.[2]

Another example of a scenario that can be tightly constrained by the Higgsstrahlung rate is the case of a composite Higgs.[33] In these models the Higgs is a pseudo-Nambu–Goldstone boson with decay constant f, and the potential has a form like

$$V(h) \sim \frac{a\lambda^2}{16\pi^2}\cos(h/f) + \frac{b\lambda^2}{16\pi^2}\sin^2(h/f)\,, \tag{4}$$

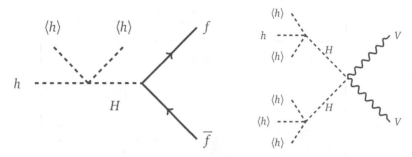

Fig. 8.   Tree-level sources of deviations in couplings of the light Higgs boson in a Two-Higgs Doublet Model. Fermion couplings have larger deviations $\sim m_Z^2/m_H^2$ whereas gauge boson couplings deviate only at order m_Z^4/m_H^4.

where λ is some spurion for shift-symmetry breaking. An obvious prediction is that $v \sim f$, so all such models are fine-tuned by adjusting $a = 2b(1 + \epsilon)$ to produce a small VEV $\langle h \rangle^2 \approx 2\epsilon f^2$. Composite Higgs models predict modified couplings

$$\frac{g_{VVh}}{g_{VVh}^{\text{SM}}} = \sqrt{1 - \frac{v^2}{f^2}}. \tag{5}$$

Thus, the Higgsstrahlung measurement has a very direct interpretation as a measure of fine-tuning. FCC-ee's projected 0.1% level measurement of the ZZh coupling would probe $f \sim 6$ TeV and tuning at the part-in-a-thousand level. This scenario is also constrained by Z pole observables, since we expect an S-parameter of parametric size Nv^2/f^2, but it is a case where the Higgs factory mode shines and sets a more stringent bound.

A very different scenario with tree-level modifications of the Higgs properties occurs in a type-II two-Higgs doublet model (2HDM). A useful way to see the leading deviations in this model is to work in a basis with one doublet h that has a VEV and a second doublet H with no VEV.[34] These are not mass eigenstates; rather, the size of the hH and h^3H terms is linked to ensure that there is no H tadpole. Because the light Higgs we have measured is approximately Standard Model-like, the misalignment between the mass and VEV bases must be small, and we can assess the modifications of the light Higgs couplings by drawing diagrams that mix h and H through VEV insertions, as shown in Fig. 8. We see that we pay a price of a propagator $1/m_H^2$ in the fermion couplings, but we pay two such factors in the couplings to gauge bosons.[35] As a result, a 2HDM model is not probed very effectively through Higgsstrahlung measurements. Rather, the main effect is to modify the fermion couplings, and in particular the largest partial width $\Gamma(h \to b\bar{b})$, through which all other Higgs branching ratios change as well. The roughly percent-level precision achievable on the $hb\bar{b}$ coupling translates into a probe of heavy Higgs boson masses up to beyond 1 TeV, comparable to the long-term reach at the LHC.[36–38]

Finally, we mention the case of stops, which when integrated out at one loop generate the Higgs couplings to photons and gluons in Eq. (3). The gluon coupling is the most sensitive measurement, which for percent-level accuracy and equal left- and right-handed stop masses translates into sensitivity to masses of about 1 TeV, as depicted in the purple curve of Fig. 5. This is comparable to the electroweak precision reach via the T-parameter, although the latter probes only the left-handed stops.

3.2. *Difficult Higgs couplings*

All of the e^+e^- colliders can measure the basic Higgs couplings to the W, Z, γ, b, c, g, τ, and μ, as well as the Higgs invisible width. But three other Higgs couplings are of interest and present much greater challenges: those to electrons, to top quarks, and to itself. The Higgs *branching ratio* to electrons is much too small to measure. On the other hand, the s-channel production of the Higgs, $e^+e^- \to h$, has a very small but not completely negligible rate.[39] This has led to interesting suggestions to build up very large luminosities (90 ab^{-1}) on the Higgs pole at FCC-ee to directly detect the coupling to electrons.[40] (I thank Oliver Fischer for bringing up this possibility at the conference.)

The Higgs self-coupling is one of the few remaining quantities in the Standard Model that we have no direct experimental information about. It is absolutely crucial that we measure it accurately in the future in order to learn about the shape of the Higgs potential and possible dynamics underlying electroweak symmetry breaking. The circular e^+e^- colliders will not be able to directly say anything, though interestingly they are sensitive to order-one deviations via their indirect loop effect on the Higgs cross-section.[41] If we want truly precise measurements, however, they will come from the circular colliders only when they operate as very high energy hadron colliders. The ILC, on the other hand, can measure the Higgs self-coupling through processes like $e^+e^- \to Zhh$ at $\sqrt{s} = 500$ GeV and $e^+e^- \to \nu\bar{\nu}hh$ at $\sqrt{s} = 1$ TeV, where a precision of 16% may be attained.[21,42,a] This again demonstrates that one of the key advantages of a linear collider over a circular collider comes from the higher energies and correspondingly wider range of physics processes that it can access. Along similar lines, we should note that the ILC would be able to directly measure the Higgs boson coupling to top quarks by operating at $\sqrt{s} \approx 500$ GeV (ideally slightly higher in energy) where the process $e^+e^- \to t\bar{t}h$ can be studied. This could give a $\sim 10\%$ direct measurement of the top Yukawa coupling,[21,43,44] which circular colliders can only indirectly constrain through its effect on the Higgs coupling to gluons.

3.3. *Exotic Higgs decays*

Most of the discussion of physics at e^+e^- colliders centers around precision Standard Model measurements, which can indirectly probe for new physics. The ILC,

---

[a]For a recent update see the talk of Masakazu Kurata at http://agenda.linearcollider.org/event/6662/session/30/; I thank Marcel Stanitzki for pointing me to this information.

by going to a center-of-mass energy of 1 TeV, has more direct discovery potential. But at a circular machine like FCC-ee or CEPC, opportunities for directly finding new particles are limited. A very exciting exception is the possibility of new light particles discovered in rare decays of the Higgs boson. Because the Higgs coupling to b-quarks is so small, there is ample room for small couplings to new physics to lead to detectable decay rates. A detailed recent survey of exotic Higgs decays discusses many different channels.[45] Possibilities include light pseudoscalars, dark photons, dark matter, and Hidden Valleys.[46]

To date there has been relatively little study of the possibilities for direct discoveries of exotic Higgs decays at future e^+e^- colliders. This will be a fruitful area to explore further. For now I just want to make one comment: these decays could involve exotic signals like particles that propagate macroscopic distances in the detector before decaying. It is important when designing detectors that opportunities to see exotic physics are not unnecessarily closed off — we want to optimize not only for Standard Model measurements, but for discovery potential. Because particles of any given lifetime have a probability distribution of decay locations, it is likely that detectors that can pick up displaced decays in the tracking volume or decays in an outer muon detector, taken together, can cover much of the parameter space, and that discoveries can be made regardless of the zeroth-order choices of detector technology (e.g. silicon trackers versus TPCs). Nonetheless, the possibilities for exotic signals should be kept in mind throughout the planning process.

4. Conclusions

As we have seen above, linear and circular colliders are to some extent complementary. If possible, it would be wonderful to have both. A linear collider has as a major advantage the capability to go to higher energies. It can improve electroweak fits by operating at the $t\bar{t}$ threshold and performing a precise top mass measurement. It could even operate at high enough energy to directly measure the top Yukawa coupling to the Higgs boson. This has great physical significance, as it is the leading interaction driving concerns about fine-tuning in the Standard Model. The linear colliders can also begin to probe the shape of the Higgs potential through self-coupling measurements. Finally, the extra discovery potential for electroweak particles is a major selling point: the ILC operating at $\sqrt{s} = 1$ TeV could discover particles with masses up to about 500 GeV, which in some cases goes well beyond the LHC's expected reach. Circular colliders have a smaller energy reach but an important advantage in energy calibration through resonant depolarization, which can be used to significantly improve our knowledge of the Z boson mass and width. The large luminosities that are proposed at circular colliders also allow extremely precise and powerful tests of the Higgs boson coupling to the Z. The most important argument in favor of circular machines is their capacity to be reused as high-energy hadron colliders with immense discovery potential. Given current uncertainties in the magnet technology that we will have in the future, it is important to build any future electron–positron collider with a large enough ring that we can be confident of its future as a hadron collider.

The LHC is a wonderful machine that will tell us a great deal about the existence or nonexistence of new colored particles at the TeV scale, but its abilities to probe particles with only electroweak interactions or even possibly colored particles that decay in ways that mimic backgrounds are limited. Higgs factories will exhaustively probe any new particles that interact with the Higgs, filling gaps that the LHC will leave behind. Electroweak precision tests are very complementary to this. For instance, the T-parameter could be the strongest constraint on the model of folded stops, providing one motivation for running at energies below $\sqrt{s} = 240$ GeV for higher accuracy.

My goal in this talk is not to advocate for any particular new collider or run plan, but to emphasize that different physics goals can be optimized by running at different energies or with different types of colliders. We must take all the options seriously. This is an exciting time for particle physics, as we lay the groundwork for future discoveries.

Acknowledgments

This work was supported in part by the NSF Grant PHY-1415548. It is partly based on work done in collaboration with JiJi Fan and Lian-Tao Wang and on portions of the CEPC pre-CDR to which Zhijun Liang and Jens Erler also made crucial contributions. I would like to thank the organizers at the Hong Kong Jockey Club Institute for Advanced Study for the invitation to speak, and the participants of the meeting for a productive and stimulating environment. I would also like to thank the organizers and participants of the earlier Workshop on Physics at the CEPC in Beijing in August 2015. I would particularly like to thank Alain Blondel, Oliver Fischer, Mike Koratzinos, Maxim Perelstein, Marcel Stanitzki, and Charlie Young for thought-provoking comments and questions at these meetings.

References

1. J. Fan, M. Reece and L.-T. Wang, *J. High Energy Phys.* **09**, 196 (2015), arXiv:1411.1054 [hep-ph].
2. J. Fan, M. Reece and L.-T. Wang, *J. High Energy Phys.* **08**, 152 (2015), arXiv:1412.3107 [hep-ph].
3. CEPC-SPPC Study Group Collab., CEPC-SPPC Preliminary Conceptual Design Report. 1. Physics and Detector, http://cepc.ihep.ac.cn/preCDR/volume.html.
4. J. Ellis and T. You, *J. High Energy Phys.* **03**, 089 (2016), arXiv:1510.04561 [hep-ph].
5. Gfitter Group Collab. (M. Baak, J. Cúth, J. Haller, A. Hoecker, R. Kogler, K. Mönig, M. Schott and J. Stelzer), *Eur. Phys. J. C* **74**, 3046 (2014), arXiv:1407.3792 [hep-ph].
6. M. Baak *et al.*, Working Group Report: Precision study of electroweak interactions, in *Community Summer Study 2013: Snowmass on the Mississippi (CSS2013)*, Minneapolis, MN, USA, 29 July–6 August 2013 (2013), arXiv:1310.6708 [hep-ph], http://www.slac.stanford.edu/econf/C1307292/docs/EnergyFrontier/Electroweak-19.pdf.

7. A. Freitas, W. Hollik, W. Walter and G. Weiglein, *Phys. Lett. B* **495**, 338 (2000), arXiv:hep-ph/0007091 [hep-ph].
8. M. Awramik and M. Czakon, *Phys. Rev. Lett.* **89**, 241801 (2002), arXiv:hep-ph/0208113 [hep-ph].
9. M. Awramik, M. Czakon and A. Freitas, *J. High Energy Phys.* **0611**, 048 (2006), arXiv:hep-ph/0608099 [hep-ph].
10. A. Freitas, *J. High Energy Phys.* **1404**, 070 (2014), arXiv:1401.2447 [hep-ph].
11. A. H. Hoang and M. Stahlhofen, *J. High Energy Phys.* **05**, 121 (2014), arXiv:1309.6323 [hep-ph].
12. SLD Electroweak Group, DELPHI, ALEPH, SLD, SLD Heavy Flavour Group, OPAL, LEP Electroweak Working Group, L3 Collabs. (S. Schael *et al.*), *Phys. Rep.* **427**, 257 (2006), arXiv:hep-ex/0509008 [hep-ex].
13. TLEP Design Study Working Group Collab. (M. Bicer *et al.*), *J. High Energy Phys.* **01**, 164 (2014), arXiv:1308.6176 [hep-ex].
14. G. P. Lepage, P. B. Mackenzie and M. E. Peskin, Expected precision of Higgs boson partial widths within the Standard Model, arXiv:1404.0319 [hep-ph].
15. M. Awramik, M. Czakon, A. Freitas and G. Weiglein, *Phys. Rev. D* **69**, 053006 (2004), arXiv:hep-ph/0311148 [hep-ph].
16. M. Koratzinos, A. Blondel, E. Gianfelice-Wendt and F. Zimmermann, FCC-ee: Energy calibration, in *Proc. 6th Int. Particle Accelerator Conference (IPAC 2015)*, Report No. TUPTY063 (2015), arXiv:1506.00933 [physics.acc-ph], http://inspirehep.net/record/1374263/files/arXiv:1506.00933.pdf.
17. P. Janot, *J. High Energy Phys.* **04**, 182 (2015), arXiv:1503.01325 [hep-ph].
18. P. Janot, *J. High Energy Phys.* **02**, 053 (2016), arXiv:1512.05544 [hep-ph].
19. N. Craig, M. Farina, M. McCullough and M. Perelstein, *J. High Energy Phys.* **03**, 146 (2015), arXiv:1411.0676 [hep-ph].
20. M. E. Peskin, Comparison of LHC and ILC capabilities for Higgs boson coupling measurements, arXiv:1207.2516 [hep-ph].
21. D. Asner *et al.*, ILC Higgs White Paper, arXiv:1310.0763 [hep-ph].
22. T. Han, Z. Liu and J. Sayre, *Phys. Rev. D* **89**, 113006 (2014), arXiv:1311.7155 [hep-ph].
23. S. Dawson *et al.*, Working Group Report: Higgs boson, in *Community Summer Study 2013: Snowmass on the Mississippi (CSS2013)*, Minneapolis, MN, USA, 29 July–6 August 2013 (2013), arXiv:1310.8361 [hep-ex], http://inspirehep.net/record/1262795/files/arXiv:1310.8361.pdf.
24. M. E. Peskin, Estimation of LHC and ILC capabilities for precision Higgs boson coupling measurements, arXiv:1312.4974 [hep-ph].
25. P. Bechtle, S. Heinemeyer, O. Stål, T. Stefaniak and G. Weiglein, Probing the Standard Model with Higgs signal rates from the Tevatron, the LHC and a future ILC, arXiv:1403.1582 [hep-ph].
26. T. Han, Z. Liu, Z. Qian and J. Sayre, *Phys. Rev. D* **91**, 113007 (2015), arXiv:1504.01399 [hep-ph].
27. N. Craig, J. Gu, Z. Liu and K. Wang, *J. High Energy Phys.* **03**, 050 (2016), arXiv:1512.06877 [hep-ph].
28. C. Englert and M. McCullough, *J. High Energy Phys.* **07**, 168 (2013), arXiv:1303.1526 [hep-ph].
29. N. Craig, C. Englert and M. McCullough, *Phys. Rev. Lett.* **111**, 121803 (2013), arXiv:1305.5251 [hep-ph].
30. B. Henning, X. Lu and H. Murayama, What do precision Higgs measurements buy us?, arXiv:1404.1058 [hep-ph].

31. Z. Chacko, H.-S. Goh and R. Harnik, *Phys. Rev. Lett.* **96**, 231802 (2006), arXiv:hep-ph/0506256 [hep-ph].

32. G. Burdman, Z. Chacko, H.-S. Goh and R. Harnik, *J. High Energy Phys.* **02**, 009 (2007), arXiv:hep-ph/0609152 [hep-ph].

33. R. Contino, The Higgs as a composite Nambu-Goldstone boson, in *Physics of the Large and the Small, TASI 09, Proc. Theoretical Advanced Study Institute in Elementary Particle Physics*, Boulder, Colorado, USA, 1–26 June 2009 (2011), pp. 235–306, arXiv:1005.4269 [hep-ph], http://inspirehep.net/record/856065/files/arXiv:1005.4269.pdf.

34. R. S. Gupta, M. Montull and F. Riva, *J. High Energy Phys.* **04**, 132 (2013), arXiv:1212.5240 [hep-ph].

35. J. F. Gunion and H. E. Haber, *Phys. Rev. D* **67**, 075019 (2003), arXiv:hep-ph/0207010 [hep-ph].

36. A. Djouadi, L. Maiani, A. Polosa, J. Quevillon and V. Riquer, *J. High Energy Phys.* **06**, 168 (2015), arXiv:1502.05653 [hep-ph].

37. N. Craig, F. D'Eramo, P. Draper, S. Thomas and H. Zhang, *J. High Energy Phys.* **06**, 137 (2015), arXiv:1504.04630 [hep-ph].

38. J. Hajer, Y.-Y. Li, T. Liu and J. F. H. Shiu, *J. High Energy Phys.* **11**, 124 (2015), arXiv:1504.07617 [hep-ph].

39. S. Jadach and R. A. Kycia, *Phys. Lett. B* **755**, 58 (2016), arXiv:1509.02406 [hep-ph].

40. D. d'Enterria, Physics at the FCC-ee, in *17th Lomonosov Conf. on Elementary Particle Physics*, Moscow, Russia, 20–26 August 2015 (2016), arXiv:1602.05043 [hep-ex], http://inspirehep.net/record/1421932/files/arXiv:1602.05043.pdf.

41. M. McCullough, *Phys. Rev. D* **90**, 015001 (2014), arXiv:1312.3322 [hep-ph] [Erratum: *ibid.* **92**, 039903 (2015)].

42. H. Abramowicz *et al.*, The International Linear Collider Technical Design Report — Vol. 4: Detectors, arXiv:1306.6329 [physics.ins-det].

43. R. Yonamine, K. Ikematsu, T. Tanabe, K. Fujii, Y. Kiyo, Y. Sumino and H. Yokoya, *Phys. Rev. D* **84**, 014033 (2011), arXiv:1104.5132 [hep-ph].

44. H. Tabassam and V. Martin, Top Higgs Yukawa coupling analysis from $e^+e^- \to ttH \to bWbWbb$, in *Int. Workshop on Future Linear Colliders (LCWS11)*, Granada, Spain, 26–30 September 2011 (2012), arXiv:1202.6013 [hep-ex], http://inspirehep.net/record/1090730/files/arXiv:1202.6013.pdf.

45. D. Curtin *et al.*, *Phys. Rev. D* **90**, 075004 (2014), arXiv:1312.4992 [hep-ph].

46. M. J. Strassler and K. M. Zurek, *Phys. Lett. B* **651**, 374 (2007), arXiv:hep-ph/0604261.

Testing Higgs Coupling Precision and New Physics Scales at Lepton Colliders

Shao-Feng Ge*

Max-Planck-Institut für Kernphysik, Heidelberg, Germany
gesf02@gmail.com

Hong-Jian He

Institute of Modern Physics and Center for High Energy Physics,
Tsinghua University, Beijing 100084, P. R. China
Center for High Energy Physics, Peking University, Beijing 100871, P. R. China
hjhe@tsinghua.edu.cn

Rui-Qing Xiao

Institute of Modern Physics and Center for High Energy Physics,
Tsinghua University, Beijing 100084, P. R. China
Lawrence Berkeley National Laboratory, Berkeley, California 94720, USA
ruiqingxiao@lbl.gov

The next-generation lepton colliders, such as CEPC, FCC-ee, and ILC will make precision measurement of the Higgs boson properties. We first extract the Higgs coupling precision from Higgs observables at CEPC to illustrate the potential of future lepton colliders. Depending on the related event rates, the precision can reach percentage level for most couplings. Then, we try to estimate the new physics scales that can be indirectly probed with Higgs and electroweak precision observables. The Higgs observables, together with the existing electroweak precision observables, can probe new physics up to 10 TeV (40 TeV for the gluon-related operator $\mathcal{O}_g$) at 95% C.L. Including the Z/W mass measurements and Z-pole observables at CEPC further pushes the limit up to 35 TeV. Although Z-pole running is originally for the purpose of machine calibration, it can be as important as the Higgs observables for probing the new physics scales indirectly. The indirect probe of new physics scales at lepton colliders can mainly cover the energy range to be explored by the following hadron colliders of pp (50–100 TeV), such as SPPC and FCC-hh.

Keywords: Lepton collider; dimension-6 operator; collider phenomenology.

*Presenter of the talk *"New Physics Scales to be Probed at Lepton Colliders"* at the IAS Program on High Energy Physics on January 11, 2016.

1. Introduction

With the discovery of Higgs boson[1] at LHC,[2] the spectrum of the SM has been completed. This culminates the searches that lasted for decades.[3] Nevertheless, there are already many motivations for making precision measurement of the Higgs coupling and testing the new physics beyond the SM.

The discovery of Higgs boson is usually claimed as a big success of particle physics and completion of the SM. Nevertheless, the SM not only requires the existence of Higgs boson, but also dictates its interactions. To claim the SM is fully tested, it is necessary to measure all interactions that the Higgs boson participates. Although the spectrum of the SM has already been completed, the SM itself still requires further experimental test. In this sense, the name "particle physics" puts too much emphasis on particle and we shall not get confused to forget the interactions between them. Even within the SM, we are well motivated to test the properties of the Higgs boson.

The existence of such a 125 GeV scalar is truly profound. It can provide masses to massive gauge bosons and fermions. Its coupling constants with the SM fermions and gauge bosons are constrained to be within 10%–20% deviation from the SM predictions.[4] In this sense, we now understand how the mechanism of mass generation can happen with a single vacuum expectation value (VEV). Nevertheless, how the Higgs acquires nonzero VEV has deep connection with vacuum stability and Higgs inflation[5] but has not been tested experimentally yet.[6] Especially, in the SM the Higgs mass receives quadratic divergence from loop corrections and hence becomes radiatively unnatural if the SM is valid up to very high energy scale. The Higgs mass is radiatively unnatural. In addition, the Yukawa couplings span several orders of magnitude and hence is hierarchically unnatural. A satisfactory model shall make this hierarchical unnaturalness in Yukawa couplings understandable.[a]

The next-generation lepton colliders, including CEPC,[8] FCC-ee,[9] and ILC,[10] are designed for making precision measurement of the Higgs properties. All three candidate machines can run at $\sqrt{s} = 250$ GeV as Higgs factory by producing Higgs boson through Higgsstrahlung $e^+e^- \to Zh$ and WW fusion $e^+e^- \to \nu\bar{\nu}h$ processes. The CEPC with 5 ab^{-1} of integrated luminosity can roughly produce 10^6 Higgs bosons. From a naive estimation, the statistical fluctuation can reach $\mathcal{O}(0.1\%)$ level for inclusive observables. Considering the fact that the Higgs can decay via various channels, the precision on its coupling can typically reach $\mathcal{O}(1\%)$ level.

In this talk, we first summarize in Sec. 2 the Higgs coupling precision that can be reached at CEPC based on the assumption that the Higgs production and

---

[a]Related to fermion masses, the quark and neutrino mixings have been experimentally established but we are still not sure how to explain. Is the mixing pattern just coincidence or consequence of some flavor symmetry? If there is any flavor symmetry dictating the mixing pattern, what is it? Is the flavor symmetry discrete or continuous? Especially, we shall keep in mind that flavor symmetry has to be broken and the mixing pattern can be determined by residual symmetry that can survive the electroweak symmetry breaking rather than the full flavor symmetry imposed on the fundamental Lagrangian.[7]

Table 1.   (Color online) The estimated 1σ precision of Higgs observables at CEPC.[8,13] The quantities in black box are experimental observables that can be directly measured while the quantities labeled in red color are the inputs to our χ^2 fit analysis.

ΔM_h	Γ_h	$\sigma(Zh)$	$\sigma(\nu\bar{\nu}h) \times \mathrm{Br}(h \to bb)$
2.6 MeV	2.8%	0.5%	2.8%
	Decay mode	$\sigma(Zh) \times \mathrm{Br}$	Br
	$h \to bb$	0.21%	0.54%
	$h \to cc$	2.5%	2.5%
	$h \to gg$	1.7%	1.8%
	$h \to \tau\tau$	1.2%	1.3%
	$h \to WW$	1.4%	1.5%
	$h \to ZZ$	4.3%	4.3%
	$h \to \gamma\gamma$	9.0%	9.0%
	$h \to \mu\mu$	17%	17%
	$h \to$ invisible	—	0.14%

decay processes can be described by rescaling the SM predictions. We then use the Higgs observables (including both production and decay rates of the Higgs boson), M_Z/M_W mass measurements, and Z-pole observables to estimate the new physics scales via dimension-6 operators in Sec. 3. Finally, our conclusion can be found in Sec. 4. Interested readers can check our full paper[11] for details.

2. Higgs Coupling Precision

Of the 10^6 Higgs bosons, most of them are produced through the Higgsstrahlung process, $e^+e^- \to Zh$. Since there are only two particles in the final state and the initial state is well defined, the Higgs boson can be reconstructed from the Z boson, $p_h = -p_Z$. This is the so-called *recoil mass reconstruction technique*[12] which allows inclusive measurement of the Higgsstrahlung cross-section $\sigma(Zh)$. Among the selected events, the Higgs decay rate $\sigma(Zh) \times \mathrm{Br}(h \to ii)$ of various channels can be measured independently. The ratio of the decay and production rates is the corresponding decay branching ratio $\mathrm{Br}(h \to ii)$. In this way, the Higgs decay branching ratios can be measured in a model-independent way at lepton colliders. For the WW fusion process, $e^+e^- \to \nu\bar{\nu}h$, only the $h \to bb$ channel has large enough rate $\sigma(\nu\bar{\nu}h) \times \mathrm{Br}(h \to bb)$. Together with the decay branching ratio $\mathrm{Br}(h \to bb)$ inferred from Higgsstrahlung, the cross-section $\sigma(\nu\bar{\nu}h)$ can also be extracted as the ratio between the directly measured $\sigma(\nu\bar{\nu}h) \times \mathrm{Br}(h \to bb)$ and the already determined branching ratio $\mathrm{Br}(h \to bb)$. In Table 1, we summarize the direct Higgs observables in black boxes and label the inputs to χ^2 fit in red color.

To extract the precision on the Higgs coupling with the SM particles, we rescale the SM prediction, $g_{hii}/g_{hii}^{\mathrm{sm}} \equiv \kappa_i$, and parametrize the deviation from the SM as

$\delta\kappa_i \equiv \kappa_i - 1$. The Higgsstrahlung and WW fusion cross-sections are then modulated by the Higgs couplings with Z and W bosons,

$$\frac{\delta\sigma(Zh)}{\sigma(Zh)} = \kappa_Z^2 - 1 \simeq 2\delta\kappa_Z, \qquad \frac{\delta\sigma(\nu\bar\nu h)}{\sigma(\nu\bar\nu h)} = \kappa_W^2 - 1 \simeq 2\delta\kappa_W, \tag{1}$$

respectively. Similarly, the decay widths are also modulated by the corresponding rescaling κ factors,

$$\frac{\delta\Gamma_{hii}}{\Gamma_{hii}^{\rm sm}} = \kappa_i^2 - 1 \simeq 2\delta\kappa_i, \qquad \frac{\Gamma_{\rm inv}}{\Gamma_{\rm tot}^{\rm sm}} = {\rm Br}(h \to \text{invisibles}) \equiv \delta\kappa_{\rm inv}. \tag{2}$$

Since no invisible decay mode is present in the SM, its contribution is parametrized directly as $\delta\kappa_{\rm inv}$ which vanishes when the SM is recovered. In contrast, for those channels already existing in the SM, the deviation $\delta\kappa_i$ is the difference from 1. The total decay width is then the sum over all decay channels, $\Gamma_{\rm tot}^{\rm sm} \equiv \sum_i \Gamma_{hii}^{\rm sm}$.

To fit the quantities labeled as red color in Table 1, the rescaled decay widths need to be expressed as decay branching ratios,

$${\rm Br}_i^{\rm th} \simeq {\rm Br}_i^{\rm sm} \left[1 + (1 - {\rm Br}_i^{\rm sm}) \frac{\delta\Gamma_i}{\Gamma_i} - \sum_{j\neq i} {\rm Br}_j^{\rm sm} \frac{\delta\Gamma_j}{\Gamma_j} \right], \tag{3}$$

where ${\rm Br}(h \to ii) \equiv \Gamma_{hii}/\Gamma_{\rm tot}$ and $\Gamma_{\rm tot} \equiv \sum_i \Gamma_{hii}$. Note that the rescaling is no longer an overall factor. For the parametrization of both channels existing in the SM and the invisible channel in (2), the theoretical prediction can be expanded as,

$${\rm Br}_i^{\rm th} \simeq {\rm Br}_i^{\rm sm} \left(1 + \sum_j A_{ij}\delta\kappa_j \right), \qquad {\rm Br}_{\rm inv}^{\rm th} \simeq \delta\kappa_{\rm inv}, \tag{4}$$

where the coefficient matrix A has elements,

$$A_{ij} = 2(\delta_{ij} - {\rm Br}_j^{\rm sm}), \qquad A_{i,{\rm inv}} = -1, \qquad A_{{\rm inv},i} = 0, \qquad A_{{\rm inv},{\rm inv}} = 1. \tag{5}$$

In the decay branching ratios, all decay channels can affect each other. For the channel with larger branching ratio ${\rm Br}_i^{\rm sm}$ in the SM, it can have larger effect ($A_{ji} = -{\rm Br}_i^{\rm sm}$) on the others but smaller effect ($A_{ii} = 1 - {\rm Br}_i^{\rm sm}$) on itself. The only exception is the invisible channel which has effect of equal size ($A_{{\rm inv},{\rm inv}} = 1$ for its own and $A_{i,{\rm inv}} = -1$ for the others) on all channels. Note that the invisible channel can be affected only by itself.

As summarized in Table 1, nine decay channels from two production modes can achieve reasonable precision. To keep the fit as general as possible when estimating the precision on measuring the deviation of Higgs couplings from the SM prediction, nine scaling κ_i ($i = b, c, g, \tau, W, Z, \gamma, \mu, {\rm inv}$) are introduced for the nine decay channels, respectively. Note that in the SM, the Higgs decay into a pair of photons or gluons is induced by triangle loops with fermion or W boson that can directly couple to Higgs and hence is not fully independent. Nevertheless, independent scaling factors κ_γ and κ_g are assigned to the $h \to \gamma\gamma$ and $h \to gg$ decay widths for

Table 2. (Color online) The 1σ precisions on measuring Higgs couplings at CEPC (250 GeV, 5 ab^{-1}), in comparison with LHC (14 TeV, 300 fb^{-1}), HL-LHC (14 TeV, 3 ab^{-1}) and ILC (250 GeV, 250 fb^{-1}) + (500 GeV, 500 fb^{-1}). The numbers for LHC, HL-LHC, and ILC are obtained from Ref. 16.

Precision (%)	CEPC		LHC	HL-LHC	ILC-250+500
κ_Z	0.249	0.249	8.5	6.3	0.50
κ_W	1.21	1.21	5.4	3.3	0.46
κ_γ	4.67	4.67	9.0	6.5	8.6
κ_g	1.55	1.55	6.9	4.8	2.0
κ_b	1.28	1.28	14.9	8.5	0.97
κ_c	1.76	1.76	—	—	2.6
κ_τ	1.39	1.39	9.5	6.5	2.0
κ_μ	—	8.59	—	—	—
Br$_{\text{inv}}$	0.135	0.135	8.0	4.0	0.52
Γ_h	2.8	2.8	—	—	—

generality. The 9-parameter fit is based on SM and can also accommodate new physics contributions.

Using the technique of analytical χ^2 fit[11,14] delivered in the BSMfitter package,[15] we estimate the precision on Higgs couplings and summarize the results in Table 2.

- Roughly speaking, the uncertainty is mainly determined by statistical fluctuations and hence the SM prediction of decay branching ratios Br$_i^{\text{sm}}$.
- Nevertheless, the precision on κ_Z is much better than the precision on κ_W although the Higgs decay $h \rightarrow WW$ (Br$_{h \rightarrow WW}^{\text{sm}} = 22.5\%$) has larger branching ratio than $h \rightarrow ZZ$ (Br$_{h \rightarrow ZZ}^{\text{sm}} = 2.77\%$). The additional constraint comes from the Higgsstrahlung cross-section $\sigma(Zh)$ which is an inclusive observable and hence has the largest event rate. Similarly, κ_W also receives constraint from both decay branching ratios and the WW fusion cross-section $\sigma(\nu\bar{\nu}h)$.
- Apart from these, the others are only constrained by decay branching ratios. Of all rescaling factors, κ_μ has the worst precision since $h \rightarrow \mu\mu$ (Br$_{h \rightarrow \mu\mu}^{\text{sm}} = 0.023\%$) has the least number of events and hence the worst statistics. For comparison, we show two fits with or without κ_μ and find that all other numbers are not affect. This demonstrates what we argued that a channel $h \rightarrow ii$ can affect other channels with weight Br$_i^{\text{sm}}$. For $h \rightarrow \mu\mu$, its branching ratio Br$_{h \rightarrow \mu\mu}^{\text{sm}} \simeq 0.023\%$ is negligibly small.
- Although the branching ratio of $h \rightarrow \gamma\gamma$ is only around 1% of $h \rightarrow WW$ and hence has 10 times larger uncertainty from naive estimation of statistical fluctuation, the precision on κ_γ is not that worse than κ_W. This is because that photon can be measured much better than the W boson. The photon can be probed directly but the W boson needs to first decay.

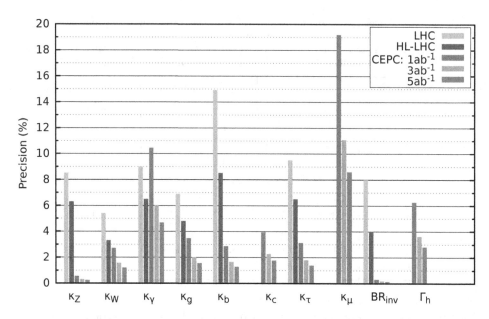

Fig. 1. (Color online) The 1σ precision on the Higgs couplings at CEPC (250 GeV) with integrated luminosity $(1, 3, 5)$ ab^{-1}, respectively, in comparison with the results at LHC (14 TeV, 300 fb^{-1}) and HL-LHC (14 TeV, 3 ab^{-1}).

- Note that the invisible decay has the best uncertainty as shown in Table 1, rendering Br$_{\text{inv}} \simeq \delta\kappa_{\text{inv}}$ to be mainly determined by this single channel $h \to$ invisible. We can see that the fitted precision 0.135% in Table 1 is roughly the same number as the original value 0.14% in Table 2 since the coefficient $A_{\text{inv,inv}}$ in (5) is 1.
- Finally, the combined precision on the total decay width Γ_h is simply the value directly from detector simulation in Table 1. This is because Γ_h is an independent variable and not entangled with the rescaling factors κ_i.

For comparison, we also show in Table 2 the precision at LHC, HL-LHC, and ILC.[16] At hadron colliders like LHC and HL-LHC, not all channels can be measured and the precision is usually much worse than at lepton colliders. To make better presentation, the results are depicted in Fig. 1. The reachable precision at CEPC can receive significant improvement from the expected measurement at LHC and HL-LHC. The largest difference appears in κ_Z and Br$_{\text{inv}}$ with improvement more than one order of magnitude. In addition, hadron colliders cannot measure the Higgs decay width due to large energy uncertainty. For ILC, it can make similar measurements as CEPC. Nevertheless, for most channels the precision is not as good as at CEPC due to smaller integrated luminosity. There are two exceptions, one is κ_W and the other is κ_b. For κ_W, the enhancement comes from the running at $\sqrt{s} = 500$ GeV where the WW production of Higgs can be significantly enhanced to introduce better constraint on $\sigma(\nu\bar{\nu}h) = \kappa_W^2 \sigma^{\text{sm}}(\nu\bar{\nu}h)$.

Table 3.   (Color online) The CP-even dimension-6 operators related to Higgs and electroweak precision observables to be measured at lepton colliders.

Higgs	EW gauge bosons	Fermions
$\mathcal{O}_H = \frac{1}{2}(\partial_\mu|\mathrm{H}|^2)^2$	$\mathcal{O}_{WW} = g^2|\mathrm{H}|^2 W_{\mu\nu}^a W^{a\mu\nu}$	$\mathcal{O}_L^{(3)} = (i\mathrm{H}^\dagger \sigma^a \overleftrightarrow{D}_\mu \mathrm{H})(\bar\Psi_L \gamma^\mu \sigma^a \Psi_L)$
$\mathcal{O}_T = \frac{1}{2}(\mathrm{H}^\dagger \overleftrightarrow{D}_\mu \mathrm{H})^2$	$\mathcal{O}_{BB} = g^2|\mathrm{H}|^2 B_{\mu\nu} B^{\mu\nu}$	$\mathcal{O}_{LL}^{(3)} = (\bar\Psi_L \gamma_\mu \sigma^a \Psi_L)(\bar\Psi_L \gamma^\mu \sigma^a \Psi_L)$
	$\mathcal{O}_{WB} = gg'\mathrm{H}^\dagger \sigma^a \mathrm{H} W_{\mu\nu}^a B^{\mu\nu}$	$\mathcal{O}_L = (i\mathrm{H}^\dagger \overleftrightarrow{D}_\mu \mathrm{H})(\bar\Psi_L \gamma^\mu \Psi_L)$
Gluon	$\mathcal{O}_{HW} = ig(D^\mu \mathrm{H})^\dagger \sigma^a (D^\nu \mathrm{H}) W_{\mu\nu}^a$	$\mathcal{O}_R = (i\mathrm{H}^\dagger \overleftrightarrow{D}_\mu \mathrm{H})(\bar\psi_R \gamma^\mu \psi_R)$
$\mathcal{O}_g = g_s^2|\mathrm{H}|^2 G_{\mu\nu}^a G^{a\mu\nu}$	$\mathcal{O}_{HB} = ig'(D^\mu \mathrm{H})^\dagger (D^\nu \mathrm{H}) B_{\mu\nu}$	

3. New Physics Scales

The purpose of building lepton colliders is not just for precision measurement of the Higgs couplings. Our final goal is new physics beyond the SM. Nevertheless, the energy of lepton colliders is probably not enough to produce the new particles which are usually expected to be heavy. Since the Higgs boson is newly discovered and other SM particles have already been precisely measured to some extent, if there is any new physics, it has large chance to appear in the Higgs couplings. It has large chance for new physics to be measured at lepton colliders indirectly.

The effect of new physics from higher energy can be parametrized in terms of dimension-6 operators. On the basis of the SM, the Lagrangian is extended by adding effective operators,

$$\mathcal{L} = \mathcal{L}_{\mathrm{SM}} + \sum_{ij} \frac{y_{ij} \sim \mathcal{O}(1)}{\Lambda \sim 10^{14}\mathrm{GeV}} \left(\bar{L}_i \tilde{H}\right)\left(\tilde{H}^\dagger L_j\right) + \sum_i \frac{c_i}{\Lambda^2}\mathcal{O}_i, \qquad (6)$$

where $\tilde{H} \equiv \tau_2 H^*$ is the CP conjugate of the Higgs doublet H. For completeness, we also show the dimension-5 operators which can provide neutrino mass matrix. For $y_{ij} \sim \mathcal{O}(1)$, the tiny neutrino masses can be produced if the cutoff $\Lambda \sim 10^{14}$ GeV is at the GUT scale. Note that the dimension-5 operators have lower suppression than the dimension-6 operators and hence are expected to have larger effect on low-energy physics. This is probably the reason why we have already measured neutrino mass and mixing in neutrino oscillation experiments but have not seen the dimension-6 operators.

Since the Higgs couplings are the thing to be measured at lepton colliders, we list all related CP-even operators in Table 3. All operators involve the Higgs doublet H with the only exception of $\mathcal{O}_{LL}^{(3)}$ which is a four-fermion operator and only affects the Fermi constant G_F. In this set, all operators are independent by removing redundant operators like $\mathcal{O}_W$ and $\mathcal{O}_B$. For simplicity, we will not elaborate the details of how these operators can modify the SM prediction of Higgs and electroweak precision observables[11] and only focus on the physical consequences.

In Table 4, we summarize the existing electroweak precision observables[17] and the Higgs observables at CEPC.[8] These observables are used to constraint the size

Table 4. (Color online) Inputs used to constrain the new physics scales of dimension-6 operators. The electroweak precision observables in the first four rows are taken from PDG2014,[17] and the 1σ precisions of Higgs measurements are taken from the CEPC detector simulations.[8] For the WW fusion cross-section $\sigma[\nu\bar{\nu}h]_{350\,\text{GeV}}$ at $\sqrt{s} = 350$ GeV, we adopt the TLEP estimate of its uncertainty[9] as an illustration.

Observables	Central value	Relative error	SM prediction
α	$7.2973525698 \times 10^{-3}$	3.29×10^{-10}	—
G_F	1.1663787×10^{-5} GeV^{-2}	5.14×10^{-7}	—
M_Z	91.1876 GeV	2.3×10^{-5}	—
M_W	80.385 GeV	1.87×10^{-4}	—
$\sigma[Zh]$	—	0.51%	—
$\sigma[\nu\bar{\nu}h]$	—	2.86%	—
$\sigma[\nu\bar{\nu}h]_{350\,\text{GeV}}$	—	0.75%	—
Br[WW]	—	1.6%	22.5%
Br[ZZ]	—	4.3%	2.77%
Br[bb]	—	0.57%	58.1%
Br[cc]	—	2.3%	2.10%
Br[gg]	—	1.7%	7.40%
Br[$\tau\tau$]	—	1.3%	6.64%
Br[$\gamma\gamma$]	—	9.0%	0.243%
Br[$\mu\mu$]	—	17%	0.023%

of dimension-6 operators with the following χ^2 function,

$$\chi^2\left(\delta\alpha, \delta G_F, \delta M_Z, \frac{c_i}{\Lambda^2}\right) = \sum_j \left[\frac{\mathcal{O}_j^{\text{th}}\left(\delta\alpha, \delta G_F, \delta M_Z, \frac{c_i}{\Lambda^2}\right) - \mathcal{O}_j^{\text{exp}}}{\Delta\mathcal{O}_j}\right]^2, \quad (7)$$

where $(\delta\alpha, \delta G_F, \delta M_Z)$ are shifts of the corresponding parameter (α, G_F, M_Z) from their reference values,

$$\alpha^{(\text{sm})} = \alpha^{(r)}\left(1 + \frac{\delta\alpha}{\alpha}\right), \quad (8a)$$

$$G_F^{(\text{sm})} = G_F^{(r)}\left(1 + \frac{\delta G_F}{G_F}\right), \quad (8b)$$

$$M_Z^{(\text{sm})} = M_Z^{(r)}\left(1 + \frac{\delta M_Z}{M_Z}\right). \quad (8c)$$

When doing χ^2 fit, both the dimension-6 operator coefficient c_j and parameter shifts $(\delta\alpha, \delta G_F, \delta M_Z)$ are treated as fitting parameters on the equal footing. For convenience, we take the reference values to be at the experimental central values, $\alpha^{(r)} = 7.2973525698 \times 10^{-3}$, $G_F^{(r)} = 1.1663787 \times 10^{-5}$ GeV^{-2}, and $M_Z^{(r)} = 91.1876$ GeV, while keeping the shifts as small deviations.

This choice of fitting parameters is different from the conventional Z-scheme by fixing the input parameters (α, G_F, M_Z) to their central values or the W-scheme by fixing (α, M_Z, M_W) instead. In these scheme-dependent approaches, only the central values of the input parameters are utilized while their uncertainties are simply discarded. The scheme-dependent approach is practically good enough if the input parameters are much more precise than the other observables. This is the case for Z-scheme with current electroweak precision measurements. As shown in Table 4, the relative errors of (α, G_F, M_Z) are negligibly small, at least one order of magnitude better than M_W. Nevertheless, the situation changes when the uncertainty on M_W is significantly improved to be comparable with the uncertainty on M_Z. This is exactly the case for CEPC.

3.1. *Sensitivity reach from Higgs observables*

Using all the observables in Table 4 (except $\sigma[\nu\bar{\nu}h]_{350\,\text{GeV}}$ which is taken from the TLEP estimation) we first estimate the effect of Higgs observables on probing new physics indirectly and show the results in Fig. 2. The 95% limit (blue) indicates the exclusion sensitivity while the 5σ value (red) is the sensitivity for discovery. The following discussions focus on the 95% limit for simplicity.

- We can see that the new physics scales for $(\mathcal{O}_T, \mathcal{O}_L^{(3)}, \mathcal{O}_L)$ can be probed up to as high as around 10 TeV. Half of the operators $(\mathcal{O}_H, \mathcal{O}_{WW}, \mathcal{O}_{BB}, \mathcal{O}_{WB}, \mathcal{O}_{HW}, \mathcal{O}_{LL}^{(3)}, \mathcal{O}_R, \mathcal{O}_{Lq}^{(3)})$ can be probed up to the scales 2–10 TeV which is already beyond the effective scale that can be probed at LHC. The remaining $(\mathcal{O}_{HB}, \mathcal{O}_{Lq}, \mathcal{O}_{Ru}, \mathcal{O}_{Rd})$ can only be probed below 1 TeV.

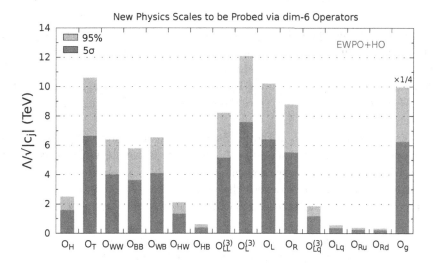

Fig. 2.   (Color online) The 95% exclusion limits (blue) and 5σ discovery sensitivities (red) to the new physics scales $\Lambda/\sqrt{|c_j|}$ by combining the current electroweak precision observables (α, G_F, M_Z, M_W)[17] and the future Higgs observables (Table 4) at the Higgs factory CEPC (250 GeV)[8] with a projected luminosity of 5 ab^{-1}. In the last column for $\mathcal{O}_g$, we have rescaled its height by a factor 1/4 to fit the plot, so its actual reach is $\Lambda/\sqrt{|c_g|} = 39.8$ TeV.

Table 5. (Color online) Impacts of adding the current electroweak precision observables (α, G_F, M_Z, M_W)[17] on probing the new physics scales $\Lambda/\sqrt{|c_j|}$ (in TeV) at 95% C.L. The limits in the first row are obtained from $\sigma(Zh)$ to be measured at the CEPC[8] only. The limits in the second row are given by combining with the current M_W measurement plus $\sigma(Zh)$. Finally, the third row presents the limits by including the current measurements of (α, G_F, M_Z) altogether. In the first two rows, (α, G_F, M_Z) are fixed to their experimental central values as in the Z-scheme, while the third row adopts the scheme-independent approach by allowing all electroweak parameters to freely vary in each fit. We label the entries of most significant improvements in red color with an underscore.

$\mathcal{O}_H$	$\mathcal{O}_T$	$\mathcal{O}_{WW}$	$\mathcal{O}_{BB}$	$\mathcal{O}_{WB}$	$\mathcal{O}_{HW}$	$\mathcal{O}_{HB}$	$\mathcal{O}_{LL}^{(3)}$	$\mathcal{O}_{L}^{(3)}$	$\mathcal{O}_L$	$\mathcal{O}_R$
2.48	2.01	4.83	0.89	1.86	2.09	0.567	5.38	11.6	10.2	8.78
2.48	10.6	4.83	0.89	5.16	2.09	0.567	8.22	12.1	10.2	8.78
2.48	10.6	4.83	0.875	5.12	2.09	0.567	8.15	12.1	10.2	8.78

- Note that the gluon involved operator $\mathcal{O}_g$ can even be probed up to 40 TeV already. The reason is that the Higgs decay into a pair of gluons, $h \to gg$, is induced by triangle loops in the SM. In comparison, the contribution of dimension-6 operators is also at the one-loop level and hence of the same order as the SM prediction $\mathrm{Br}_{h \to gg}^{\mathrm{sm}}$. Deviation in this channel then has magnified effect.

3.2. *The improvement from electroweak precision observables*

Among the most sensitive operators, the limit for $\mathcal{O}_T$ is mainly provided by M_W. In Table 5, we show the effect of imposing the existing electroweak precision observables. The first row shows the effect of only $\sigma(Zh)$. On this basis, M_W is added in the second row and then all of (α, G_F, M_Z, M_W) in the third row. We can see that the most significant enhancement from M_W appears in $\mathcal{O}_T$ by promoting its limit from 2 TeV up to 10 TeV. In addition, the limits on $\mathcal{O}_{WB}$ and $\mathcal{O}_{LL}^{(3)}$ can also be increased by a factor of 2–3. Nevertheless, further imposing the measurements of (α, G_F, M_Z) has no effective help since their effect is basically fixing the electroweak parameters (g, g', v).

The situation changes when the precision on M_W is significantly increased to be comparable with the precision on M_Z. In Table 6, we show the projected 1σ precision of Z and W mass measurements at CEPC.[8,18] Comparing with the existing measurements in Table 4, where the precision on M_W is almost an order of magnitude worse than the precision on M_Z, the M_W measurement at CEPC is

Table 6. The projected precision (1σ) of Z and W mass measurements to be achieved at the CEPC.[8,18]

Observables	Relative error	Absolute error
M_Z	$(0.55 - 1.1) \times 10^{-5}$	$(0.5 - 1)$ MeV
M_W	$(3.7 - 6.2) \times 10^{-5}$	$(3 - 5)$ MeV

Table 7.   (Color online) Impacts of the projected M_Z and M_W measurements at CEPC[8,18] on the reach of new physics scale $\Lambda/\sqrt{|c_j|}$ (in TeV) at 95% C.L. The Higgs observables (including $\sigma(\nu\bar{\nu}h)$ at 350 GeV) and the existing electroweak precision observables (Table 4) are always included in each row. The differences among the four rows arise from whether taking into account the measurements of M_Z and M_W (Table 6) or not. The second (third) row contains the measurement of M_Z (M_W) alone, while the first (last) row contains none (both) of them. We mark the entries of the most significant improvements from M_Z and/or M_W measurements in red color with an underscore.

$\mathcal{O}_H$	$\mathcal{O}_T$	$\mathcal{O}_{WW}$	$\mathcal{O}_{BB}$	$\mathcal{O}_{WB}$	$\mathcal{O}_{HW}$	$\mathcal{O}_{HB}$	$\mathcal{O}_{LL}^{(3)}$	$\mathcal{O}_L^{(3)}$	$\mathcal{O}_L$	$\mathcal{O}_R$	$\mathcal{O}_{L,q}^{(3)}$	$\mathcal{O}_{L,q}$	$\mathcal{O}_{R,u}$	$\mathcal{O}_{R,d}$	$\mathcal{O}_g$
2.74	10.6	6.38	5.78	6.53	2.15	0.603	8.57	12.1	10.2	8.78	1.85	0.565	0.391	0.337	39.8
2.74	10.7	6.38	5.78	6.54	2.15	0.603	8.61	12.1	10.2	8.78	1.85	0.565	0.391	0.337	39.8
2.74	21.0	6.38	5.78	10.4	2.15	0.603	15.5	16.4	10.2	8.78	1.85	0.565	0.391	0.337	39.8
2.74	23.7	6.38	5.78	11.6	2.15	0.603	17.4	18.1	10.2	8.78	1.85	0.565	0.391	0.337	39.8

relatively improved more than M_Z. Fixing M_Z to the experimental central value may disguise its interplay with M_W as summarized in Table 7. On the basis of fitting Higgs and existing electroweak precision observables in Table 4 (first row), adding better measurement on M_Z (second row) does not improve the reach on new physics scales while the M_W measurement at CEPC (third row) can significantly improve the results. The enhancement from M_W can be as large as a factor of 2 for ($\mathcal{O}_T$, $\mathcal{O}_{WB}$, $\mathcal{O}_{LL}^{(3)}$). It is interesting to see that further imposing the M_Z measurement at CEPC, after already imposing M_W, the sensitivity can be increased by another 10%.

Another feature is the scale of $\mathcal{O}_H$ increases from 2.5 TeV in Table 5 to 2.74 TeV in Table 7 by 10%. This is because of the WW fusion at $\sqrt{s} = 350$ GeV which has been added into the χ^2 fit for Table 7. From the Higgsstrahlung peak at $\sqrt{s} = 250$ GeV to $t\bar{t}$ threshold $\sqrt{s} = 350$ GeV leads to significant increase in $\sigma(\nu\bar{\nu}h)$ but decrease in $\sigma(Zh)$. Consequently, we can expect the reduced uncertainty in $\sigma(\nu\bar{\nu}h)$, estimated by TLEP,[9] is the major gain from increasing the lepton collider energy. Nevertheless, the benefit for constraining new physics scales is just 10%.

3.3. *The improvement from Z-pole observables*

Finally, we further include the Z-pole measurements listed in Table 8 and show the results in Fig. 3. Now the new physics scales can reach as high as 35 TeV, in addition to the 40 TeV of $\mathcal{O}_g$. Three operators ($\mathcal{O}_{LL}^{(3)}$, $\mathcal{O}_L^{(3)}$, $\mathcal{O}_g$) can be probed up to above 30 TeV. Among them, the scales of $\mathcal{O}_{LL}^{(3)}$ and $\mathcal{O}_L^{(3)}$ are enhanced by the Z-pole observables by almost another factor of 2. Most of the operators shown in

Table 8.   Projected Z-pole measurements at CEPC[8,18] with integrated luminosity of 5 ab^{-1}.

N_ν	$A_{FB}(b)$	R^b	R^μ	R^τ	$\sin^2\theta_w$
1.8×10^{-3}	1.5×10^{-3}	8×10^{-4}	5×10^{-4}	5×10^{-4}	1×10^{-4}

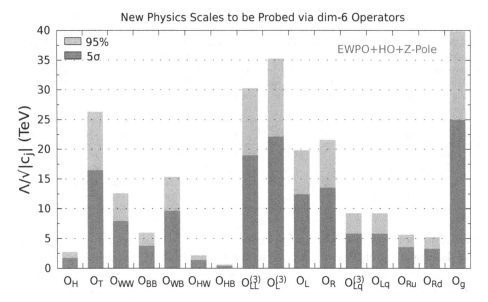

Fig. 3. (Color online) The 95% exclusion (blue) and 5σ discovery (red) sensitivities to the new physics scales $\Lambda/\sqrt{|c_j|}$ by combining the current electroweak precision measurements (α, G_F, M_Z, M_W)[17] with the future Higgs observables at the Higgs factory CEPC (Table 1) and Z-pole measurements (Table 6) under a projected luminosity of 5 ab^{-1}.[8]

Fig. 3 can be probed above 5 TeV, including ($\mathcal{O}_T$, $\mathcal{O}_{WW}$, $\mathcal{O}_{BB}$, $\mathcal{O}_{WB}$, $\mathcal{O}_L$, $\mathcal{O}_R$, $\mathcal{O}_{Lq}^{(3)}$, $\mathcal{O}_{Lq}$, $\mathcal{O}_{Ru}$, $\mathcal{O}_{Rd}$).

Note that the most significant improvement comes from quark related operators ($\mathcal{O}_{Lq}^{(3)}$, $\mathcal{O}_{Lq}$, $\mathcal{O}_{Ru}$, $\mathcal{O}_{Rd}$). Roughly speaking, the scale of $\mathcal{O}_{Lq}^{(3)}$ is increased by a factor of 5 while ($\mathcal{O}_{Lq}$, $\mathcal{O}_{Ru}$, $\mathcal{O}_{Rd}$) are increased by a factor of 10. This is because the quark related operators cannot enter the Higgs production cross-sections $\sigma(Zh)$ or $\sigma(\nu\bar{\nu}h)$ but can affect the $h \to ZZ$ and $h \to WW$ decays indirectly to some extent. Since the Higgs boson mass at 125 GeV is not large enough to put the two Z/W bosons on shell, the decay width has to be calculated with at least one off-shell Z/W boson. Actually, the contribution of two off-shell Z/W can be as large as 25%. Then, the decay width has to be evaluated for the complete chains $h \to ZZ \to f_1\bar{f}_i f_j\bar{f}_j$ (f denoting fermions) and $h \to WW \to u_i\bar{d}_j d_k\bar{u}_l$ (u for up-type fermion and d for the down-type). The quark related operators then enters as correction to the $Zf\bar{f}$ and $W^- u_i\bar{d}_j$ vertices. Since the $h \to ZZ$ and $h \to WW$ channels do not have dominating branching ratios, see Table 4, and the precision on decay branching ratios is not as good as the inclusive measurement of $\sigma(Zh)$, the new physics scales probed by Higgs observables are very low. For ($\mathcal{O}_{L,q}$, $\mathcal{O}_{R,u}$, $\mathcal{O}_{R,d}$), the sensitivity can only reach 300–500 GeV while the scale for $\mathcal{O}_{L,q}^{(3)}$ is 1.85 TeV. With Z-pole measurements, where the $Zf\bar{f}$ and $W^- u_i\bar{d}_j$ vertices dominate, the sensitivities are significantly enhanced to 5–9 TeV.

4. Conclusion

The discovery of the Higgs boson at LHC completes the particle spectrum of the SM. Nevertheless, the SM as a whole can be claimed as complete only after fully testing all the interactions that it dictates. The particle physics is not just about particle but also interactions between them. Especially, the scalar Higgs boson mediates new type of interactions. Our current understanding of the Higgs coupling with fermions and Higgs self-interactions is not as good as the gauge interactions. New physics may enter by modifying the Higgs coupling with the SM particles. From the point of view of testing the SM or going beyond, a lepton collider (Higgs factory) for precision measurement is necessary.

The next-generation lepton colliders, such as CEPC, FCC-ee, and ILC, are motivated by the discovery of Higgs boson at LHC for precision measurement of its properties. We present the extracted Higgs coupling precision from Higgs observables at CEPC to show its physics potential. With one million events, the Higgs couplings can be measured to $\mathcal{O}(1\%)$ level. In particular, the Higgs coupling with Z boson can be as good as 0.25% due to the inclusive cross-section of the Higgsstrahlung process $e^+e^- \to Zh$.

Although CEPC runs at energy $\sqrt{s} = 250$ GeV and Z-pole, it can probe the new physics beyond the SM indirectly. We use dimension-6 operators to estimate the new physics scales that can be reached at CEPC. The result shows that the Higgs observables together with existing electroweak precision observables can constrain the new physics up to 10 TeV at 95% C.L. (40 TeV for the gluon related operator $\mathcal{O}_g$). If the M_Z/M_W mass measurements and Z-pole observables are also utilized, the new physics scale can be further pushed up to 35 TeV. The Z-pole observables are as important as the Higgs observables and the Z-pole running is useful not only for the purpose of calibration but also for probing the new physics scales. It is beneficial to assign more time for Z-pole running at $\sqrt{s} \approx 90$ GeV before switching to the Higgs factory mode $\sqrt{s} = 250$ GeV or returning to Z-pole after finishing the Higgs observable measurements. Since CEPC is a circular collider, the 50–100 km tunnel can host a hadron (pp) collider SPPC with $\sqrt{s} = 50$–100 TeV of energy after CEPC. The energy scale (10–40 TeV) indirectly reachable at CEPC basically covers the energy range to be effectively explored at SPPC. The lepton collider CEPC can guide and pave the road for the sequent hadron collider SPPC.

Although our study takes CEPC for illustration, the conclusion can apply to other candidate lepton colliders, such as FCC-ee and ILC. The technical details for obtaining the results presented here can be found in our formal paper.[11]

Acknowledgments

We thank Matthew McCullough, Manqi Ruan, and Tevong You for many valuable discussions. We are grateful to Michael Peskin for discussing the analysis of Ref. 16. We also thank Timothy Barklow, Tao Han, Zhijun Liang and Matthew Strassler for discussions. S.-F. Ge is grateful to the Jockey Club Institute for Advanced Study at the Hong Kong University of Science and Technology, especially Henry Tye and Tao Liu, for kind invitation to attend the IAS Program/Conference on High Energy Physics and hospitality during the stay.

References

1. F. Englert and R. Brout, *Phys. Rev. Lett.* **13**, 321 (1964). P. W. Higgs, *Phys. Lett.* **12**, 132 (1964); *Phys. Rev. Lett.* **13**, 508 (1964); G. S. Guralnik, C. R. Hagen and T. Kibble, *ibid.* **13**, 585 (1965); T. Kibble, *Phys. Rev.* **155**, 1554 (1967).
2. ATLAS Collab. (G. Aad *et al.*), *Phys. Lett. B* **716**, 1 (2012), arXiv:1207.7214 [hep-ex]; CMS Collab. (S. Chatrchyan *et al.*), *ibid.* **716**, 30 (2012), arXiv:1207.7235 [hep-ex].
3. J. Ellis, M. K. Gaillard and D. V. Nanopoulos, An updated historical profile of the Higgs boson, arXiv:1504.07217 [hep-ph].
4. J. Ellis and T. You, *J. High Energy Phys.* **1306**, 103 (2013), arXiv:1303.3879 [hep-ph]; C. Englert, R. Kogler, H. Schulz and M. Spannowsky, arXiv:1511.05170 [hep-ph].
5. S. F. Ge, H. J. He, J. Ren and Z. Z. Xianyu, *Phys. Lett. B* **757**, 480 (2016), arXiv:1602.01801; J. Ellis, H. J. He and Z. Z. Xianyu, *J. Cosmol. Astropart. Phys.* **1608**, 068 (2016), arXiv:1606.02202.
6. W. Yao, *Proc. Snowmass Community Summer Study (CSS 2013)*, July 29–August 6, 2013, Minneapolis, USA, arXiv:1308.6302 [hep-ph]; H. J. He, J. Ren and W. Yao, *Phys. Rev. D* **93**, 015003 (2015), arXiv:1506.03302 [hep-ph]; A. J. Barr, M. J. Dolan, C. Englert, D. E. Ferreira de Lima and M. Spannowsky, *J. High Energy Phys.* **1502**, 016 (2015), arXiv:1412.7154 [hep-ph]; C. R. Chen and I. Low, *Phys. Rev. D* **90**, 013018 (2014), arXiv:1405.7040 [hep-ph]; D. Curtin, P. Meade and C. T. Yu, *J. High Energy Phys.* **1411**, 127 (2014), arXiv:1409.0005 [hep-ph]; A. Azatov, R. Contino, G. Panico and M. Son, *Phys. Rev. D* **92**, 035001 (2015), arXiv:1502.00539 [hep-ph]; Q. Li, Z. Li, Q.-S. Yan and X. Zhao, *ibid.* **92**, 014015 (2015), arXiv:1503.07611 [hep-ph]; A. V. Kotwal, S. Chekanov and M. Low, *ibid.* **91**, 114018 (2015), arXiv:1504.08042 [hep-ph]; M. Dall'Osso, T. Dorigo, C. A. Gottardo, A. Oliveira, M. Tosi and F. Goertz, arXiv:1507.02245 [hep-ph]; B. Batell, M. McCullough, D. Stolarski and C. B. Verhaaren, *J. High Energy Phys.* **1509**, 216 (2015), arXiv:1508.01208 [hep-ph]; L. C. Lv, C. Du, Y. Fang, H. J. He and H. Zhang, *Phys. Lett. B* **755**, 509 (2016), arXiv:1507.02644; A. Papaefstathiou and K. Sakurai, *ibid.* **1602**, 006 (2016), arXiv:1508.06524 [hep-ph]; D. Curtin and P. Saraswat, arXiv:1509.04284 [hep-ph]; C.-Y. Chen, Q.-S. Yan, X. Zhao, Z. Zhao and Y.-M. Zhong, *Phys. Rev. D* **93**, 013007 (2016), arXiv:1510.04013 [hep-ph]; Q.-H. Cao, Y. Liu and B. Yan, arXiv:1511.03311 [hep-ph]; R. Grober, M. Muhlleitner and M. Spira, arXiv:1602.05851 [hep-ph].
7. S. F. Ge, H. J. He and F. R. Yin, *J. Cosmol. Astropart. Phys.* **1005**, 017 (2010), arXiv:1001.0940 [hep-ph]; D. A. Dicus, S. F. Ge and W. W. Repko, *Phys. Rev. D* **83**, 093007 (2011), arXiv:1012.2571 [hep-ph]; S. F. Ge, D. A. Dicus and W. W. Repko, *Phys. Lett. B* **702**, 220 (2011), arXiv:1104.0602 [hep-ph]; H. J. He and F. R. Yin, *Phys. Rev. D* **84**, 033009 (2011), arXiv:1104.2654; H. J. He and X. J. Xu, *Phys. Rev. D* **86**, 111301(R) (2012), arXiv:1203.2908; S. F. Ge, D. A. Dicus and W. W. Repko, *Phys. Rev. Lett.* **108**, 041801 (2012), arXiv:1108.0964 [hep-ph]; A. D. Hanlon, S. F. Ge and W. W. Repko, *Phys. Lett. B* **729**, 185 (2014), arXiv:1308.6522 [hep-ph]; S. F. Ge, arXiv:1406.1985 [hep-ph].
8. CEPC Collab., CEPC-SPPC Preliminary Conceptual Design Report, http://cepc. ihep.ac.cn; M. Ruan, Higgs measurement at e^+e^- circular colliders, in *37th Int. Conf. on High Energy Physics (ICHEP-2014)*, July 2–9 2014, Valencia, Spain, arXiv:1411.5606 [hep-ex].

9. FCC Collab., http://cern.ch/FCC-ee; M. Bicer *et al.*, *J. High Energy Phys.* **1401**, 164 (2014), arXiv:1308.6176 [hep-ex]; D. d'Enterria, arXiv:1602.05043 [hep-ex].
10. H. Baer *et al.*, arXiv:1306.6352 [hep-ph]; G. Moortgat-Pick *et al.*, *Eur. Phys. J. C* **75**, 371 (2015), arXiv:1504.01726 [hep-ph]; K. Fujii *et al.*, arXiv:1506.05992 [hep-ex].
11. S. F. Ge, H. J. He and R. Q. Xiao, *J. High Energy Phys.* **1610**, 007 (2016), arXiv:1603.03385 [hep-ph].
12. ILD Design Study Group Collab. (H. Li *et al.*), HZ recoil mass and cross-section analysis in ILD, arXiv:1202.1439 [hep-ex].
13. Results presented at *CEPC Physics Software Meeting*, IHEP, China, March 26–27, 2016; *CEPC-SPPC Workshop*, IHEP, China, April 8–9, 2016.
14. S. F. Ge, K. Hagiwara, N. Okamura and Y. Takaesu, *J. High Energy Phys.* **1305**, 131 (2013), arXiv:1210.8141 [hep-ph].
15. S. F. Ge, BSMfitter – Beyond standard model fitter, http://bsmfitter.hepforge.org.
16. M. E. Peskin, Estimation of LHC and ILC capabilities for precision Higgs boson coupling measurements, in *Proc. Snowmass 2013*, arXiv:1312.4974v3 [hep-ph].
17. Particle Data Group Collab., (K. A. Olive *et al.*), *Chin. Phys. C* **38**, 090001 (2014).
18. H. Yang, H. Li, Q. Li, J. Guo, M. Ruan, Y. Wu and Z. Liang, Presentation of the CEPC Detector Working Group, *Z* and *W* Physics at CEPC, http://indico.ihep.ac.cn/event/4338/session/2/material/slides/1?contribId=32.

Probing the Higgs with Angular Observables at Future e^+e^- Colliders

Zhen Liu

Theoretical Physics Department,
Fermi National Accelerator Laboratory,
Batavia, IL 60510, USA
zliu2@fnal.gov

I summarize our recent works on using differential observables to explore the physics potential of future e^+e^- colliders in the framework of Higgs effective field theory. This proceeding is based upon Refs. 1 and 2. We study angular observables in the $e^+e^- \to ZH\ell^+\ell^- b\bar{b}$ channel at future circular e^+e^- colliders such as CEPC and FCC-ee. Taking into account the impact of realistic cut acceptance and detector effects, we forecast the precision of six angular asymmetries at CEPC (FCC-ee) with center-of-mass energy $\sqrt{s} = 240$ GeV and 5 (30) ab^{-1} integrated luminosity. We then determine the projected sensitivity to a range of operators relevant for the Higgsstrahlung process in the dimension-6 Higgs EFT. Our results show that angular observables provide complementary sensitivity to rate measurements when constraining various tensor structures arising from new physics. We further find that angular asymmetries provide a novel means of constraining the "blind spot" in indirect limits on supersymmetric scalar top partners. We also discuss the possibility of using ZZ-fusion at e^+e^- machines at different energies to probe new operators.

Keywords: Asymmetries; Higgs boson; collider; ILC; CEPC; FCC; EFT; higher-dimensional operators.

1. Introduction

Following the discovery of a Standard Model-like Higgs at the LHC,[3,4] the study of Higgs properties has become one of the highest priorities for current and future colliders. High-luminosity electron–positron colliders are particularly well suited to this end, promising a large sample of relatively clean Higgs production events and the ability to directly probe Higgs properties in a model-independent fashion. Such precision tests of Higgs couplings will provide a window into physics beyond the Standard Model (BSM) well above the weak scale.

Thus far, much attention has focused on the potential of future e^+e^- colliders to probe deviations in Higgs properties in terms of a re-scaling of Standard Model couplings,[5–7] with sensitivity exceeding the percent level in some channels. However, in general deviations in Higgs properties may encode additional information, for example in the form of operators with different tensor structure in the Higgs Effective Field Theory (EFT). Disentangling contributions from these different operators provide a further handle on BSM physics by both increasing the effective reach of e^+e^- colliders and distinguishing different BSM scenarios in the event of deviations from the Standard Model.

2. Angular Observables at CEPC and FCC-ee

Given the apparent parametric separation of scales between the Higgs boson and new physics, the Higgs EFT provides a useful framework for characterizing deviations in Higgs properties from their Standard Model (SM) predictions.

Here we will work in terms of a minimal operator basis given in Ref. 8; for a comparable choice of basis, see Ref. 9. The relevant operators defining our operator basis are given in Table 1 of Ref. 1. After electroweak symmetry breaking, these dimension-6 operators give rise to a variety of interaction terms relevant for $e^+e^- \rightarrow ZH$ of the form

$$\mathcal{L}_{\text{eff}} \supset c_{ZZ}^{(1)} h Z_\mu Z^\mu + c_{ZZ}^{(2)} h Z_{\mu\nu} Z^{\mu\nu} + c_{Z\tilde{Z}} h Z_{\mu\nu} \tilde{Z}^{\mu\nu} + c_{AZ} h Z_{\mu\nu} A^{\mu\nu} + c_{A\tilde{Z}} h Z_{\mu\nu} \tilde{A}^{\mu\nu}$$

$$+ h Z_\mu \bar{\ell} \gamma^\mu (c_V + c_A \gamma_5) \ell + Z_\mu \bar{\ell} \gamma^\mu (g_V - g_A \gamma_5) \ell - g_{\text{em}} Q_\ell A_\mu \bar{\ell} \gamma^\mu \ell, \qquad (1)$$

where h is the real CP-even Higgs scalar, $Z_{\mu\nu}$ and $A_{\mu\nu}$ are the Z boson and photon gauge field strengths, and $\tilde{V}^{\mu\nu} = \epsilon^{\mu\nu\alpha\beta} V_{\alpha\beta}$. Here we again use the notation of Ref. 8 for clarity. The couplings in this broken-phase effective Lagrangian may be straightforwardly expressed in terms of coefficients in the dimension-6 Higgs EFT and is outlined in details in Ref. 1.

2.1. *Angular observables in ZH production*

In general, these effective operators contribute to a shift in the cross-section for $e^+e^- \rightarrow ZH$, so that a linear combination of Wilson coefficients can be constrained to high precision by future e^+e^- colliders.[9] However, there is additional information available in Higgsstrahlung events that allows us to constrain independent linear combinations of Wilson coefficients. This independent information can be effectively parametrized in terms of angular observables. In this paper, we will work in terms of the parametrization in Ref. 8, although other definitions of angular observables are possible and in principle may prove more efficient in isolating specific Wilson coefficients.

We define the angles $\cos\theta_1, \cos\theta_2$ and ϕ as follows: the z direction is defined by the momentum of the on-shell Z boson in the rest frame of the incoming e^+e^- pair. The xz plane is the plane defined by the momentum of the outgoing Z boson and its ℓ^- decay product. Then θ_1 is the angle between the momentum of the outgoing ℓ^+ and the z-axis. θ_2 is the angle between the momentum of the incoming e^+ and the momentum of the outgoing h along the z-axis. Finally, the angle ϕ corresponds to the angle in the xy plane between the planes defined by the incoming e^+e^- and the outgoing $\ell^+\ell^-$, respectively. These angles are illustrated in Fig. 1.

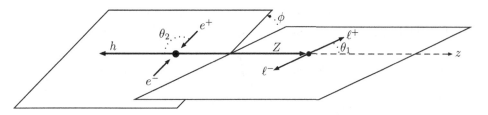

Fig. 1.   Definition of the angles θ_1, θ_2, and ϕ in the $e^+e^- \to ZH$ production process relevant for the construction of angular observables. See text for further details.

In terms of these angles, we parametrize the triple differential cross-section for $e^+e^- \to Z(\to \ell^+\ell^-)h$ as

$$\frac{d\sigma}{d\cos\theta_1 d\cos\theta_2 d\phi} = \frac{1}{2^{10}(2\pi)^3} \frac{1}{\sqrt{r}\gamma_Z} \frac{\sqrt{\lambda(1,s,r)}}{s^2} \frac{1}{m_h^2} \mathcal{J}(q^2,\theta_1,\theta_2,\phi), \quad (2)$$

where $r = m_Z^2/m_h^2 \approx 0.53$, $\gamma_Z = \Gamma_Z/m_h \approx 0.020$, $s = q^2/m_h^2$, $\lambda(a,b,c) = a^2 + b^2 + c^2 - 2ab - 2ac - 2bc$, and the function $\mathcal{J}$ contains nine independent angular structures with coefficients $J_1,\ldots,J_9$ decomposed as

$$\mathcal{J}(q^2,\theta_1,\theta_2,\phi) = J_1(1 + \cos^2\theta_1 \cos^2\theta_2 + \cos^2\theta_1 + \cos^2\theta_2) + J_2 \sin^2\theta_1 \sin^2\theta_2$$
$$+ J_3 \cos\theta_1 \cos\theta_2 + J_4 \sin\theta_1 \sin\theta_2 \sin\phi + J_5 \sin 2\theta_1 \sin 2\theta_2 \sin\phi.$$
$$(3)$$

The explicit form of the J_i in terms of the EFT coefficients and Standard Model parameters was computed by Ref. 8 and for convenience is given in Appendix of Ref. 1. The total integrated cross-section for $e^+e^- \to ZH$ is given in terms of the J_i simply by

$$\sigma(s) = \frac{32\pi}{9} \frac{1}{2^{10}(2\pi)^3} \frac{1}{\sqrt{r}\gamma_Z} \frac{\sqrt{\lambda(1,s,r)}}{s^2} \frac{1}{m_h^2} (4J_1 + J_2). \quad (4)$$

It is useful to isolate various combinations of terms in the differential cross-section through the following angular observables $\mathcal{A}_i$, normalized to σ:

$$\mathcal{A}_{\theta_1} = \frac{1}{\sigma} \int_{-1}^{1} d\cos\theta_1 \, \mathrm{sgn}(\cos(2\theta_1)) \frac{d\sigma}{d\cos\theta_1} = 1 - \frac{5}{2\sqrt{2}} + \frac{3J_1}{\sqrt{2}(4J_1 + J_2)}, \quad (5)$$

$$\mathcal{A}_\phi^{(1)} = \frac{1}{\sigma} \int_0^{2\pi} d\phi \, \mathrm{sgn}(\sin\phi) \frac{d\sigma}{d\phi} = \frac{9\pi}{32} \frac{J_4}{4J_1 + J_2}, \quad (6)$$

$$\mathcal{A}_\phi^{(2)} = \frac{1}{\sigma} \int_0^{2\pi} d\phi \, \mathrm{sgn}(\sin(2\phi)) \frac{d\sigma}{d\phi} = \frac{2}{\pi} \frac{J_8}{4J_1 + J_2}, \quad (7)$$

$$\mathcal{A}_\phi^{(3)} = \frac{1}{\sigma} \int_0^{2\pi} d\phi \, \mathrm{sgn}(\cos\phi) \frac{d\sigma}{d\phi} = \frac{9\pi}{32} \frac{J_6}{4J_1 + J_2}, \quad (8)$$

$$\mathcal{A}_\phi^{(4)} = \frac{1}{\sigma} \int_0^{2\pi} d\phi \, \mathrm{sgn}(\cos(2\phi)) \frac{d\sigma}{d\phi} = \frac{2}{\pi} \frac{J_9}{4J_1 + J_2}. \quad (9)$$

Here $\text{sgn}(\pm|x|) = \pm 1$. In addition to these five angular observables, it is also useful to define the forward–backward asymmetry

$$
\begin{aligned}
\mathcal{A}_{c\theta_1, c\theta_2} &= \frac{1}{\sigma} \int_{-1}^{1} d\cos\theta_1 \, \text{sgn}(\cos\theta_1) \int_{-1}^{1} d\cos\theta_2 \, \text{sgn}(\cos\theta_2) \frac{d^2\sigma}{d\cos\theta_1 d\cos\theta_2} \\
&= \frac{9}{16} \frac{J_3}{4J_1 + J_2}.
\end{aligned}
\tag{10}
$$

Although there are nine J_i, only six are independent, leading to six independent angular observables corresponding to the six independent form factors in the $e^+e^- \to ZH$ amplitude. Each of the angular observables is sensitive to a different linear combination of coefficients in the dimension-6 Higgs EFT.

The numerical values of the input parameters are chosen carefully and discussed in details in Ref. 1. Working only to linear order in the Wilson coefficients, at $\sqrt{s} = 240$ GeV the dependence of the total cross-section on the hatted coefficients $\hat{\alpha}_i$ in the unbroken-phase HEFT and similarly, the dependence of the angular observables on the hatted coefficients $\hat{\alpha}_i$ in the unbroken-phase HEFT are both numerically expressed in Ref. 1. The inclusive cross-section σ is unsurprisingly sensitive to all CP-even operators, with particular sensitivity to operators that shift the couplings between gauge bosons and leptons, as well as those that generate new $hZ\ell\ell$ contact terms. The asymmetry variables $\mathcal{A}_{\theta_1}$ and $\mathcal{A}_\phi^{(4)}$ provide independent sensitivity to the operators $\mathcal{O}_{\Phi W}, \mathcal{O}_{\Phi B}, \mathcal{O}_{WB}$, which are also the operators in this basis that are constrained by measurements of $h \to \gamma\gamma$. The forward–backward asymmetry $\mathcal{A}_{c\theta_1, c\theta_2}$ and the angular asymmetry $\mathcal{A}_\phi^{(3)}$ are sensitive to independent linear combinations of the CP-even operators (excepting $\mathcal{O}_{\Phi\Box}$, whose contribution has been eliminated by construction in the angular asymmetries). Finally, the asymmetries $\mathcal{A}_\phi^{(1)}$ and $\mathcal{A}_\phi^{(2)}$ are sensitive to independent linear combinations of CP-odd operators.

2.2. *Expected precision and statistical uncertainty*

Having defined the set of angular variables relevant for probing the Higgs EFT in $e^+e^- \to ZH$, we now develop projections for the sensitivity attainable at various proposed Higgs factories. In particular, we study the reach in angular observables at two proposed future e^+e^- colliders: the Circular Electron-Positron Collider (CEPC) and the e^+e^- mode of the CERN Future Circular Collider (FCC-ee). Both of these colliders are designed to produce large numbers of $e^+e^- \to ZH$ events at the center-of-mass energy $\sqrt{s} = 240$ GeV. With a proposed luminosity of 2×10^{34} cm^{-2} s^{-1} per Interaction Point (IP), the integrated luminosity at CEPC will be 5 ab^{-1} over a running time of 10 years with 2 IPs.[7,10] The machine parameters of FCC-ee[11] project that its luminosity can reach 6×10^{34} cm^{-2} s^{-1} at $\sqrt{s} = 240$ GeV, which is three times that of CEPC. In addition, there is a factor of 2 increase in luminosity on account of the projected 4 IPs at FCC-ee, bringing the total FCC-ee luminosity

Table 1. The SM expectation at $\sqrt{s} = 240$ GeV for the asymmetry observables and the standard deviation (σ_A) for different sample sizes. We consider the process with $Z \to \mu^+\mu^-/e^+e^-$ and $H \to b\bar{b}$, which is almost entirely background-free. According to Subsec. 3.3.3.1 in the CEPC pre-CDR,[7] the number of events after basic cuts is 22100 for 5 ab^{-1}. We use this number here and also scale it up with luminosity for 30 ab^{-1} and the full statistics scenario detailed in text.

| Observable | SM expectation | Precision σ_A | | |
| | | 5 ab^{-1} | 30 ab^{-1} | |
		CEPC	FCC-ee	Full Stat.
$\mathcal{A}_{\theta_1}$	-0.448	0.0060	0.0025	0.00078
$\mathcal{A}_\phi^{(1)}$	0	0.0067	0.0027	0.00087
$\mathcal{A}_\phi^{(2)}$	0	0.0067	0.0027	0.00087
$\mathcal{A}_\phi^{(3)}$	0.0136	0.0067	0.0027	0.00087
$\mathcal{A}_\phi^{(4)}$	0.0959	0.0067	0.0027	0.00086
$\mathcal{A}_{c\theta_1, c\theta_2}$	-0.0075	0.0067	0.0027	0.00087

to six times that of CEPC. Considering the same running time of 10 years, we therefore take the integrated luminosity at FCC-ee to be 30 ab^{-1} for the purpose of our projections.

In Table 1, we list the theoretical expectations for all the relevant asymmetry observables assuming only Standard Model contributions, as well as the 1σ errors (σ_A) for various integrated luminosity benchmarks. We consider only the process with $Z \to \mu^+\mu^-/e^+e^-$ and $H \to b\bar{b}$, which is almost entirely background-free. According to Subsec. 3.3.3.1 in the pre-CDR of CEPC,[7] the number of events after basic cuts in both $\mu^+\mu^-$ and e^+e^- channels is $11067 + 11033 = 22100$ for 5 ab^{-1}. We assume for simplicity that FCC-ee will conduct a very similar study on this channel, and consequently scale the statistics up directly to 30 ab^{-1} for FCC-ee.

For the purposes of forecasting, we assume the experimental results are SM-like and obtain the expected constraints on new physics using a simple χ^2 fit. For the sake of concreteness, we focus on the channel $e^+e^- \to ZH \to \ell^+\ell^- b\bar{b}$ at CEPC with $\sqrt{s} = 240$ GeV and 5 ab^{-1} integrated luminosity, although we also forecast sensitivity for several scenarios at FCC-ee. To compensate the omission of systematics, we judiciously apply a universal 5% penalty factor for the sensitivities of asymmetry observables to Wilson coefficients, as mentioned in Sec. 2. For the uncertainty in the cross-section, we adopt the values in the preCDR,[7] which are 0.9% for the $\mu^+\mu^- b\bar{b}$ channel and 1.1% for the $e^+e^- b\bar{b}$ channel. The combined precision for the $e^+e^- \to ZH \to \ell^+\ell^- b\bar{b}$ channel is therefore 0.7%, assuming statistical uncertainties dominate.

2.3. Constraining Wilson coefficients

In this section, we present the model-independent constraints on the Wilson coefficients in the Higgs effective Lagrangian, Eq. (1), parametrized the nine Wilson coefficients,

$$\hat{\alpha}_{ZZ}, \quad \hat{\alpha}_{ZZ}^{(1)}, \quad \hat{\alpha}_{\Phi\ell}^{V}, \quad \hat{\alpha}_{\Phi\ell}^{A}, \quad \hat{\alpha}_{AZ}, \quad \delta g_{V}, \quad \delta g_{A}, \quad \hat{\alpha}_{Z\tilde{Z}}, \quad \hat{\alpha}_{A\tilde{Z}}.$$

Treating the nine coefficients as independent parameters, there are totally seven constraints from the rate and the six asymmetry observables, less than the number of unknowns. Therefore, one cannot obtain independent constraints on the Wilson coefficients without making further assumptions. However, with a reduced set of coefficients the angular observables can break the degeneracy of the rate measurement, which by itself could only constrain one linear combination of the Wilson coefficients. To illustrate this point, we focus on two coefficients at a time while setting the rest to zero. One of the coefficients is always chosen to be $\hat{\alpha}_{ZZ}^{(1)}$, which parametrizes a modification of the SM $HZ^{\mu}Z_{\mu}$ interaction and is most strongly constrained by the rate measurement. The angular observables, being normalized to the total cross-section, are independent of $\hat{\alpha}_{ZZ}^{(1)}$ by construction. We provide in Table 2 the constraints on individual Wilson coefficients with the assumption that all other coefficients are zero. Table 2 shows the 1σ uncertainties for each Wilson coefficient (setting others to zero) from the rate measurements only, the angular observables measurements only, and the combination of the two. We use "∞" to denote coefficients for which no constraint can be derived within our procedure. In particular, the angular observables are insensitive to $\hat{\alpha}_{ZZ}^{(1)}$ by construction, while the rate measurements are independent of the CP-odd operators at leading order in the Wilson coefficients.

As discussed in earlier, with the same running time FCC-ee is able to deliver a sample size six times larger than that of CEPC. It is also reasonable to expect that statistical uncertainties dominate for the $e^{+}e^{-} \to ZH \to \ell^{+}\ell^{-} b\bar{b}$ process at FCC-ee as they do at CEPC. Furthermore, the inclusion of additional decay modes of H and Z would increase the statistics and could potentially significantly increase the constraining power. While the reaches of other channels would require further

Table 2. 1σ uncertainties for individual Wilson coefficients, with the assumption that all other coefficients are zero. The second row shows the constraints from the rate measurements only, the third row shows the constraints from measurements of angular observables (combined) only, and the last row shows the total combined constraints from both rate and angular measurements. If no constraint could be derived within our procedure, a ∞ is shown.

	$\hat{\alpha}_{ZZ}$	$\hat{\alpha}_{ZZ}^{(1)}$	$\hat{\alpha}_{\Phi\ell}^{V}$	$\hat{\alpha}_{\Phi\ell}^{A}$	$\hat{\alpha}_{AZ}$	δg_{V}	δg_{A}	$\hat{\alpha}_{Z\tilde{Z}}$	$\hat{\alpha}_{A\tilde{Z}}$
Rate	0.00064	0.0035	0.0079	0.00059	0.012	0.023	0.0018	∞	∞
Angles	0.016	∞	0.0058	0.078	0.0087	0.017	0.23	0.012	0.036
Total	0.00064	0.0035	0.0047	0.00059	0.0070	0.014	0.0018	0.012	0.036

study, to illustrate their potential usefulness we perform a naive scaling of statistics from the FCC-ee $e^+e^- \to ZH \to \ell^+\ell^- b\bar{b}$ process by another factor of 10, and denote this scenario as FCC-ee FS (full statistics).

2.4. *Constraining stops*

As a final example of the discriminating power of angular observables, we consider a concrete weakly-coupled model that may be constrained with precision measurements at e^+e^- colliders: scalar top partners (stops) in supersymmetric extensions of the Standard Model. For simplicity, we will consider stops with degenerate stop soft masses $m_{\tilde{t}}^2 = m_{\tilde{Q}_3}^2 = m_{\tilde{t}_R}^2$ plus mixing terms of the form $X_t = A_t - \mu \cot \beta$. The mass scale of the effective operators is $\Lambda = m_{\tilde{t}}$. Wilson coefficients for this scenario were computed in Ref. 12, while the constraint on the stop parameter space due to rate measurements at e^+e^- colliders was determined in Ref. 9. Here we include the additional sensitivity contributed by angular observables by translating the results of Refs. 9 and 12 into our preferred basis of Wilson coefficients and applying the results of the previous section.

In Fig. 2, we show the sensitivity provided by rate measurements and the inclusion of angular observables in the plane of the two stop mass eigenvalues M_1 and M_2, which are functions of $m_{\tilde{t}}^2$ and X_t given by the stop mass mixing matrix. For definiteness we have set $\tan \beta = 10$, while the results are insensitive to $\tan \beta$ as long as $\tan \beta \gtrsim$ few.

The features of the exclusion provided by rate measurements were discussed extensively in Ref. 9. The most noteworthy feature of the rate measurements is the so-called "blind spot" along the line $M_2 = M_1 + \sqrt{2}m_t$ where the shift in the hZZ rate is zero. Such blind spots arise more generally in stop corrections to various Higgs properties such as hgg, $h\gamma\gamma$ couplings and precision electroweak observables.

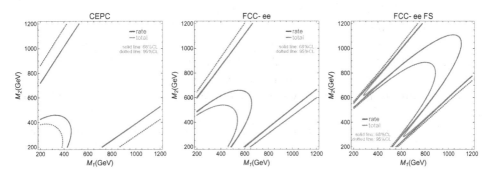

Fig. 2. (Color online) Expected constraints in the (M_1, M_2) plane for different collider scenarios, assuming SM-like results. M_1 and M_2 are the two mass eigenvalues of the left- and right-handed stops. The three scenarios, CEPC, FCC-ee and FCC-ee FS are described in Subsec. 2.3. We set $\tan \beta = 10$. The blue contours show the constraints from the rate measurements only and the red contours show the total combined constraints from the measurements of rate and the angular observables. The solid (dotted) lines corresponds to 68% (95%) CL. The region in the upper-right part of each plot is allowed by projected coupling measurements.

Each blind spot corresponds to a zero in physical linear combinations of Wilson coefficients where the contributions from two stops cancel. While the exact zeroes in observables arise in different places in the $M_1 - M_2$ plane, they are collected around the line in which the coupling of the lightest stop mass eigenstate to the Higgs goes to zero.

In general, the addition of angular observables does not lead to immense improvements over the rate measurement in generic regions of parameter space. This is not surprising, since the relevant Wilson coefficients well-constrained by angular observables are generated at one loop and thus are small for all values of the stop masses. However, it is apparent in Fig. 2 that the addition of angular observables provides significantly improved sensitivity in the blind spot of the hZZ rate measurement. This is simply because the Wilson coefficients contributing to angular observables are suppressed but nonzero along the line where the ZH cross-section shift is zero, and so provide complementary sensitivity at small M_1, M_2 provided sufficient statistics. This demonstrates the value of angular observables even in the case of BSM scenarios that are generally well-constrained by rate measurements.

3. Physics Potential from Measuring Higgs Inclusive Rate at Different Energies

BSM physics could give rise to modifications of the Higgs couplings. The proper framework to describe such possibilities in a model-independent manner is the effective field theory approach. With respect to the SM gauge symmetry, such effects are expressed by dimension-six Higgs operators after integrating out heavy particles or loop functions.[13–16,a] The operators modifying Higgs to ZZ couplings are naturally of particular interest in our case. This is partly because it will be one of the most precisely determined quantities through a recoil-mass measurement and partly because it is one of the key couplings that could help reveal the underlying dynamics of electroweak symmetry breaking. Certain operators may have different momentum dependence and thus measurements of differential cross-sections may be more sensitive to the new effects.[b] The ILC is expected to have several operational stages with different center-of-mass energies, and the high-precision measurement achievable from ZZ fusion will contribute to our knowledge of these different operators.[c]

To demonstrate this important feature, we consider the following two representative operators

$$\mathcal{O}_H = \partial^\mu(\phi^\dagger\phi)\partial_\mu(\phi^\dagger\phi)\,, \quad \mathcal{O}_{HB} = g'D^\mu\phi^\dagger D^\nu\phi B_{\mu\nu}\,, \tag{11}$$

[a]For recent reviews of these operators, see e.g. Refs. 17–20. Many of these operators not only contribute to Higgs physics, but also modify electroweak precision tests simultaneously.[21–24]
[b]For discussions of the effects on Higgs decays due to these operators, see Ref. 8.
[c]Assuming existence of a single operator at a time, limits can be derived, see, e.g. Ref. 9.

with

$$\mathcal{L}^{\text{dim-6}} \supset \frac{c_H}{2\Lambda^2}\mathcal{O}_H + \frac{c_{HB}}{\Lambda^2}\mathcal{O}_{HB}\,, \tag{12}$$

where ϕ is the SM SU(2)$_L$ doublet and Λ is the new physics scale. The coefficients c_H and c_{HB} are generically of order unity. We adopt the scaled coefficients $\bar{c}_H = \frac{v^2}{\Lambda^2}c_H$ and $\bar{c}_{HB} = \frac{m_W^2}{\Lambda^2}c_{HB}$. This translates to generic values of $\bar{c}_H \approx 0.06$ and $\bar{c}_{HB} \approx 0.006$ for $\Lambda = 1$ TeV.

The operator $\mathcal{O}_H$ modifies the Higgs-ZZ coupling in a momentum-independent way at lowest order. This operator renormalizes the Higgs kinetic term and thus modifies the Higgs coupling to any particles universally.[25,26] Equivalently, one may think of rescaling the standard model coupling constant. In contrast, the operator $\mathcal{O}_{HB}$ generates a momentum-dependent Higgs-ZZ coupling. This leads to a larger variation of the production rate versus c.m. energy for the Zh process than the ZZ fusion because of the energy difference in the intermediate Z bosons.

Such operators receive direct constraints from the LHC from similar production processes,[21,22] off-shell Higgs-to-ZZ measurement,[27] etc., all of which lack desirable sensitivities due to the challenging hadron collider environment. Based on an analysis of current data the coefficient $\bar{c}_{HB}$ is excluded for values outside the window $(-0.045, 0.075)^{\text{d}}$ and $\bar{c}_H$ is far less constrained.[21,22]

We only list above the cross-sections which can be precisely measured at different ILC stages, with corresponding polarizations taken into account. The distinction between ZZ fusion (e^-e^+h) and Zh-associated production with Z decaying to electron–positron pairs is easily made by applying a minimal m_{ee} cut above m_Z.

In Fig. 3, we plot the expected constraints on the constants $\bar{c}_H$ and $\bar{c}_{HB}$ from the Zh and ZZ processes measured at the ILC, assuming only these two constants among the six-dimensional terms are nonzero. We show the 95% CL contours for different measurements. The dashed (dot–dashed) blue line represents the contour from Zh-associated measurement at ILC 250 GeV (500 GeV). The red line represents the contour from combined ZZ fusion measurements at ILC 500 GeV and 1 TeV. One can see that at a given energy for a simple production mode only a linear combination of the two operators is constrained, resulting in a flat direction in the contours. However, measurements of Zh at two different energies would allow us to measure both simultaneously, as shown in the gray contour. Moreover, the addition of the ZZ information at 1 TeV would offer *significant* improvements as shown in the yellow contour. This allows us to measure $\bar{c}_H$ and $\bar{c}_{HB}$ at the level of 0.04 and 0.004, respectively. Much of the improvement comes from the fact that in ZZ fusion, in contrast to Zh-associate production, the $\mathcal{O}_{HB}$ operator contributes with the opposite sign of the $\mathcal{O}_H$ operator.

---

^{d}The window is $(-0.053, 0.044)$ for single-operator analysis. This smallness of the difference between the marginalized analysis and single-operator analysis illustrates that this operator mainly affects Higgs physics and thus other electroweak precision observables do not provide much information.

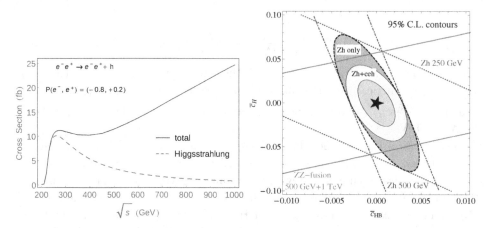

Fig. 3. (Color online) Left panel: Total cross-section (in fb) for $e^-e^+ \to e^-e^+ + h$ at ILC versus $\sqrt{s}$. The dashed curve is for Higgsstrahlung mode only. Right panel: Constraints on coefficients of dimension-six operators $\bar{c}_H$ and $\bar{c}_{HB}$ with and without the inclusion of the ZZ fusion channel. The dashed and dot–dashed lines represent 2σ deviations from zero in the Zh channel at 250 GeV and 500 GeV (blue lines), respectively. The solid (red) lines indicates the constraint from ZZ fusion for 500 GeV plus 1 TeV. The outer (black-dashed) contour shows the constraint from combined Zh measurements and the middle (yellow) and inner (green) contours show the combined 2σ and 1σ results with ZZ fusion included.

4. Conclusions

Future e^+e^- provide unprecedented opportunities to explore the Higgs sector. The large sample size of clean Higgs events may be used to constrain not only deviations in Higgs couplings, but also nonstandard tensor structures arising from BSM physics. While the former are readily probed by rate measurements, the latter may be effectively probed using appropriately-constructed angular asymmetries. In this work, we have initiated the study of angular observables at future e^+e^- colliders such as CEPC and FCC-ee. We have taken particular care to account for the impact of realistic cut acceptance and detector effects on angular asymmetries.

Our primary result is a forecast of the precision with which angular asymmetries may be measured at future e^+e^- colliders. We have translated this forecast into projected sensitivity to a range of operators in the dimension-6 EFT, where angular measurements provide complementary sensitivity to rate measurements. Among other things, we have found that angular asymmetries provide a novel means of probing BSM corrections to the $hZ\gamma$ coupling beyond direct measurement of $e^+e^- \to h\gamma$. We also apply our results to a complete model of BSM physics, namely scalar top partners in supersymmetric extensions of the Standard Model, where angular observables help to constrain the well-known "blind spot" in rate measurements.

There are wide range of interesting future directions. In this work, we have focused on ZH events with $Z \to \ell^+\ell^-$ and $h \to b\bar{b}$ in order to obtain a relatively pure sample of signal events without significant background contamination.

Of course, there will be far more events involving alternate decays of the Z and Higgs which, while not background-free, could add considerable discriminating power. It would be useful to conduct a realistic study of these additional channels to determine the maximum possible sensitivity of angular asymmetries. Although we have taken care to account for the impact of cut acceptance and detector effects on angular asymmetries, our work has neglected the possible impact of theory uncertainties in the Standard Model prediction for angular asymmetries. A detailed estimate of current and projected theory uncertainties in the Standard Model prediction for Higgsstrahlung differential distributions would be broadly useful to future studies. More generally, this work serves as a starting point for investigating the full set of Higgs properties accessible at future e^+e^- colliders.

We also show sensitivities on the inclusive cross-section σ_Z^{inc} at multiple energies offer the possibility to distinguish contributions from different higher-dimensional operators induced by BSM physics. We demonstrate the ability to simultaneously constrain two operators whose effects are difficult to observe at the LHC, as shown in Sec. 3. Including the ZZ fusion channel provides as large as 50% relative improvement for the constraint on the chosen operators compared to the Zh-associated production channel alone.

Acknowledgment

Z. Liu acknowledges the hospitality of the HKIAS.

References

1. N. Craig, J. Gu, Z. Liu and K. Wang, *J. High Energy Phys.* **03**, 050 (2016), arXiv:1512.06877.
2. T. Han, Z. Liu, Z. Qian and J. Sayre, *Phys. Rev. D* **91**, 113007 (2015), arXiv:1504.01399.
3. ATLAS Collab. (G. Aad *et al.*), *Phys. Lett. B* **716**, 1 (2012), arXiv:1207.7214.
4. CMS Collab. (S. Chatrchyan *et al.*), *Phys. Lett. B* **716**, 30 (2012), arXiv:1207.7235.
5. D. Asner *et al.*, ILC Higgs White Paper, arXiv:1310.0763.
6. TLEP Design Study Working Group Collab. (M. Bicer *et al.*), *J. High Energy Phys.* **01**, 164 (2014), arXiv:1308.6176.
7. M. Ahmad *et al.*, CEPC-SPPC Preliminary Conceptual Design Report, Volume I: Physics and Detector, http://cepc.ihep.ac.cn/preCDR/volume.html (2015).
8. M. Beneke, D. Boito and Y.-M. Wang, *J. High Energy Phys.* **1411**, 028 (2014), arXiv:1406.1361.
9. N. Craig, M. Farina, M. McCullough and M. Perelstein, *J. High Energy Phys.* **1503**, 146 (2015), arXiv:1411.0676.
10. A. Apyan *et al.*, CEPC-SPPC Preliminary Conceptual Design Report, Volume II: Accelerator, http://cepc.ihep.ac.cn/preCDR/volume.html (2015).
11. http://tlep.web.cern.ch/content/machine-parameters.
12. B. Henning, X. Lu and H. Murayama, What do precision Higgs measurements buy us?, arXiv:1404.1058.
13. S. Weinberg, *Physica A* **96**, 327 (1979).
14. C. N. Leung, S. Love and S. Rao, *Z. Phys. C* **31**, 433 (1986).

15. W. Buchmuller and D. Wyler, *Nucl. Phys. B* **268**, 621 (1986).
16. K. Hagiwara, R. Szalapski and D. Zeppenfeld, *Phys. Lett. B* **318**, 155 (1993), arXiv:hep-ph/9308347.
17. G. Giudice, C. Grojean, A. Pomarol and R. Rattazzi, *J. High Energy Phys.* **0706**, 045 (2007), arXiv:hep-ph/0703164.
18. B. Grzadkowski, M. Iskrzynski, M. Misiak and J. Rosiek, *J. High Energy Phys.* **1010**, 085 (2010), arXiv:1008.4884.
19. R. Contino, M. Ghezzi, C. Grojean, M. Muhlleitner and M. Spira, *J. High Energy Phys.* **1307**, 035 (2013), arXiv:1303.3876.
20. Particle Data Group Collab. (K. Olive *et al.*), *Chin. Phys. C* **38**, 090001 (2014).
21. J. Ellis, V. Sanz and T. You, *J. High Energy Phys.* **1407**, 036 (2014), arXiv:1404.3667.
22. J. Ellis, V. Sanz and T. You, *J. High Energy Phys.* **1503**, 157 (2015), arXiv:1410.7703.
23. A. Biekötter, A. Knochel, M. Krämer, D. Liu and F. Riva, *Phys. Rev. D* **91**, 055029 (2015), arXiv:1406.7320.
24. A. Falkowski and F. Riva, *J. High Energy Phys.* **1502**, 039 (2015), arXiv:1411.0669.
25. V. Barger, T. Han, P. Langacker, B. McElrath and P. Zerwas, *Phys. Rev. D* **67**, 115001 (2003), arXiv:hep-ph/0301097.
26. N. Craig, S. Knapen and P. Longhi, *Phys. Rev. Lett.* **114**, 061803 (2015), arXiv:1410.6808.
27. A. Azatov, C. Grojean, A. Paul and E. Salvioni, *Zh. Eksp. Teor. Fiz.* **147**, 410 (2015), arXiv:1406.6338.

Probing the Nonunitarity of the
Leptonic Mixing Matrix at the CEPC

Stefan Antusch[*,†,‡] and Oliver Fischer[*,§]

*Department of Physics, University of Basel,
Klingelbergstr. 82, CH-4056 Basel, Switzerland
† Max-Planck Institut für Physik (Werner-Heisenberg-Institut),
Föhringer Ring 6, D-80805 München, Germany
‡ stefan.antusch@unibas.ch
§ oliver.fischer@unibas.ch

The nonunitarity of the leptonic mixing matrix is a generic signal of new physics aiming at the generation of the observed neutrino masses. We discuss the Minimal Unitarity Violation (MUV) scheme, an effective field theory framework which represents the class of extensions of the Standard Model (SM) by heavy neutral leptons, and discuss the present bounds on the nonunitarity parameters as well as estimates for the sensitivity of the CEPC, based on the performance parameters from the preCDR.

Keywords: Neutrino masses; nonunitarity of the leptonic mixing matrix; CEPC.

1. Introduction

The experimental results on neutrino oscillations have unambiguously shown that *at least* two of the three known Standard Model (SM) neutrinos are massive. However, the mass generating mechanism is still unknown and constitutes one of the great open questions in particle physics. At present, no experimental evidence exists to tell us which among the various possible extensions of the SM to realize the observed neutrino masses is realized in nature.

One of the best-motivated extensions of the SM to generate the masses for the SM neutrinos consists in adding sterile neutrinos to the SM particle content, which are often also referred to as "gauge singlet fermions," "heavy neutral leptons" or simply "heavy" or "right-handed" neutrinos (see e.g. Ref. 1 and references therein).

When the sterile neutrinos are much heavier than the energy scale of a given experiment, only the light neutrinos propagate, which implies that the experiments are sensitive to an "effective leptonic mixing matrix." This effective leptonic mixing matrix is given by a submatrix of the full unitary leptonic mixing matrix, and thus is in general not unitary. The new physics effects within this class of SM extensions can be described in a model independent way by the effective theory framework of the Minimal Unitarity Violation (MUV) scheme.[2]

In this paper, we discuss the expected sensitivity of the Circular Electron Positron Collider (CEPC) and other present and future experiments for probing the nonunitarity of the leptonic mixing matrix in the MUV scheme, updating some of our results[3] to account for the performance parameters from the CEPC preCDR.

2. Origin of Nonunitarity of the Leptonic Mixing Matrix

A very general effective description of leptonic nonunitarity in scenarios with an arbitrary number of sterile neutrinos is given by the MUV scheme.[2] It describes the generic situation that the left-handed neutrinos mix with other neutral fermionic fields that are much heavier than the energy scale where the considered experiments are performed.

Adding n sterile neutrinos to the SM, the resulting generalized "seesaw extension of the SM" is given by

$$\mathscr{L} = \mathscr{L}_{\text{SM}} - \frac{1}{2}\overline{N_{\text{R}}^I}M_{IJ}^N N_{\text{R}}^{c\,J} - (y_{\nu_\alpha})_{I\alpha}\overline{N_{\text{R}}^I}\tilde{\phi}^\dagger L^\alpha + \text{H.c.}, \tag{1}$$

where the N_{R}^I ($I = 1,\ldots,n$) are sterile neutrinos, $\tilde{\phi} \equiv i\tau_2\phi^*$ and L^α ($\alpha = 1, 2, 3$) are the lepton doublets. ϕ denotes the SM Higgs doublet field, which breaks the electroweak (EW) symmetry spontaneously by acquiring a vacuum expectation value $v_{\text{EW}} = 246.44$ GeV in its neutral component.

When the mass scale of the sterile neutrinos, denoted by M in the following, is much larger than v_{EW}, the N_{R}^I can be integrated out of the theory below M and an effective theory emerges which contains the effective dimension 5 and dimension 6 operators $\delta\mathcal{L}^{d=5}$ and $\delta\mathcal{L}^{d=6}$. The SM extended by these two effective operators defines the MUV scheme.

The operator $\delta\mathcal{L}^{d=5}$ generates the masses of the light neutrinos after electroweak symmetry breaking, and is given by

$$\delta\mathcal{L}^{d=5} = \frac{1}{2}c_{\alpha\beta}^{d=5}(\overline{L^c}_\alpha\tilde{\phi}^*)(\tilde{\phi}^\dagger L_\beta) + \text{H.c.} \tag{2}$$

The second effective operator

$$\delta\mathcal{L}^{d=6} = c_{\alpha\beta}^{d=6}(\bar{L}_\alpha\tilde{\phi})i\slashed{\partial}(\tilde{\phi}^\dagger L_\beta) \tag{3}$$

generates a contribution to the kinetic terms of the neutrinos (and only the neutrinos) after EW symmetry breaking. Canonically normalizing the kinetic terms in general involves a transformation of the neutrino fields by a nonunitary matrix and leads to a nonunitary (effective) leptonic mixing matrix (see, e.g. Refs. 2, 4 and 5).

The coefficient matrix in the definition of $\delta\mathcal{L}^{d=5}$, (cf. Eq. (2)) can be connected to the parameters of the seesaw extension of the SM by

$$c_{\alpha\beta}^{d=5} = (y_\nu^T)_{\alpha I}(M_N)_{IJ}^{-1}(y_\nu)_{J\beta}. \tag{4}$$

After EW symmetry breaking, $\delta\mathcal{L}^{d=5}$ generates the mass matrix of the light neutrinos:

$$(m_\nu)_{\alpha\beta} = -\frac{v_{\text{EW}}^2}{2} c_{\alpha\beta}^{d=5} \,, \tag{5}$$

which corresponds to the usual (generalized) seesaw formula. The coefficient matrix for the dimension 6 operator, i.e. $c_{\alpha\beta}^{d=6}$ in Eq. (3), is related to the sterile neutrino parameters by (cf. Ref. 6)

$$c_{\alpha\beta}^{d=6} = \frac{y_{\nu_\alpha}^* y_{\nu_\beta}}{M^2} \,. \tag{6}$$

After EW symmetry breaking and the canonical normalization of the neutrino kinetic terms, the unitary mixing matrix in the lepton sector is modified to the (effective) nonunitary leptonic mixing matrix $\mathcal{N}$. $\mathcal{N}$ is related to $c_{\alpha\beta}^{d=6}$ via:

$$(\mathcal{N}\mathcal{N}^\dagger)_{\alpha\beta}^{-1} - 1_{\alpha\beta} = \frac{v_{\text{EW}}^2}{2} c_{\alpha\beta}^{d=6} \,. \tag{7}$$

One can use the Hermitean matrix ε (with small entries) to parametrize the deviation of the leptonic mixing matrix $\mathcal{N}$ from unitarity by defining

$$(\mathcal{N}\mathcal{N}^\dagger)_{\alpha\beta} = 1_{\alpha\beta} + \varepsilon_{\alpha\beta} \,. \tag{8}$$

The matrix elements $\varepsilon_{\alpha\beta}$ are related to $c_{\alpha\beta}^{d=6}$ by

$$\varepsilon_{\alpha\beta} = -\frac{v_{\text{EW}}^2}{2} c_{\alpha\beta}^{d=6} \,, \tag{9}$$

up to higher order terms in the small elements $\varepsilon_{\alpha\beta}$.

In the MUV scheme, the charged and neutral EW currents are modified as:

$$j_\mu^\pm = \bar{\ell}_\alpha \gamma_\mu \mathcal{N}_{\alpha j} \nu_j + \text{H.c.} \,, \quad j_\mu^0 = \bar{\nu}_i (\mathcal{N}^\dagger \mathcal{N})_{ij} \gamma_\mu \nu_j \,, \tag{10}$$

where ℓ_α denote the charged leptons and the ν_i, ν_j denote the light neutrino mass eigenstates ($i, j \in \{1, 2, 3\}$). These modifications can lead to various observable effects, which allow to test the nonunitarity of the leptonic mixing matrix experimentally, as we will now discuss.

3. Electroweak Precision Observables

Due to the modifications of the EW interactions, the presence of sterile neutrinos changes the theory predictions for the Electroweak Precision Observables (EWPOs). For calculating the prediction in the MUV scheme we make use of the high precision of the most accurately measured[7] parameters:

$$\alpha(m_z)^{-1} = 127.944(14) \,,$$

$$G_F = 1.1663787(6) \times 10^{-5} \text{ GeV}^{-2} \,, \tag{11}$$

$$m_Z = 91.1875(21) \,,$$

where m_Z denotes the Z pole mass, α the fine structure constant and G_F the Fermi constant.

The parameters α and m_Z are not modified in the presence of sterile neutrinos (i.e. within MUV). The Fermi constant G_F, however, is measured from muon decay, which is sensitive to the charged current interactions (cf. Eq. (10)). The experimentally measured quantity is the muon decay constant G_μ, which, in the MUV scheme, has the following tree level relation to G_F:

$$G_\mu^2 = G_F^2 (\mathcal{N}\mathcal{N}^\dagger)_{\mu\mu} (\mathcal{N}\mathcal{N}^\dagger)_{ee} , \tag{12}$$

where we have assumed that $M > m_\mu$ and $\mathcal{N}$ is the nonunitary leptonic (PMNS) mixing matrix. This is very important since the Fermi constant enters the theory predictions for most EWPOs. For a list of modifications of the EWPOs, see Ref. 3 and references therein.

4. Universality Tests and Low Energy Precision Observables

In addition to the EWPOs, also the lepton universality tests from W decays at the CEPC will be very powerful probes of leptonic nonunitarity. Furthermore, various low energy precision observables are very relevant:[3]

- The lepton universality observables are given by the ratios of decay rates: $R^X_{\alpha\beta} = \Gamma^X_\alpha / \Gamma^X_\beta$, where Γ^X_α denotes a decay width including a charged lepton ℓ_α and a neutrino. In the SM, $R^X_{\alpha\beta} = 1$ holds for all α, β, X, which is generally called lepton universality. In our analysis, we consider the low energy measurements of π, μ, τ, K and W decays from experiments, which are very precise probes of the charged current as defined in Eq. (10).
- Strong constraints on the nonunitarity of leptonic mixing can also be obtained from measurements of rare lepton flavor violating processes, in particular from the searches for charged lepton decays $\ell_\rho \to \ell_\sigma \gamma$. The presently strongest constraint comes from the MEG collaboration[8] with the upper bound $\text{Br}[\mu \to e\gamma] \leq 5.7 \times 10^{-13}$. Such experiments however probe only the off-diagonal nonunitarity parameters.
- The very precise measurements of meson decays are used to determine the unitarity of the first row of the CKM matrix. The unitarity condition on the CKM matrix implies strong bounds on the modification of the weak currents in Eq. (10) and thus on the nonunitarity parameters.
- The modified theory prediction of the weak mixing angle can be constrained by the results from the NuTeV experiment, which measured deep inelastic scattering of a neutrino beam off a nucleon. The initially significant deviation[9] from the LEP measurement has been reinvestigated (cf. Ref. 10) with the effect that the tension with other precision data was removed.
- In our analysis, we furthermore include low energy measurements of parity violation from the weak currents at energies far below the Z boson mass. This approach can allow to measure the weak mixing angle with a precision below

the percent level, see e.g. Refs. 11 and 12 for an overview. Currently, the best measured values for the weak mixing angle at low energies come from the determination of the weak charge of Caesium[13] and of the proton,[14] and the parity violating asymmetry in Møller scattering.[12]

4.1. *Present constraints from a global fit*

We performed a Markov Chain Monte Carlo fit of the relevant six nonunitarity parameters in the MUV, including the correlations between the observables.[a] Our global analysis yields the following highest posterior density (HPD) intervals at 68% (1σ) Bayesian confidence level (CL):[3]

$$\begin{aligned}
\epsilon_{ee} &= -0.0012 \pm 0.0006 & |\epsilon_{e\mu}| &< 0.7 \times 10^{-5}\,, \\
|\epsilon_{\mu\mu}| &< 0.00023 & |\epsilon_{e\tau}| &< 0.00135\,, \\
\epsilon_{\tau\tau} &= -0.0025 \pm 0.0017 & |\epsilon_{\mu\tau}| &< 0.00048\,.
\end{aligned} \tag{13}$$

We remark that the moduli of the off-diagonal elements are generally restricted by the triangle inequality which implies:[17]

$$|\varepsilon_{\alpha\beta}| \le \sqrt{|\varepsilon_{\alpha\alpha}||\varepsilon_{\beta\beta}|}\,. \tag{14}$$

We found that the best fit points for the off-diagonal $\varepsilon_{\alpha\beta}$ and for $\varepsilon_{\mu\mu}$ are at zero. At 90% CL, the constraints on the nonunitary PMNS matrix are:

$$\left|\mathcal{N}\mathcal{N}^\dagger\right| = \begin{pmatrix} 0.9979 - 0.9998 & < 10^{-5} & < 0.0021 \\ < 10^{-5} & 0.9996 - 1.0 & < 0.0008 \\ < 0.0021 & < 0.0008 & 0.9947 - 1.0 \end{pmatrix}\,. \tag{15}$$

Our result show that nonzero ε_{ee} and $\varepsilon_{\tau\tau}$ improve the fit to the data, while the best-fit value for $\varepsilon_{\mu\mu}$ is zero. We note, that negative $\varepsilon_{\alpha\alpha}$ were imposed as prior in the analysis (as required in the MUV scheme).

In our fit, it turned out that the experimental bounds on $\varepsilon_{e\tau}$ and $\varepsilon_{\mu\tau}$ are comparable to those from Eq. (14). It is interesting to remark that a future observation of rare tau decays beyond the level allowed by the triangle inequality, or a strong experimental indication of a positive $\varepsilon_{\alpha\alpha}$, has the potential to rule out the MUV scheme.

5. Improved Sensitivity to Leptonic Nonunitarity at the CEPC

As we will now discuss, the CEPC would provide excellent sensitivity to the nonunitarity of the leptonic mixing matrix, in particular via improved measurements of the Electroweak Precision Observables (EWPOs) and via universality tests using the leptonic decays of the large W boson sample.

---

[a]We note that the phases of the complex off-diagonal nonunitarity parameters can be probed by future neutrino oscillation experiments.[15,16] Also note that within the MUV scheme, the diagonal nonunitarity parameters are real and ≤ 0.

Table 1. Estimated precision for the EWPOs, based on the CEPC preCDR.[18]

Observable	LEP precision	From CEPC preCDR
M_W [MeV]	33	3
$\sin^2 \theta_W^{\text{eff}}$	0.07%	0.01%
R_b	0.3%	0.08%
R_c	0.3%	0.07%
R_{inv}	0.27%	8.9×10^{-4}
R_ℓ	0.1%	0.1%
Γ_ℓ	0.1%	0.1%
σ_h^0 [nb]	8.9×10^{-4}	1×10^{-4}

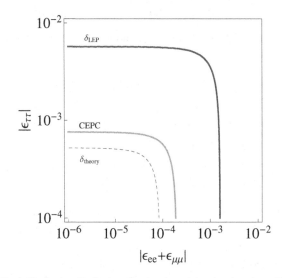

Fig. 1. (Color online) Exclusion limits on the $|\varepsilon_{\tau\tau}|$ and $|\varepsilon_{ee}| + |\varepsilon_{\mu\mu}|$ at 90% confidence level (updated version of the figure from Ref. 3). The solid blue line denotes the present experimental precision of the EWPOs from LEP, while the dashed blue line denotes the present theory uncertainty. The solid yellow line corresponds to the estimated precision of the CEPC.

5.1. Sensitivity via EWPOs

The modifications of the theory predictions for the EWPOs depend on the sum $|\varepsilon_{ee}| + |\varepsilon_{\mu\mu}|$ and on $|\varepsilon_{\tau\tau}|$. In order to estimate the possible sensitivity of the CEPC, we use the estimated precision for the EWPOs given in Table 1, based on the preCDR. Note that we consider the projected systematic uncertainties rather than the statistical ones, which could be much smaller. The theory uncertainties are set to zero unless stated otherwise.

The estimated sensitivity for the (combination of) nonunitarity parameters $|\varepsilon_{ee}| + |\varepsilon_{\mu\mu}|$ and $|\varepsilon_{\tau\tau}|$ is shown in Fig. 1, at the 90% confidence level, by the solid yellow line. The solid blue line denotes the present constraints from LEP and the dashed blue line denotes the limit from the present theory uncertainty. The sensitivity of the CEPC to $|\varepsilon_{ee}| + |\varepsilon_{\mu\mu}|$ of $\sim 2 \times 10^{-4}$ can be translated via Eq. (6) into a mass for the "sterile" or "right-handed" neutrino of ~ 12 TeV, for Yukawa couplings of order one.

5.2. *Sensitivities via lepton universality tests*

In order to estimate the sensitivities of the CEPC and other planned future experiments to leptonic nonunitarity via universality tests, we use the performance parameters given in Table 2.

The prediction for the universality observables in the MUV scheme depends on the differences between the diagonal nonunitarity parameters. We choose the following differences as parameters:

$$\Delta_{\tau\mu} \equiv \varepsilon_{\tau\tau} - \varepsilon_{\mu\mu} \quad \text{and} \quad \Delta_{\mu e} \equiv \varepsilon_{\mu\mu} - \varepsilon_{ee}. \tag{16}$$

The exclusion sensitivity contours for the two combinations of nonunitarity parameters at 90% CL is shown in Fig. 2. The blue line represents the current constraints,

Table 2.   Performance parameters of future experiments for testing lepton universality.

Observable	R^{ℓ}	R^{π}	R^{K}	$\text{Br}(W \to \ell\nu)$
Precision	0.001	0.001	0.004	0.0003
Experiment	Tau factories[19]	TRIUMF,[20] PSI[21]	NA62[22]	CEPC[18]

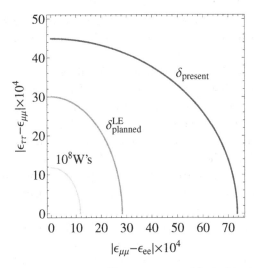

Fig. 2.   (Color online) Exclusion limits at 90% confidence level.[3] The blue line denotes the present experimental exclusion limits, the orange line represents the estimated improvements[19–22,26] in the low energy sector. The green line includes the estimated improvement from $W \to \ell_{\alpha}$ decays at the CEPC.

while the improvements by the planned low energy experiments are shown by the orange line. The estimated sensitivity of the CEPC (with 10^8 W decays), combined with the future low energy experiments, is represented by the green line.

5.3. *Expected improvements from charged lepton flavor violation and neutrino oscillation experiments*

In the fit to the present data, the experimental constraint on the charged lepton flavor violating (cLFV) decay $\mu \to e\gamma$ from MEG[8] drives the strong bound on the nonunitarity parameter $|\varepsilon_{e\mu}|$. Future tests of cLFV will provide a substantial improvement in precision. For example, the sensitivity of the PRISM/PRIME project[23] and a Mu2e upgrade[24] could achieve a sensitivity of $|\varepsilon_{e\mu}| \sim 3.6 \times 10^{-7}$.

Furthermore, the precision to the branching ratio for the cLFV tau decay $\tau \to e\gamma$ is expected to improve 10^{-9} at SuperKEKB,[25] which leads to an improved sensitivity of $|\varepsilon_{e\tau}| \sim 1.5 \times 10^{-3}$. Finally, we remark that a sensitivity to the phases of the parameters could be achieved in neutrino oscillation experiments, as discussed in Refs. 15 and 16 within the MUV scheme.

6. Discussion and Conclusions

In this paper, we have reviewed and partly updated the results from our previous work[3] where we have investigated the potential of future lepton colliders, such as the CEPC, for testing the nonunitarity of the leptonic mixing matrix. As framework we considered the MUV scheme, which is an effective field theory description of extensions of the SM by "heavy" sterile neutrinos.

For "heavy" sterile neutrinos, i.e. with masses much larger than the EW scale, we presented the results for the present constraints on the nonunitarity parameters from a global fit in the MUV scheme. Although the present electroweak precision data favors sterile neutrinos with active-sterile mixing $|\theta_e|^2 = \mathcal{O}(10^{-3})$ at the 2σ level, which might be a first hint at the existence of sterile neutrinos, we rather view the constraints on the nonunitarity parameters at the 90% Bayesian confidence level as the main conclusion from the fit to the present data.

Future experiments have the potential to push the search for new physics (such as sterile neutrinos) via the nonunitarity of the leptonic mixing matrix to a new level. A very high sensitivity to $|\varepsilon_{ee}| + |\varepsilon_{\mu\mu}|$ and $|\varepsilon_{\tau\tau}|$ down to $\sim 2 \times 10^{-4}$ and $\sim 10^{-3}$, respectively, can be achieved via the precision measurements of the EWPOs at the CEPC. Also the universality tests in W decays at the CEPC would be very powerful, and can yield a sensitivity to the differences of the diagonal nonunitarity parameters down to $\sim 10^{-3}$. Furthermore, complementary experiments, for instance on cLFV and on neutrino oscillations, could probe very sensitively the off-diagonal flavor-violating nonunitarity parameters.

In summary, the CEPC would be a powerful experiment to probe the nonunitarity of the leptonic mixing matrix and thereby search for new physics beyond the

SM (such as sterile neutrinos) needed to explain the observed masses of the light neutrinos.

Acknowledgments

This work was supported by the Swiss National Science Foundation. We thank the organizers of the IAS Program on the Future of High Energy Physics in Hong Kong for their hospitality.

References

1. K. N. Abazajian *et al.*, arXiv:1204.5379 [hep-ph].
2. S. Antusch, C. Biggio, E. Fernandez-Martinez, M. B. Gavela and J. Lopez-Pavon, *J. High Energy Phys.* **0610**, 084 (2006), doi:10.1088/1126-6708/2006/10/084, arXiv:hep-ph/0607020.
3. S. Antusch and O. Fischer, *J. High Energy Phys.* **1410**, 094 (2014), arXiv:1407.6607 [hep-ph].
4. A. De Gouvea, G. F. Giudice, A. Strumia and K. Tobe, *Nucl. Phys. B* **623**, 395 (2002), arXiv:hep-ph/0107156.
5. A. Broncano, M. B. Gavela and E. E. Jenkins, *Phys. Lett. B* **552**, 177 (2003) [Erratum: *ibid.* **636**, 332 (2006)], arXiv:hep-ph/0210271.
6. A. Broncano, M. B. Gavela and E. E. Jenkins, *Nucl. Phys. B* **672**, 163 (2003), arXiv:hep-ph/0307058.
7. Particle Data Group Collab. (K. A. Olive *et al.*), *Chin. Phys. C* **38**, 090001 (2014).
8. MEG Collab. (J. Adam *et al.*), *Phys. Rev. Lett.* **110**, 201801 (2013), arXiv:1303.0754 [hep-ex].
9. NuTeV Collab. (G. P. Zeller *et al.*), *Phys. Rev. Lett.* **88**, 091802 (2002) [Erratum: *ibid.* **90**, 239902 (2003)], arXiv:hep-ex/0110059.
10. W. Bentz, I. C. Cloet, J. T. Londergan and A. W. Thomas, *Phys. Lett. B* **693**, 462 (2010), arXiv:0908.3198 [nucl-th].
11. J. Erler and M. J. Ramsey-Musolf, *Prog. Part. Nucl. Phys.* **54**, 351 (2005), arXiv:hep-ph/0404291.
12. K. S. Kumar, S. Mantry, W. J. Marciano and P. A. Souder, *Annu. Rev. Nucl. Part. Sci.* **63**, 237 (2013), arXiv:1302.6263 [hep-ex].
13. V. A. Dzuba, J. C. Berengut, V. V. Flambaum and B. Roberts, *Phys. Rev. Lett.* **109**, 203003 (2012), arXiv:1207.5864 [hep-ph].
14. Qweak Collab. (Nuruzzaman), *EPJ Web Conf.* **71**, 00100 (2014), arXiv:1312.6009 [nucl-ex].
15. E. Fernandez-Martinez, M. B. Gavela, J. Lopez-Pavon and O. Yasuda, *Phys. Lett. B* **649**, 427 (2007), arXiv:hep-ph/0703098.
16. S. Antusch, M. Blennow, E. Fernandez-Martinez and J. Lopez-Pavon, *Phys. Rev. D* **80**, 033002 (2009), arXiv:0903.3986 [hep-ph].
17. S. Antusch, J. P. Baumann and E. Fernandez-Martinez, *Nucl. Phys. B* **810**, 369 (2009), arXiv:0807.1003 [hep-ph].
18. http://cepc.ihep.ac.cn/preCDR/.
19. M. E. Biagini *et al.*, arXiv:1310.6944 [physics.acc-ph].
20. A. Aguilar-Arevalo *et al.*, *Nucl. Instrum. Methods A* **609**, 102 (2009).
21. D. Pocanic *et al.*, *Phys. Rev. Lett.* **93**, 181803 (2004), arXiv:hep-ex/0312030.

22. NA62 Collab. (E. Goudzovski), *Nucl. Phys. B (Proc. Suppl.)* **210-211**, 163 (2011).
23. R. J. Barlow, *Nucl. Phys. B (Proc. Suppl.)* **218**, 44 (2011).
24. mu2e Collab. (K. Knoepfel *et al.*), arXiv:1307.1168.
25. SuperKEKB Physics Working Group Collab. (A. G. Akeroyd *et al.*), arXiv:hep-ex/0406071.
26. MOLLER Collab. (J. Mammei), *Nuovo Cimento C* **035**, 203 (2012), arXiv:1208.1260 [hep-ex].

Higgs Production Through Sterile Neutrinos

Stefan Antusch,[*,†,‡] Eros Cazzato[*,§] and Oliver Fischer[*,¶]

*Department of Physics, University of Basel,
Klingelbergstr. 82, CH-4056 Basel, Switzerland
†Max-Planck-Institut für Physik (Werner-Heisenberg-Institut),
Föhringer Ring 6, D-80805 München, Germany
‡stefan.antusch@unibas.ch
§e.cazzato@unibas.ch
¶oliver.fischer@unibas.ch

In scenarios with sterile (right-handed) neutrinos with an approximate "lepton-number-like" symmetry, the heavy neutrinos (the mass eigenstates) can have masses around the electroweak scale and couple to the Higgs boson with, in principle, unsuppressed Yukawa couplings, while the smallness of the light neutrinos' masses is guaranteed by the approximate symmetry. The on-shell production of the heavy neutrinos at lepton colliders, together with their subsequent decays into a light neutrino and a Higgs boson, constitutes a resonant contribution to the Higgs production mechanism. This resonant mono-Higgs production mechanism can contribute significantly to the mono-Higgs observables at future lepton colliders. A dedicated search for the heavy neutrinos in this channel exhibits sensitivities for the electron neutrino Yukawa coupling as small as $\sim 5 \times 10^{-3}$. Furthermore, the sensitivity is enhanced for higher center-of-mass energies, when identical integrated luminosities are considered.

Keywords: Neutrino physics; sterile neutrinos; Higgs physics; future lepton colliders; CEPC.

1. Introduction

Convincing evidence from neutrino oscillation experiments requires that at least two of the known neutrinos are massive. The absolute scale of the light neutrinos' masses is bounded to lie below about 0.2 eV, see for instance Ref. 1 for a recent review. A renormalizable term for the observed neutrino masses with the particle content of the SM is possible, when for instance sterile (right-handed) neutrinos are introduced, as is suggestively depicted in Fig. 1 by the missing right-chiral component of the active neutrinos. Sterile neutrinos can have Majorana masses, and Yukawa couplings to the three active neutrinos and to the Higgs field, which

¶Corresponding author.

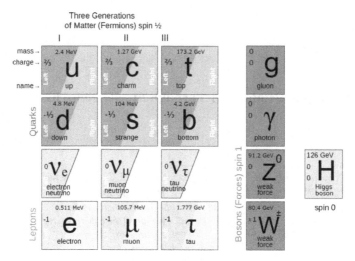

Fig. 1. Summary of the field content of the Standard Model: while the charged fermions have left- and right-chiral components, the right-chiral component of the neutrinos is missing.

leads to a mixing of the neutral fermions. Diagonalization of the resulting mass matrix leads to the masses of the light neutrinos via the so-called seesaw formula:

$$(m_\nu)_{\alpha\beta} = -\frac{1}{2}v_{EW}^2 \left(Y_\nu^T \cdot M^{-1} Y_\nu\right)_{\alpha\beta},$$ (1)

with the Yukawa matrices Y, the Majorana matrix M, and with $v_{EW} = 246.22$ GeV being the vacuum expectation value (VEV) of the neutral component of the Higgs $SU(2)_L$-doublet. In the simplistic case of one active and one sterile neutrino, with a large Majorana mass M_R and a small Yukawa coupling y_ν, such that $M_R \gg y_\nu v_{EW}$, the mass of the light neutrino m_ν is simply given by the naive seesaw formula:

$$m_\nu = \frac{1}{2}\frac{v_{EW}^2 |y_\nu|^2}{M_R}.$$ (2)

The experimental constraint of $m_\nu < 0.2$ eV makes it very unlikely to observe the associated heavy neutrino in an experiment directly, as it either has a mass of the order of the Grand Unification scale (for order one neutrino Yukawa couplings), or tiny Yukawa couplings (for a Majorana mass $\sim v_{EW}$), both of which strongly suppresses the active-sterile mixing.

The naive seesaw mechanism, for a scenario with two active neutrinos and two sterile neutrinos, is given by the following neutrino Yukawa and mass matrices:

$$Y_\nu = \begin{pmatrix} \mathcal{O}(y_\nu) & 0 \\ 0 & \mathcal{O}(y_\nu) \end{pmatrix}, \quad \begin{pmatrix} M_R & 0 \\ 0 & M_R + \varepsilon \end{pmatrix}.$$ (3)

Here, the light neutrinos' masses are given by

$$m_{\nu_i} = \frac{v_{EW}^2 \mathcal{O}(y_\nu^2)}{M_R}(1 + \delta_{i2}\varepsilon),$$ (4)

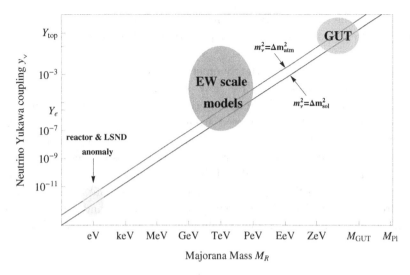

Fig. 2. (Color online) Schematic illustration of the parameter space of type-I seesaw extensions of the SM with right-handed neutrinos. The diagonal blue lines denote the naive seesaw relation from Eq. (2) for the two observed mass differences. The yellow area represents the reactor anomaly and the LSND findings, and the green and blue areas denote the domains of symmetry protected and GUT motivated seesaw models, respectively.

and the mass splitting between the two light neutrinos can be controlled with perturbations ($\sim \varepsilon$) in the mass matrix. Also in this scenario, the knowledge of m_{ν_i} implies a relation between y_ν and M_R.

In scenarios of the theory featuring a protective "lepton-number-like" symmetry, a specific structure of the Yukawa and mass matrices can be present, that can lead to a cancellation of the lepton-number violating masses of the light neutrinos. An example of such a structure, for two active and two sterile neutrinos, is given by

$$Y_\nu = \begin{pmatrix} \mathcal{O}(y_\nu) & 0 \\ \mathcal{O}(y_\nu) & 0 \end{pmatrix}, \quad \begin{pmatrix} 0 & M_R \\ M_R & \varepsilon \end{pmatrix}. \tag{5}$$

This specific structure of the mass matrix leads to the light neutrinos' masses:

$$m_{\nu_i} = 0 + \varepsilon \frac{v_{\mathrm{EW}}^2 \mathcal{O}(y_\nu^2)}{M_R^2}. \tag{6}$$

In these scenarios, the light neutrinos' mass scale is controlled by a perturbation of the mass matrix ($\sim \varepsilon$).

Therefore the constraints on the light neutrinos' mass scale do not affect y_ν and M_R, such that simultaneous $M_R \sim v_{\mathrm{EW}}$ and $y_\nu \sim 1$ are theoretically allowed. Various models of this type are known in the literature,[2-7] one well-known example being the so-called "inverse seesaw."[2,3]

In Fig. 2 we show a schematic illustration of the sterile neutrino parameter space. In the figure, two blue lines denote the naive seesaw relation between mass and Yukawa coupling for the case of one active and one sterile neutrino. The green

area represents the domain of the symmetry protected seesaw models with masses
on the electroweak scale, which could be probed by past, present and future particle
colliders.

A theoretical formulation of the Symmetry Protected Seesaw Scenarios (SPSS)
with heavy neutrino masses on the EW scale was given in Ref. 8. Therein,
the present constraints on the active–sterile mixing for heavy neutrino masses
above ~ 10 GeV have been extracted from a combination of precision experiments (including electroweak precision observables, charged lepton flavor violating decays,[9] lepton universality, decays of the Higgs boson) and direct searches
at the large electron positron collider (LEP) by the collaborations L3,[10] Delphi,[11]
Aleph[12] and Opal.[13] Furthermore, the prospects for direct and indirect heavy neutrino searches at the CEPC have been presented. Present and future constraints
on sterile neutrinos with masses on the electroweak scale have also been studied in
Ref. 14.

2. The Symmetry Protected Seesaw Scenario

The Symmetry Protected Seesaw Scenario (SPSS)[8] is a benchmark model that
captures the relevant features of seesaw models with a protective symmetry, which
allows for sterile (right-handed) neutrinos with masses around the electroweak (EW)
scale and up to $\mathcal{O}(1)$ Yukawa couplings.

The SPSS introduces one pair of sterile neutrinos N_R^I ($I = 1, 2$) and a suitable
"lepton-number-like" symmetry, where N_R^1 (N_R^2) has the same (opposite) charge
as the left-handed $SU(2)_L$ doublets L^α, $\alpha = e, \mu, \tau$. Possibly suppressed lepton-
number-violating effects such as the masses of the light neutrinos arise, when this
symmetry gets slightly broken. For the discussion in the following it is not necessary
to consider the effects from the small breaking of the protective symmetry.

The Lagrangian density of a generic seesaw model with two sterile neutrinos
with an exact "lepton-number-like" symmetry is given by

$$\mathscr{L} \supset \mathscr{L}_{\text{SM}} - \overline{N_R^1} M N_R^{2\,c} - y_{\nu_\alpha} \overline{N_R^1} \tilde{\phi}^\dagger L^\alpha + \text{H.c.}, \tag{7}$$

where the kinetic terms of the sterile neutrinos were omitted, $\mathscr{L}_{\text{SM}}$ contains the
known SM fields, and L^α and ϕ denote the lepton and Higgs doublets, respectively.
The y_{ν_α} are complex-valued neutrino Yukawa couplings and the sterile neutrino
mass parameter M can be chosen real without loss of generality.

Note that in the benchmark scenario two right-handed neutrinos are assumed
to dominate the collider phenomenology beyond the SM. This captures the general
features of symmetry protected seesaw scenarios with more than two right-handed
neutrinos, provided that the effects of the additional right-handed neutrinos can
be neglected, for example when additional sterile neutrinos have large masses, or,
alternatively, are uncharged under the "lepton-number-like" symmetry. In the limit
of exact symmetry, the additional sterile neutrinos indeed decouple from the other
particles, since no Yukawa couplings to the lepton doublets are allowed and they
cannot mix with the other sterile states.

In the SPSS, the mass matrix of the two sterile neutrinos and the neutrino Yukawa matrix take the form

$$
M_N = \frac{1}{2}\begin{pmatrix} 0 & M \\ M & 0 \end{pmatrix}, \quad Y_\nu = \begin{pmatrix} y_{\nu_e} & 0 \\ y_{\nu_\mu} & 0 \\ y_{\nu_\tau} & 0 \end{pmatrix}, \tag{8}
$$

where the zeroes correspond to the case of the "lepton-number-like" symmetry being exactly realized and are replaced with small quantities when the symmetry is slightly broken.

After EW symmetry breaking, the 5×5 mass matrix of the electrically neutral leptons emerges:

$$
\mathscr{L}_{\text{mass}} = -\frac{1}{2}
\begin{pmatrix} \overline{\nu^c_{eL}} \\ \overline{\nu^c_{\mu L}} \\ \overline{\nu^c_{\tau L}} \\ \overline{N^1_R} \\ \overline{N^2_R} \end{pmatrix}^T
\begin{pmatrix}
0 & 0 & 0 & m_e & 0 \\
0 & 0 & 0 & m_\mu & 0 \\
0 & 0 & 0 & m_\tau & 0 \\
m_e & m_\mu & m_\tau & 0 & M \\
0 & 0 & 0 & M & 0
\end{pmatrix}
\begin{pmatrix} \nu_{eL} \\ \nu_{\mu L} \\ \nu_{\tau L} \\ (N^1_R)^c \\ (N^2_R)^c \end{pmatrix}
+ \text{H.c.}, \tag{9}
$$

with the Dirac masses $m_\alpha = y_{\nu_\alpha} v_{\text{EW}}/\sqrt{2}$. The diagonalization of the mass matrix in Eq. (9), referred to as $\mathcal{M}$ in the following, with a unitary matrix U, results in the physical mass matrix, neglecting $\mathcal{O}(\theta^2)$ corrections to the heavy neutrinos:

$$
U^T \mathcal{M} U = \text{Diag}(0, 0, 0, M, M). \tag{10}
$$

The matrix U is the leptonic 5×5 mixing matrix. When the protective symmetry is broken, two of the light neutrinos obtain nonzero masses, and e.g. a third sterile neutrino could explain a nonzero mass for the third light neutrino. The mixing of the active and sterile neutrinos can be quantified by the mixing angles, defined as

$$
\theta_\alpha = \frac{y^*_{\nu_\alpha}}{\sqrt{2}} \frac{v_{\text{EW}}}{M}. \tag{11}
$$

With the leptonic mixing angles the leptonic mixing matrix U in Eq. (10), in the limit of exact symmetry, can be written as:

$$
U = \begin{pmatrix}
\mathcal{N}_{e1} & \mathcal{N}_{e2} & \mathcal{N}_{e3} & -\frac{i}{\sqrt{2}}\theta_e & \frac{1}{\sqrt{2}}\theta_e \\
\mathcal{N}_{\mu 1} & \mathcal{N}_{\mu 2} & \mathcal{N}_{\mu 3} & -\frac{i}{\sqrt{2}}\theta_\mu & \frac{1}{\sqrt{2}}\theta_\mu \\
\mathcal{N}_{\tau 1} & \mathcal{N}_{\tau 2} & \mathcal{N}_{\tau 3} & -\frac{i}{\sqrt{2}}\theta_\tau & \frac{1}{\sqrt{2}}\theta_\tau \\
0 & 0 & 0 & \frac{i}{\sqrt{2}} & \frac{1}{\sqrt{2}} \\
-\theta^*_e & -\theta^*_\mu & -\theta^*_\tau & \frac{-i}{\sqrt{2}}(1-\frac{1}{2}\theta^2) & \frac{1}{\sqrt{2}}(1-\frac{1}{2}\theta^2)
\end{pmatrix}. \tag{12}
$$

The mixing matrix shown above is unitary up to second-order in θ_α. The elements of the nonunitary 3×3 submatrix $\mathcal{N}$, which is the effective mixing matrix of the three active neutrinos, i.e. the Pontecorvo–Maki–Nakagawa–Sakata (PMNS) matrix, are given as

$$\mathcal{N}_{\alpha i} = (\delta_{\alpha\beta} - \tfrac{1}{2}\theta_\alpha\theta_\beta^*)(U_\ell)_{\beta i}, \tag{13}$$

with U_ℓ being a unitary 3×3 matrix. In the limit of the protective symmetry being exact, the SPSS introduces four parameters (beyond the SM) that are relevant for collider phenomenology: the moduli of the neutrino Yukawa couplings ($|y_{\nu_e}|$, $|y_{\nu_\mu}|$, $|y_{\nu_\tau}|$) and the Majorana mass M.

2.1. Weak interactions of the light and heavy neutrinos

Leptonic mixing leads to light and heavy neutrino mass eigenstates that both interact with the weak gauge bosons, which can be expressed in the mass basis as

$$j_\mu^\pm = \sum_{i=1}^5 \sum_{\alpha=e,\mu,\tau} \frac{g}{\sqrt{2}} \bar{\ell}_\alpha \gamma_\mu P_L U_{\alpha i} \tilde{n}_i + \text{H.c.}, \tag{14a}$$

$$j_\mu^0 = \sum_{i,j=1}^5 \sum_{\alpha=e,\mu,\tau} \frac{g}{2c_W} \bar{\tilde{n}}_j U_{j\alpha}^\dagger \gamma_\mu P_L U_{\alpha i} \tilde{n}_i, \tag{14b}$$

where g is the weak coupling constant, c_W is the cosine of the Weinberg angle and $P_L = \tfrac{1}{2}(1-\gamma^5)$ is the left-chiral projection operator, and where the mass eigenstates $\tilde{n}_j$ of the light and heavy neutrinos are defined via the active and sterile neutrinos:

$$\tilde{n}_j = (\nu_1, \nu_2, \nu_3, N_4, N_5)_j^T = U_{j\alpha}^\dagger n_\alpha, \quad n = \left(\nu_{e_L}, \nu_{\mu_L}, \nu_{\tau_L}, (N_R^1)^c, (N_R^2)^c\right)^T. \tag{15}$$

The neutrino mass eigenstates interact with the Higgs boson via the Yukawa interaction in the Lagrangian density in Eq. (7).

The weak interactions of the heavy neutrinos lead to the following partial decay widths (if kinematically allowed):

$$\Gamma(N_j \to W^\pm \ell_\alpha^\mp) = \frac{|\theta_\alpha|^2}{2} \frac{G_F M^3}{4\sqrt{2}\pi} \Pi_{(1+1)}(\mu_W), \tag{16a}$$

$$\Gamma(N_j \to Z\nu_i) = |\vartheta_{ij}|^2 \frac{G_F M^3}{4\sqrt{2}\pi} \Pi_{(1+1)}(\mu_Z), \tag{16b}$$

$$\Gamma(N_j \to h\nu_i) = |\vartheta_{ij}|^2 \frac{M^3}{8\pi v_{\text{EW}}^2} \left(1 - \mu_h^2\right)^2, \tag{16c}$$

where we introduced $\mu_X = m_X/M$, G_F as the Fermi constant, and the kinematic factor

$$\Pi_{(1+1)}(\mu_X) = \frac{1}{2}(1 - \mu_X^2)^2(2 + \mu_X^2). \tag{17}$$

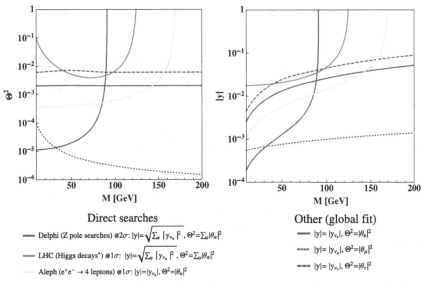

Fig. 3.   Summary of present bounds from direct searches for sterile neutrinos and indirect searches via a global fit of precision data. See Ref. 8 for details.

2.2. *Input parameters*

For the theory prediction we use the set of input parameters[15] with the highest experimental precision, which are summarized in the table below.

Input parameter	m_Z [GeV]	$\alpha(m_Z)^{-1}$	$G_F^{\rm SM}$ [GeV^{-2}]
Value	91.1875(21)	127.944(14)	$1.1663787(6) \times 10^{-5}$

The value of the Fermi constant has been inferred in the context of the SM, hence the superscript. It can be related to the Fermi constant in the context of the SPSS via the definition of the charged current interactions. This effect, and those due to the nonunitarity of the PMNS matrix from Eq. (13) leads to a modification of the theory prediction for a number of other theory parameters, which we generally refer to as "nonunitarity effects." For the details, we refer the reader to Ref. 8 which includes a detailed discussion and up-to-date constraints on the model parameters. The summarized constraints on the active-sterile mixing parameters as a function of the heavy neutrino mass are shown in Fig. 3.

3. Higgs Boson Production at Future Lepton Colliders

In the following we consider mono-Higgs events, where a Higgs boson is produced together with two light neutrinos. In the SM the two main mechanisms for Higgs production in leptonic collisions, at a circular electron–positron collider, such as the FCC-ee or the CEPC, are given by Higgsstrahlung and WW fusion. In the SPSS,

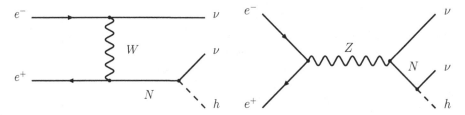

Fig. 4.   The dominating Feynman diagrams that give rise to the resonant mono-Higgs production cross-section $\sigma_{h\nu\nu}^{\text{Direct}}$ including the heavy neutrinos. The contribution from these diagrams leads to a resonant enhancement of the mono-Higgs production cross-section.

the corresponding cross-sections are modified because of the nonunitarity effects mentioned in Sec. 2. Furthermore, it is possible to produce Higgs bosons from the decays of on-shell heavy neutrinos. The total mono-Higgs production cross-section can therefore be decomposed as

$$\sigma_{h\nu\nu} = \sigma_{h\nu\nu}^{\text{SM}} + \sigma_{h\nu\nu}^{\text{Non-U}} + \sigma_{h\nu\nu}^{\text{Direct}}, \tag{18}$$

with $\sigma_{h\nu\nu}^{\text{SM}}$ being the SM prediction and $\sigma_{h\nu\nu}^{\text{Non-U}}$ the induced modification of the SM prediction due to the nonunitarity effects. The cross-section $\sigma_{h\nu\nu}^{\text{Direct}}$ denotes the direct production of heavy neutrinos on the mass shell and their subsequent decays into a Higgs boson and a light neutrino. We show the Feynman diagrams that correspond to $\sigma_{h\nu\nu}^{\text{Direct}}$ in Fig. 4 and note in passing, that the interference between direct production and the SM Higgs production mechanism is strongly suppressed.

The relevance of the individual contributions from the W- and Z-exchange diagrams to the combined cross-section $\sigma_{h\nu\nu}^{\text{Direct}}$ depends on the center-of-mass energy. At lower energies, in particular at the Z-pole, the s-channel production of heavy neutrinos is dominant, which renders $\sigma_{h\nu\nu}^{\text{Direct}}$ sensitive to the neutrino Yukawa couplings of all flavors. At higher center-of-mass energies, the W-exchange processes numerically dominate the heavy neutrino-production cross-section, which renders $\sigma_{h\nu\nu}^{\text{Direct}}$ sensitive only to y_{ν_e}. The heavy-neutrino-production cross-section is shown in Fig. 5 as a function of the heavy neutrino mass, for different center-of-mass energies. In the left panel, the exemplary values $|\theta_e|^2 = 0.0018$, $|\theta_{\mu,\tau}|^2 = 0$ are used for the W exchange diagram, while in the right panel $|\theta_\tau|^2 = 0.0042$, $|\theta_{e,\mu}|^2 = 0$. We notice, that the contribution from the W exchange diagram to the heavy-neutrino-production cross-section grows logarithmically with the center-of-mass energy, such that the relative amount of Higgs bosons produced from heavy neutrinos over the contribution produced from the SM mechanism increases for larger energies.

3.1. *Numerical analysis*

The contribution to the mono-Higgs events from the decays of heavy neutrinos is considered the signal, and we included all four-fermion processes without non-unitarity (i.e. all $|y_{\nu_\alpha}| = 0$) as the SM background. All processes are evaluated at

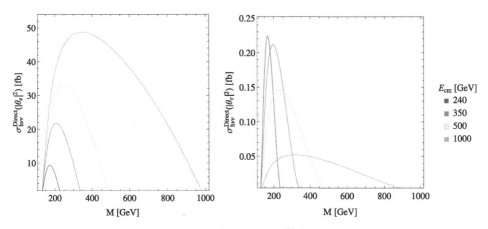

Fig. 5.   Resonant mono-Higgs production cross-section $\sigma_{h\nu\nu}^{\text{Direct}}$ as a function of the heavy neutrino mass. Left: The active–sterile mixing of $|\theta_e|^2 = 0.0018$ and $|\theta_{\mu,\tau}| = 0$ is used. Right: Active–sterile mixing of $|\theta_\tau|^2 = 0.0042$ and $|\theta_{e,\mu}| = 0$ is used. In this figure, the analytic formula from Buchmüller *et al.*[16] has been used for $\sigma(e^+e^- \to \nu N)$.

the tree level, which is sufficient in the SPSS (cf. Ref. 18), and the signal processes include all effects to order θ^2 in the active–sterile mixing parameters.

The SPSS was implemented via Feynrules[19] into the Monte Carlo event generator WHIZARD 2.2.7,[20,21] which allows for the appropriate simulation of leptonic collisions including initial state radiation. We remark that the effects from beamstrahlung are negligible for the considered center-of-mass energies and will be neglected in the following. The parton showering and hadronization has been carried out with PYTHIA 6.427[22] and the events were reconstructed with the ILD detector card using Delphes 3.2.0.[23] We considered the circular lepton collider FCC-ee, with the modi operandi of $\sqrt{s} = 240$, 350 and 500 GeV with integrated luminosities of 10, 3.5 and 1 ab^{-1}. Our results are valid also for the CEPC, provided the same integrated luminosities and energies are achieved. The results are only indicative for the ILC, where the beam polarization requires a systematic re-evaluation.

In order to enhance the statistics we focused on the Higgs decays into two hadronic jets (di-jet) which have a very large combined branching ratio of $\sim 70\%$. The di-jet plus missing energy signature comprises our mono-Higgs search channel such that we select events with two hadronic jets with an invariant mass of 100 GeV $\leq M_{jj} \leq 140$ GeV.

We simulated and reconstructed 10^5 events for each final state, with the exception for di-b-jet plus missing energy, where 3×10^6 events have been simulated and reconstructed. The di-jet invariant mass distribution in the mono-Higgs search channel in Fig. 6 shows the signal at the center-of-mass energy of 240 GeV and an integrated luminosity of 10 ab^{-1}, for a heavy neutrino mass of $M = 152$ GeV and with neutrino Yukawa coupling $|y_{\nu_e}| = 0.036$ and $|y_{\nu_{\mu,\tau}}| = 0$, which is compatible

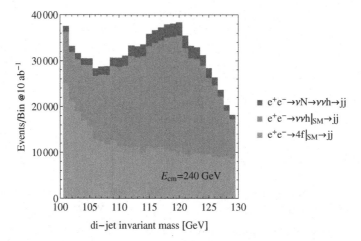

Fig. 6. (Color online) The di-jet invariant mass spectrum in the mono-Higgs search channel, for $\sqrt{s} = 240$ GeV, and a total integrated luminosity of 10 ab^{-1}. The red, green, and blue area denotes the dominant background, subdominant background, and the signal, respectively. Parameters used are $M = 152$ GeV, and y_{ν_e} given by the present upper bound[8,17] and $|y_{\nu_{\mu,\tau}}| = 0$.

with the present upper bounds. In the figure, the following not fully optimized cuts have been used: $N_j = 2$, $N_\ell = 0$, $110 < M_{jj} < 125$ GeV, $P_{jj} > 70$, $\not{E}_T > 15$ GeV.

In the analysis[24] the effects from all three neutrino Yukawa couplings $|y_{\nu_\alpha}|$, $\alpha = e$, μ, τ were studied, but only the electron flavor neutrino Yukawa coupling was found to create an observable effect. We will therefore focus on the case of $|y_{\nu_e}| \neq 0$ and $|y_{\nu_{\mu,\tau}}| = 0$.

We emphasize that, despite our kinematic cuts not being optimal, this example leads to a significance, i.e. signal/$\sqrt{\text{signal} + \text{background}}$, of 30, which implies that it would be hard to miss the heavy neutrino induced additional mono-Higgs events.

3.2. Contamination of SM parameters and the direct search

For the analysis of the Higgs boson at future lepton colliders so-called "standard cuts" have been defined,[25] which are designed to improve the ratio of mono-Higgs events over SM background events:

$\sqrt{s}$	240 GeV	350 GeV
Missing Mass [GeV]	$80 \leq M_{\text{miss}} \leq 140$	$50 \leq M_{\text{miss}} \leq 240$
Transverse P [GeV]	$20 \leq P_T \leq 70$	$10 \leq P_T \leq 140$
Longitudinal P [GeV]	$|P_L| < 60$	$|P_L| < 130$
Maximum P [GeV]	$|P| < 30$	$|P| < 60$
Di-jet Mass [GeV]	$100 \leq M_{jj} \leq 130$	$100 \leq M_{jj} \leq 130$
Angle (jets) [Rad]	$\alpha > 1.38$	$\alpha > 1.38$

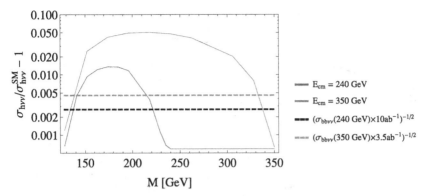

Fig. 7.   Heavy neutrino induced deviation of the mono-Higgs production cross-section after application of standard cuts, for y_{ν_e} at the 1σ upper bound. The horizontal dashed lines denote the relative statistical precision of the SM predicted events N, given by $1/\sqrt{N}$.

The cuts are not very efficient in filtering out the heavy neutrino events, such that a contamination of the sample of mono-Higgs events can occur. This can lead to a deviation from the theory prediction when interpreted in the context of the SM.

The heavy neutrino induced deviation of the mono-Higgs production cross-section is shown in Fig. 7, for $|y_{\nu_e}|$ at the present upper bound. The black and gray dashed lines denote the statistical accuracy at 1σ for 240 GeV and 350 GeV, respectively. As the mono-Higgs production cross-section can deviate significantly this can lead to a discrepancy when comparing the mono-Higgs with the other Higgs channels at 240 GeV, as can be inferred from their respective forecast[26] accuracy.

It is worth emphasizing that the above deviations of the mono-Higgs-production cross-section from the SM prediction is compatible with present constraints on the active–sterile mixing. Furthermore, provided that the present nonzero best-fit value for $|\theta_e|$ as reported[8,17,27] recently, get confirmed, this deviation in the mono-Higgs channel would be a prediction.

Apart from deviations of the SM parameters, a direct search in the mono-Higgs events for heavy neutrino signals is possible. We refer the interested reader to Ref. 24 for the details on the analysis. Here we show the resulting sensitivity of the mono-Higgs channel to active–sterile neutrino mixing $|y_{\nu_e}|$ in Fig. 8. In the figure, we show the sensitivities of the center-of-mass energies of 240, 350 and 500 GeV, with the total integrated luminosities of 10, 3.5, 1 ab^{-1}, respectively. The solid lines denote the reconstructed level, while the dashed lines denote the sensitivity from a parton level analysis.

Values of the neutrino Yukawa coupling y_{ν_e} above the solid lines give rise to a signal that can be distinguished from the SM background with a significance larger than 1σ. The sensitivity at 500 GeV is competitive with 240 GeV and 350 GeV, even for heavy neutrino masses $M \leq 250$ GeV and despite the lower luminosity. It is worth noting that also for values of the heavy neutrino mass above the kinematic threshold (i.e. $M > \sqrt{s}$) a signal is generated due to the nonunitarity effects in

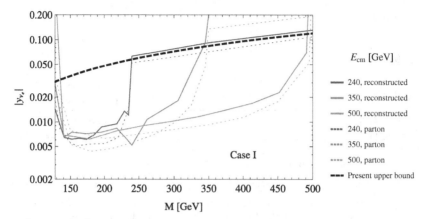

Fig. 8. (Color online) Future lepton collider sensitivity of the mono-Higgs search channel to the neutrino Yukawa coupling y_{ν_e} at 1σ. In this figure, for the center-of-mass energies of 240, 350 and 500 GeV, we considered total integrated luminosities of 10, 3.5 and 1 ab^{-1}, respectively.

mono-Higgs production, as is visible in the blue line in Fig. 8, above 240 GeV. It turns out, however, that the corresponding sensitivity above the present bound.

Mono-Higgs events from heavy neutrino decays can have a larger amount of missing energy compared to the SM, which provides an unambiguous signal. However, the considered target luminosities results in $\mathcal{O}(1)$ and $\mathcal{O}(10)$ signal events at 240 GeV and 350 GeV, respectively, such that they do not provide an improvement of the sensitivity. We note, that at 500 GeV, the considered luminosity results in less than $\mathcal{O}(1)$ events of this kind.

Altogether a future circular lepton collider has very promising prospects for discovering heavy neutrino signals in the mono-Higgs channel.

4. Summary and Conclusions

Sterile neutrinos are a well motivated extension of the SM, and in scenarios with protective symmetries the mass eigenstates can have masses around the electroweak scale and couple "strongly" (i.e. with Yukawa couplings comparable to the b-quark) to the Higgs boson while remaining compatible with the smallness of the light neutrinos' masses. The interactions with the Higgs boson can give rise to an efficient *resonant mono-Higgs production* mechanism, where a heavy neutrino is produced on shell and decays into a light neutrino and a Higgs boson. Therefore, future lepton colliders provide a promising environment to search for resonantly produced Higgs bosons as a signal for heavy neutrinos.

We have shown that the resonantly produced mono-Higgs events can effectively contaminate the SM analysis of the mono-Higgs channel. With $|\theta_e|$ within the present 1σ upper bounds, this can lead to deviations from the SM prediction at the percent level, which is larger than the estimated future accuracy.

We found further that a future circular lepton collider (where we considered the FCC-ee as an example) would be sensitive to the neutrino Yukawa coupling $|y_{\nu_e}|$ (respectively to the active–sterile mixing parameter $|\theta_e|$) down to $\sim 5 \times 10^{-3}$ via searches in the mono-Higgs channel. It is interesting to note that higher $\sqrt{s}$ allows for an increased range of testable heavy neutrino masses M and it also increases the signal-to-background ratio, such that a comparable sensitivity can be achieved with less integrated luminosity. In particular, the operation at $\sqrt{s} = 350$ GeV turns out to be even more sensitive than the one at 240 GeV.

We conclude that the mono-Higgs channel constitutes a sensitive probe for sterile neutrinos at future lepton colliders due to the possible sizeable deviations from the SM prediction. It possesses a large degree of complementarity to the other search channels for sterile neutrinos, outlined in Ref. 8 with estimated sensitivities.

Acknowledgments

This work was supported by the Swiss National Science Foundation. We thank the organizers of the IAS Program on the Future of High Energy Physics in Hong Kong, 2016, for their hospitality.

References

1. S. Gariazzo, C. Giunti, M. Laveder, Y. F. Li and E. M. Zavanin, *J. Phys. G* **43**, 033001 (2016), arXiv:1507.08204 [hep-ph].
2. D. Wyler and L. Wolfenstein, *Nucl. Phys. B* **218**, 205 (1983).
3. R. N. Mohapatra and J. W. F. Valle, *Phys. Rev. D* **34**, 1642 (1986).
4. M. Shaposhnikov, *Nucl. Phys. B* **763**, 49 (2007), arXiv:hep-ph/0605047.
5. J. Kersten and A. Y. Smirnov, *Phys. Rev. D* **76**, 073005 (2007), arXiv:0705.3221 [hep-ph].
6. M. B. Gavela, T. Hambye, D. Hernandez and P. Hernandez, *J. High Energy Phys.* **0909**, 038 (2009), arXiv:0906.1461 [hep-ph].
7. M. Malinsky, J. C. Romao and J. W. F. Valle, *Phys. Rev. Lett.* **95**, 161801 (2005), arXiv:hep-ph/0506296.
8. S. Antusch and O. Fischer, *J. High Energy Phys.* **1505**, 053 (2015), arXiv:1502.05915 [hep-ph].
9. MEG Collab. (J. Adam *et al.*), *Phys. Rev. Lett.* **110**, 201801 (2013), arXiv:1303.0754 [hep-ex].
10. L3 Collab. (O. Adriani *et al.*), *Phys. Rept.* **236**, 1 (1993).
11. DELPHI Collab. (P. Abreu *et al.*), *Z. Phys. C* **74**, 57 (1997) [Erratum: *ibid.* **75**, 580 (1997)].
12. ALEPH Collab. (D. Decamp *et al.*), *Phys. Rept.* **216**, 253 (1992).
13. OPAL Collab. (M. Z. Akrawy *et al.*), *Phys. Lett. B* **247**, 448 (1990).
14. F. del Aguila and J. A. Aguilar-Saavedra, *Phys. Lett. B* **672**, 158 (2009), arXiv:0809.2096 [hep-ph]; F. del Aguila, J. A. Aguilar-Saavedra and J. de Blas, *Acta Phys. Pol. B* **40**, 2901 (2009), arXiv:0910.2720 [hep-ph]; A. Das and N. Okada, *Phys. Rev. D* **88**, 113001 (2013), arXiv:1207.3734 [hep-ph]; K. N. Abazajian *et al.*, arXiv:1204.5379 [hep-ph]; P. S. Bhupal Dev, R. Franceschini and R. N. Mohapatra, *Phys. Rev. D* **86**, 093010 (2012), arXiv:1207.2756 [hep-ph]; M. Drewes, *Int. J.*

Mod. Phys. E **22**, 1330019 (2013), arXiv:1303.6912 [hep-ph]; C. H. Lee, P. S. Bhupal Dev and R. N. Mohapatra, *Phys. Rev. D* **88**, 093010 (2013), arXiv:1309.0774 [hep-ph]; J. C. Helo, M. Hirsch and S. Kovalenko, *Phys. Rev. D* **89**, 073005 (2014), arXiv:1312.2900 [hep-ph]; A. Das, P. S. Bhupal Dev and N. Okada, *Phys. Lett. B* **735**, 364 (2014), arXiv:1405.0177 [hep-ph]; FCC-ee study team Collab. (A. Blondel *et al.*), arXiv:1411.5230 [hep-ex]; A. Abada, V. De Romeri, S. Monteil, J. Orloff and A. M. Teixeira, *J. High Energy Phys.* **1504**, 051 (2015), arXiv:1412.6322 [hep-ph]; M. Drewes and B. Garbrecht, arXiv:1502.00477 [hep-ph]; F. F. Deppisch, P. S. Bhupal Dev and A. Pilaftsis, *New J. Phys.* **17**, 075019 (2015), arXiv:1502.06541 [hep-ph]; P. Humbert, M. Lindner and J. Smirnov, *J. High Energy Phys.* **1506**, 035 (2015), arXiv:1503.03066 [hep-ph]; S. Banerjee, P. S. B. Dev, A. Ibarra, T. Mandal and M. Mitra, *Phys. Rev. D* **92**, 075002 (2015), arXiv:1503.05491 [hep-ph]; S. Antusch and O. Fischer, *Int. J. Mod. Phys. A* **30**, 1544004 (2015); L. Duarte, J. Peressutti and O. A. Sampayo, *Phys. Rev. D* **92**, 093002 (2015), arXiv:1508.01588 [hep-ph]; N. Bizot and M. Frigerio, arXiv:1508.01645 [hep-ph]; C. O. Dib and C. S. Kim, *Phys. Rev. D* **92**, 093009 (2015), arXiv:1509.05981 [hep-ph]; A. Das and N. Okada, arXiv:1510.04790 [hep-ph]; A. Abada, V. De Romeri and A. M. Teixeira, arXiv:1510.06657 [hep-ph]; A. de Goueva and A. Kobach, *Phys. Rev. D* **93**, 033005 (2016), arXiv:1511.00683 [hep-ph]; L. Basso, arXiv:1512.06381 [hep-ph].

15. Particle Data Group Collab. (K. A. Olive *et al.*), *Chin. Phys. C* **38**, 090001 (2014).
16. W. Buchmuller and C. Greub, *Nucl. Phys. B* **363**, 345 (1991).
17. S. Antusch and O. Fischer, *J. High Energy Phys.* **1410**, 094 (2014), arXiv:1407.6607 [hep-ph].
18. E. Fernandez-Martinez, J. Hernandez-Garcia, J. Lopez-Pavon and M. Lucente, *J. High Energy Phys.* **1510**, 130 (2015), arXiv:1508.03051 [hep-ph].
19. A. Alloul, N. D. Christensen, C. Degrande, C. Duhr and B. Fuks, *Comput. Phys. Commun.* **185**, 2250 (2014), arXiv:1310.1921 [hep-ph].
20. W. Kilian, T. Ohl and J. Reuter, *Eur. Phys. J. C* **71**, 1742 (2011), arXiv:0708.4233 [hep-ph].
21. M. Moretti, T. Ohl and J. Reuter, arXiv:hep-ph/0102195.
22. T. Sjostrand, S. Mrenna and P. Z. Skands, *J. High Energy Phys.* **0605**, 026 (2006), arXiv:hep-ph/0603175.
23. M. Cacciari, G. P. Salam and G. Soyez, *Eur. Phys. J. C* **72**, 1896 (2012), arXiv:1111.6097 [hep-ph].
24. S. Antusch, E. Cazzato and O. Fischer, arXiv:1512.06035 [hep-ph].
25. H. Ono and A. Miyamoto, *Eur. Phys. J. C* **73**, 2343 (2013), arXiv:1207.0300 [hep-ex].
26. M. Ruan, arXiv:1411.5606 [hep-ex].
27. L. Basso, O. Fischer and J. J. van der Bij, *Europhys. Lett.* **105**, 11001 (2014), arXiv:1310.2057 [hep-ph].

Triple Gauge Couplings at Future Hadron and Lepton Colliders

Ligong Bian, Jing Shu and Yongchao Zhang

State Key Laboratory of Theoretical Physics and
Kavli Institute for Theoretical Physics China (KITPC),
Institute of Theoretical Physics, Chinese Academy of Sciences, Beijing 100190, P. R. China

The WW production is the primary channel to directly probe the triple gauge couplings (TGCs). We analyze the $e^+e^- \to W^+W^-$ process at the proposed circular electron–positron collider (CEPC), and find that the anomalous TGCs and relevant dimension six operators can be probed up to the order of 10^{-4}. We also estimate constraints at the 14 TeV (LHC), with both 300 fb^{-1} and 3000 fb^{-1} integrated luminosity from the leading lepton p_T and azimuthal angle difference $\Delta\phi_{ll}$ in the di-lepton channel. The constrain is somewhat weaker, up to the order of 10^{-3}. The limits on the TGCs are complementary to those on the electroweak precision observables and Higgs couplings.

Keywords: CEPC; LHC; TGCs.

1. Introduction: Anomalous Triple Gauge Couplings Beyond the SM

Given the 125 GeV Higgs has been found at the large hadron collider (LHC), the standard model (SM) is in some sense complete. However, it might suffer from new physics, e.g. in the gauge sector, which can be examined with an unprecedented precision at future lepton colliders such as the circular electron–positron collider (CEPC) and LHC Run II. In this short note, we summarize some of the key points in one recent paper by us,[1] regarding the future prospects of anomalous charged triple gauge couplings (TGCs) beyond SM at future lepton and hadron colliders.

Due to non-Abelian nature of the weak interaction, there exist triple and quartic couplings among the EW gauge bosons in the SM. Here we focus only on the charged TGCs, i.e. those involving the couplings $WW\gamma$ and WWZ. With the anomalous contributions beyond SM, the charged TGCs can be generally parametrized as[2]

$$\mathcal{L}_{\text{TGC}}/g_{WWV} = ig_{1,V}(W^+_{\mu\nu}W^-_\mu V_\nu - W^-_{\mu\nu}W^+_\mu V_\nu) + i\kappa_V W^+_\mu W^-_\nu V_{\mu\nu}$$

$$+ \frac{i\lambda_V}{M_W^2} W^+_{\lambda\mu}W^-_{\mu\nu}V_{\nu\lambda} + g_5^V \varepsilon_{\mu\nu\rho\sigma}(W^+_\mu \overset{\leftrightarrow}{\partial_\rho} W_\nu)V_\sigma$$

$$- g_4^V W^+_\mu W^-_\nu (\partial_\mu V_\nu + \partial_\nu V_\mu) + i\tilde\kappa_V W^+_\mu W^-_\nu \tilde V_{\mu\nu} + \frac{i\tilde\lambda_V}{M_W^2} W^+_{\lambda\mu}W^-_{\mu\nu}\tilde V_{\nu\lambda},$$

$$(1)$$

where $V = \gamma, Z$, the gauge couplings $g_{WW\gamma} = -e$, $g_{WWZ} = -e \cot \theta_W$ with $\cos \theta_W$ the weak mixing angle, the field strength tensor $F_{\mu\nu} \equiv \partial_\mu A_\nu - \partial_\nu A_\mu$ with $A = W$, γ, Z, and the conjugate tensor $\tilde{V}_{\mu\nu} = \frac{1}{2} \varepsilon_{\mu\nu\rho\sigma} V_{\rho\sigma}$, and $A \overset{\leftrightarrow}{\partial_\mu} B \equiv A(\partial_\mu B) - (\partial_\mu A)B$. Besides the SM TGCs, the Lagrangian equation (1) contains 14 anomalous TGCs up to dimension six. The parity (P) and charge conjugate (C) conservative couplings beyond SM $\Delta g_{1,Z} \equiv g_{1,Z} - 1$, $\Delta \kappa_{\gamma,Z} \equiv \kappa_{\gamma,Z} - 1$ and $\lambda_{\gamma,Z}$ can be related to the effective field theory (EFT) beyond SM, e.g. the dimension-6 operators in the SILH basis[3,4]

$$\Delta \mathcal{L} = \frac{i c_W \, g}{2 M_W^2} (H^\dagger \sigma^i \overset{\leftrightarrow}{D^\mu} H)(D^\nu W_{\mu\nu})^i + \frac{i c_{HW} \, g}{M_W^2} (D^\mu H)^\dagger \sigma^i (D^\nu H) W_{\mu\nu}^i$$

$$+ \frac{i c_{HB} \, g'}{M_W^2} (D^\mu H)^\dagger (D^\nu H) B_{\mu\nu} + \frac{c_{3W} \, g}{6 M_W^2} \epsilon^{ijk} W_\mu^{i\,\nu} W_\nu^{j\,\rho} W_\rho^{k\,\mu}. \qquad (2)$$

The c_W operator is strictly constrained by precision measurements and can be neglected as a first-order approximation, with only three operators left at the Dim-6 level[5-7]

$$\Delta g_{1,Z} = -\frac{c_{HW}}{\cos^2 \theta_W},$$

$$\Delta \kappa_\gamma = -(c_{fHW} + c_{HB}), \qquad (3)$$

$$\lambda_\gamma = -c_{3W}$$

with $\Delta \kappa_Z = \Delta g_{1,Z} - \tan^2 \theta_W \Delta \kappa_\gamma$ and $\lambda_\gamma = \lambda_Z$. Under such circumstance, the anomalous TGCs are related by the EW $SU(2)_L \times U(1)_Y$ gauge symmetry, and there is only three independent couplings in the C and P conserving sector.

2. CEPC Constraints

At both lepton and hadron colliders, the TGCs can be directly probed in the WW pair process. When the information of W boson decay is taken into consideration, the kinematics of $e^+ e^- \to W^+ W^- \to f_1 \bar{f}_2 \bar{f}_3 f_4$ is dictated by five angles in the narrow W width approximation: the scattering angle θ between e^- and W^- and the polar angles $\theta_{1,2}^*$ and the azimuthal angles $\phi_{1,2}^*$ for the decay products in the rest frame of $W^\mp$. The polarization of W bosons can be described by the spin density matrix (SDM) which contains the full helicity information of the W pairs. We resort alternatively to the differential cross-sections with regard to the five kinematic angles, which are more physically intuitive, and examine the effects of anomalous TGCs on the distributions of final fermions. It is found that all the distributions of the five angles contribute significantly to the sensitives of TGCs in most of the channels.

To examine the response of the angular distributions to the TGCs, we expand the differential cross-sections in terms of the aTGCs,

$$\frac{d\sigma}{d\Omega_k} = \frac{d\sigma_0}{d\Omega_k} [1 + \omega_i(\Omega_i)\alpha_i + \omega_{ij}(\Omega_k)\alpha_i\alpha_j], \qquad (4)$$

where $\Omega = \cos\theta$, $\cos\theta_{1,2}^*$, $\phi_{1,2}^*$ (or alternatively $\Omega = \cos\theta$, $\cos\theta_{\ell,\jmath}^*$, $\phi_{\ell,\jmath}^*$ with ℓ and $\jmath$ leptons and quark jets). It is straightforward to obtain analytically the linear coefficient functions ω_i from the differential cross-sections, which are used by the LEP experimental groups to extract constraints on the anomalous couplings.[8]

With a huge luminosity of 5 ab^{-1} at CEPC at a center-of-mass energy of $\sqrt{s} = 240$ GeV, we can collect a total number of 8.6×10^7 events of W pairs with 45%, 44% and 11% decaying respectively in the hadronic, semileptonic and leptonic channels. With such a huge statistics, these anomalous TGCs are expected to be severely constrained. Here for simplicity, we assume that the CEPC is optimistically designed such that the systematic errors are comparatively small and the TGC sensitivities are dominated by the statistical uncertainties. The SM radiative corrections are expected to be of the same order as the Dim-6 operators and cannot be simply ignored. Nevertheless, the high order corrections can be treated as a constant term and affect only the best fit values but not the relative errors. Since there is no measurement from CEPC yet, the radiative correction can be omitted for the time being.

For large numbers of events at CEPC, the statistical errors can be estimated to be $\sqrt{N_i}$ with N_i the number of events. Then it is straightforward to estimate the sensitivities of the anomalous TGCs and the relevant Dim-6 operators in Eq. (3). For the sake of simplicity, we concentrate only on the semileptonic channel in this short note, where the ambiguities involving quark jets have been taken into account. More information from the pure leptonic and hadronic channels and the combined sensitivities can be found in Ref. 1.

The expected semileptonic channel sensitivities of the aTGCs and Dim-6 operators at CEPC are presented in Table 1, where we assume an integrated luminosity of 5 ab^{-1}. All the sensitivities in this table are one parameter constraint where all other couplings or coefficients are fixed to zero. It is transparent that the limits on the TGCs and Dim-6 operators can reach up to the level of 10^{-4} in the semileptonic channel. When all the available channels are combined together, the sensitivities can be further improved. When two of the three anomalous couplings (or operator coefficients) are allowed to vary, we obtain the two-dimensional sensitivity plots presented in Fig. 1. The correlations between the three TGCs and Dim-6 operators are, respectively,

$$\rho^{\text{TGC}} = \begin{pmatrix} 1 & & \\ 0.793 & 1 & \\ 0.969 & 0.795 & 1 \end{pmatrix}, \quad \rho^{\text{operator}} = \begin{pmatrix} 1 & & \\ 0.906 & 1 & \\ 0.956 & 0.795 & 1 \end{pmatrix}. \quad (5)$$

Obviously, some of the TGCs (such as $\Delta g_{1,Z}$ and λ_γ) and operators are strongly correlated and there exist blind directions (such as $\Delta g_{1,Z} + \lambda_\gamma$) which are much less severely constrained. Such blind directions might be effected by experimental data and can be removed to some extent by incorporating the helicity information of $e^\pm$ and $W^\pm$.[9,10]

Table 1. Estimations of the 1σ prospects (in units of 10^{-4}) for the aTGCs and Dim-6 operators in the semileptonic decay channel of WW process at CEPC with $\sqrt{s} = 240$ GeV and an integrated luminosity of 5 ab^{-1}.

$\Delta g_{1,Z}$	$\Delta \kappa_\gamma$	λ_γ	c_{HW}	c_{HB}	c_{3W}
2.19	3.33	2.35	1.18	3.34	2.35

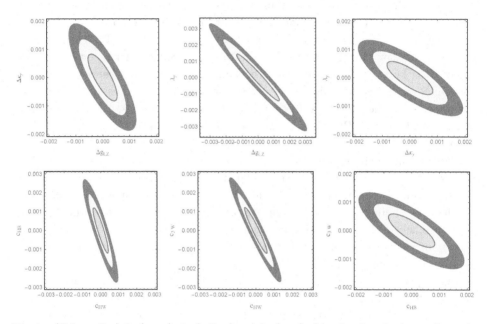

Fig. 1.   (Color online) 1σ (green), 2σ (yellow) and 3σ (gray) allowed regions in the semileptonic channel for the aTGCs (upper panels) and Dim-6 operators (lower panels) at CEPC with $\sqrt{s} = 240$ GeV and an integrated luminosity of 5 ab^{-1}. In drawing the plots, two of the three couplings (or coefficients) are allowed to vary and the third one is fixed to zero.

3. Constraints at LHC

The TGCs can also be probed directly at hadron colliders in a similar process $q\bar{q} \rightarrow W^+W^-$. These measurements are complementary to the EW precision tests, the accurate Higgs coupling probes, and all of these can be combined together to constrain the beyond SM physics.[11,12] As a direct comparison, we consider simply the WW production at the forthcoming LHC running at 14 TeV as an illustration. To suppress the huge QCD backgrounds, we focus on the purely leptonic decay channels $W \rightarrow e\nu, \mu\nu$. Though the neutrino events cannot be fully reconstructed, the p_T of charged leptons is widely used to study the TGCs. The azimuthal angle difference $\Delta\phi_{\ell\ell}$ projected onto the transverse plane in the lab frame, analogous to the azimuthal angles at lepton colliders, can be used to further improve the sensitivities. We do detector level simulations and apply the following cuts: for

Table 2. 1σ constraints on the TGCs and Dim-6 operators (in unit of 10^{-4}) from the same flavor leptonic decay channels of $pp \to W^+W^-$ at 14 TeV LHC with a luminosity of 300 fb^{-1} and 3000 fb^{-1}.

	$\Delta g_{1,Z}$	$\Delta \kappa_\gamma$	λ_γ	c_{HW}	c_{HB}	c_{3W}
300 fb^{-1}	23	73	17	14	73	17
3000 fb^{-1}	11	30	5.7	6.3	30	5.7

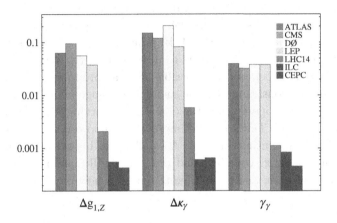

Fig. 2.   Current and future 95% confidence level constraints on the aTGCs. See text for details.

the charged leptons $l = e,\ \mu$, leading $p_T > 25$ GeV and subleading $p_T > 20$ GeV, $|\eta| < 2.5$, $\Delta R_{ll} > 0.4$, $m_{ll} > 15(10)$ GeV, $E_T > 45(15)$ GeV for the same (different) flavor channels, with the additional cut $|m_{ll} - M_Z| > 15$ GeV for the same flavor channels. It is found that the anomalous couplings tend to generate large p_T events. To optimize the constraints, we set further the leading $p_T > 300$ GeV and > 500 GeV, respectively for a luminosity of 300 fb^{-1} and 3000 fb^{-1} and apply the cuts $\Delta \phi_{ll} > 170°$,[a] with the constrains collected in Table 2. For a luminosity of 300 fb^{-1}, the limits are of the order of magnitude of 10^{-3}. When the luminosity is 10 times larger, the constraints go two or three times stronger.

We collect in Fig. 2 the current 95% confidence level constraints on the anomalous TGCs $\Delta g_{1,Z}$, $\Delta \kappa_\gamma$, λ_γ from LEP, Tevatron and LHC and that from 14 TeV LHC and the future lepton colliders CEPC and ILC. The current lepton and hadron collider bounds are from Ref. 13, the data for LHC 14 TeV assume a luminosity of 3000 fb^{-1}, the limits for both CEPC and ILC use only the semileptonic channel, and for ILC they are the limits at $\sqrt{s} = 500$ GeV with a luminosity of 500 fb^{-1} from Refs. 14 and 15. It should be noted that at ILC the injected beams are polarized leptons, which is different from the circular collider CEPC. Comparing naïvely the

---

[a]When more data are collected, to improve the signal significance and suppress the SM background, the cuts can be made in a more aggressive manner.

limits in this figure, the 14 TeV LHC and future lepton colliders can improve the limits on the aTGCs by one to two orders of magnitude. At future lepton colliders, using more decay channels, higher energies and larger luminosity can improve further the constraints in this figure.

4. Conclusion

In the precision era, the triple couplings among the SM electroweak gauge bosons is an essential part to test the SM in the gauge sector and set constraints on precision electroweak and Higgs physics, which can give us a powerful guidance on searching for new physics beyond the SM. *WW* process is one of the most important channels at hadron and lepton colliders to measure directly the charged TGCs. In this work, we use the kinematical observable of decay products at CEPC and LHC to study the future constraints on the anomalous gauge couplings and the relevant three dimension-6 operators in the C and P conserving sector. It is promising that constraints on the TGCs can be improved by two orders of magnitude and reach the order of magnitude of 10^{-4}. A smaller gap between electroweak precision measurements, Higgs couplings and the TGCs will push us to reconsider the complementarity of them and the shrinking space for new physics.

References

1. L. Bian, J. Shu and Y. Zhang, *J. High Energy Phys.* **1509**, 206 (2015), doi:10.1007/JHEP09(2015)206, arXiv:1507.02238 [hep-ph].
2. K. Hagiwara, R. D. Peccei, D. Zeppenfeld and K. Hikasa, *Nucl. Phys. B* **282**, 253 (1987).
3. G. F. Giudice, C. Grojean, A. Pomarol and R. Rattazzi, *J. High Energy Phys.* **0706**, 045 (2007), arXiv:hep-ph/0703164.
4. R. Contino, M. Ghezzi, C. Grojean, M. Muhlleitner and M. Spira, *J. High Energy Phys.* **1307**, 035 (2013), arXiv:1303.3876 [hep-ph].
5. A. De Rujula, M. B. Gavela, P. Hernandez and E. Masso, *Nucl. Phys. B* **384**, 3 (1992).
6. K. Hagiwara, S. Ishihara, R. Szalapski and D. Zeppenfeld, *Phys. Lett. B* **283**, 353 (1992).
7. K. Hagiwara, S. Ishihara, R. Szalapski and D. Zeppenfeld, *Phys. Rev. D* **48**, 2182 (1993).
8. M. Diehl and O. Nachtmann, *Z. Phys. C* **62**, 397 (1994).
9. G. Brooijmans *et al.*, arXiv:1405.1617 [hep-ph].
10. A. Falkowski and F. Riva, *J. High Energy Phys.* **1502**, 039 (2015), arXiv:1411.0669 [hep-ph].
11. T. Corbett, O. J. P. Éboli, J. Gonzalez-Fraile and M. C. Gonzalez-Garcia, *Phys. Rev. Lett.* **111**, 011801 (2013), arXiv:1304.1151 [hep-ph].
12. J. Ellis, V. Sanz and T. You, *J. High Energy Phys.* **1503**, 157 (2015), arXiv:1410.7703 [hep-ph].
13. ATLAS Collab. (G. Aad *et al.*), *J. High Energy Phys.* **1501**, 049 (2015), arXiv:1410.7238 [hep-ex].
14. H. Baer *et al.*, arXiv:1306.6352 [hep-ph].
15. ECFA/DESY LC Physics Working Group Collab. (J. A. Aguilar-Saavedra *et al.*), arXiv:hep-ph/0106315.

The LHC Searches for Heavy Neutral Higgs Bosons by Jet Substructure Analysis

Ning Chen

Department of Modern Physics,
University of Science and Technology of China,
Hefei, Anhui 230026, China
chenning@ustc.edu.cn

The two-Higgs-doublet model contains extra Higgs bosons with mass ranges spanning from several hundred GeV to about 1 TeV. We study the possible experimental searches for the neutral Higgs bosons of A and H at the future high-luminosity LHC runs. Besides of the conventional search modes that are inspired by the supersymmetric models, we discuss two search modes which were not quite addressed previously. They are the decay modes of $A \to hZ$ and $A/H \to t\bar{t}$. Thanks to the technique of tagging boosted objects of SM-like Higgs bosons and top quarks, we show the improved mass reaches for heavy neutral Higgs bosons with masses up to $\sim \mathcal{O}(1)$ TeV. The modes proposed here are complementary to the conventional experimental searches motivated by the MSSM.

Keywords: LHC; Higgs boson; jet substructure.

1. Introduction

In many of new physics models beyond the SM (BSM), the Higgs sector is extended with several scalar multiplets. Examples include the minimal supersymmetric standard model (MSSM),[1] the left–right symmetric models,[2] and the composite Higgs models.[3] There are several Higgs bosons in these models with one of them to be identified as the 125 GeV Higgs boson discovered at LHC. Therefore, extra heavy Higgs bosons are yet to be searched for by the future LHC experiments and the future high-energy pp colliders running at $\sqrt{s} = 50$–100 TeV.[4,5]

A very widely studied scenario beyond the minimal one-doublet setup is the two-Higgs-doublet model (2HDM), which is the low-energy description of the scalar sectors in various new physics models. A recent review of the phenomenology in the context of the general 2HDM can be found in Ref. 6. References 7–23, 44 studied the 2HDM phenomenology at the LHC in light of the 125 GeV Higgs discovery. The scalar spectrum in the 2HDM contains five states, namely, two neutral CP-even Higgs bosons (h, H), one neutral CP-odd Higgs boson A and two charged Higgs bosons $H^\pm$. In the context of the general 2HDM, each Higgs boson mass is actually a free parameter before applying any constraint. By including the perturbative unitarity and stability constraints to the general 2HDM potential,[19] the masses of the heavy Higgs bosons in the spectra are generally bounded from above as $(M_A, M_H, M_\pm) \lesssim 1$ TeV. Therefore, it becomes evident that the upcoming LHC runs at 14 TeV would search for these heavy states in the mass range of several hundred GeV to $\sim \mathcal{O}(1)$ TeV.

Within the framework of the 2HDM, we study the high-luminosity (HL) LHC searches for the heavy neutral Higgs bosons A and H at 14 TeV run. The previous experimental searches often focus on the benchmark models in the MSSM, which has type-II 2HDM Yukawa couplings. Thus, the interesting final states to be looked for are the $A/H \to \bar{b}b$ [24,25] and $A/H \to \tau^+\tau^-$ [26–32] due to the significant enhancements to the Yukawa couplings. Different from the existing experimental search modes, we consider the decay modes of $A/H \to t\bar{t}$ and $A \to hZ$. The $A/H \to t\bar{t}$ decay mode can be the most dominant one with the low-t_β inputs, for both 2HDM-I and 2HDM-II setups. Due to the large SM background of $t\bar{t}$, the searches for the $t\bar{t}$ final states from the heavy Higgs boson decays are thought to be very challenging. In addition, it is known that the signal channel of $gg \to A/H \to t\bar{t}$ strongly interferes with the SM background[33–35] and results in a peak-dip structure. Therefore, one can only rely on the heavy-quark associated production channels to study the $A/H \to t\bar{t}$ decays. For the $A \to hZ$ decay modes, we study the $\bar{b}b\ell^+\ell^-$ final state searches, where an SM-like Higgs boson with mass of 125 GeV is involved.

A common feature is that both decay modes involve heavy states. With heavy mother particles of A/H, it is natural to expect large boosts to the SM-like Higgs boson h and/or top quarks in our study. For highly boosted top quarks and/or h, the jets in their hadronic decay modes may lie close together and may not be independently resolved. As a result, the top quarks and/or SM-like Higgs boson h in the boosted region may appear as single jets with three or two subjets in a small region of the calorimeter. The separations by angular scales of subjets are of order $2m_h/p_T$ or $2m_t/p_T$. The method of tagging the boosted SM-like Higgs jets was suggested in Refs. 36 and 37, where one is likely to use the dominant decay mode of $h_{\rm SM} \to b\bar{b}$. This is dubbed "BDRS" algorithm. The method of reconstructing the boosted top jet uses the hadronic decay mode of $t_h \to bW_h \to b + jj$. Specifically, there are two classes of methods of tagging the boosted top quarks. One algorithm is called the JHUTopTagger,[38] which requires the summation of the transverse momenta of the decayed particles to be larger than 1 TeV. This is very challenging when one is interested in mother particles with masses of several hundred GeV to $\mathcal{O}(1)$ TeV. Alternatively, we study the LHC searches for the $A/H \to t\bar{t}$ decays by using the HEPTopTagger method,[39–42] which is efficient in tagging the top jets with intermediate transverse momenta of $\mathcal{O}(100)$ GeV.

The rest of the paper is organized as follows. In Sec. 2, we have a brief review of the heavy neutral Higgs bosons in the framework of the general CP-conserving 2HDM. In the precise alignment limit of $c_{\beta-\alpha} \to 0$, the decay modes of $A/H \to t\bar{t}$ are always the most dominant ones for the 2HDM-I, and also dominant ones for the 2HDM-II with low-t_β inputs. The inclusive production cross-sections of $\sigma[pp \to t\bar{t} + (A/H \to t\bar{t})]$ at the LHC 14 TeV runs are evaluated. The relaxation of the alignment limit leads to possible exotic decay modes, such as $A \to hZ$. We show that this decay mode can also become significant, especially for the 2HDM-I case. In Sec. 3, we search for the $t\bar{t} + (A/H \to t\bar{t})$ channel[43] at the HL-LHC. We focus on the $t\bar{t} + A/H$ production channel, with the sequential decay of $A/H \to t\bar{t}$. This process is always controlled by the top quark Yukawa coupling of the heavy Higgs bosons. By applying the HEPTOPTAGGER method for reconstructing one boosted top quark, plus selecting two additional same-sign-dilepton (SSDL) events, we obtain a signal reach for $M_{A/H} \sim \mathcal{O}(1)$ TeV with low-t_β inputs at the HL-LHC runs. The results in this part are mainly based on Ref. 43. In Sec. 4, the analysis of LHC searches for the CP-odd Higgs boson via the $A \to hZ$ final states[44] is provided. In order to eliminate the SM background sufficiently, we apply the BDRS algorithm in Ref. 36 to reconstruct the boosted Higgs. The LHC search potential to the $A \to hZ$ decay channel at different phases of the upcoming runs at 14 TeV is shown. The results in this part are mainly based on Ref. 44. Finally, we make conclusion in Sec. 5.

2. The Heavy Neutral Higgs Bosons in the 2HDM

In this section, we briefly discuss the productions and decays of the heavy neutral Higgs bosons A and H in the context of the general CP-conserving 2HDM.

2.1. *The general 2HDM setup and couplings*

The most general 2HDM Higgs potential is composed of all gauge-invariant and renormalizable terms by two Higgs doublets $(\Phi_1, \Phi_2) \in 2_{+1}$ of the $SU(2)_L \times U(1)_Y$ electroweak gauge symmetries. For the CP-conserving case, there can be two mass terms plus seven quartic coupling terms with real parameters. For simplicity, we consider the soft breaking of a discrete $\mathbb{Z}_2$ symmetry, under which two Higgs doublets transform as $(\Phi_1, \Phi_2) \to (-\Phi_1, \Phi_2)$. The corresponding Lagrangian is expressed as

$$\mathcal{L} = \sum_{i=1,2} |D\Phi_i|^2 - V(\Phi_1, \Phi_2), \tag{1}$$

$$V(\Phi_1, \Phi_2) = m_{11}^2 |\Phi_1|^2 + m_{22}^2 |\Phi_2|^2 - m_{12}^2 (\Phi_1^\dagger \Phi_2 + \text{H.c.})$$

$$+ \frac{1}{2}\lambda_1 |\Phi_1|^4 + \frac{1}{2}\lambda_2 |\Phi_2|^4 + \lambda_3 |\Phi_1|^2 |\Phi_2|^2 + \lambda_4 |\Phi_1^\dagger \Phi_2|^2$$

$$+ \frac{1}{2}\lambda_5 [(\Phi_1^\dagger \Phi_2)(\Phi_1^\dagger \Phi_2) + \text{H.c.}]. \tag{2}$$

Two Higgs doublets Φ_1 and Φ_2 pick up VEVs to trigger the EWSB, and one parametrizes the ratio of the two Higgs VEVs as

$$t_\beta \equiv \tan\beta = \frac{v_2}{v_1}. \tag{3}$$

The perturbative bounds of the heavy Higgs boson Yukawa couplings constrain the choices of t_β, which should be neither as small as $\mathcal{O}(0.1)$ nor as large as $\mathcal{O}(50)$. Three of the eight real components correspond to the Nambu–Goldstone bosons giving rise to the electroweak gauge boson masses, with the remaining five as the physical Higgs bosons, namely, two CP-even Higgs bosons h and H, one CP-odd Higgs boson A and the charged Higgs bosons $H^\pm$. The light CP-even Higgs boson h is taken as the only state in the 2HDM spectra with mass of 125 GeV and its couplings with SM fermions and gauge bosons are controlled by two parameters of (α, β). A more convenient choice of 2HDM parameter set is $(c_{\beta-\alpha}, t_\beta)$. The current global fits[9,10,45,46] by using the LHC 7 $\oplus$ 8 TeV runs to the 2HDM parameters point to the alignment limit of $c_{\beta-\alpha} \to 0$.

In the general 2HDM, SM fermions with the same quantum numbers couple to the same Higgs doublet, which will avoid the tree-level flavor-changing neutral currents. For the 2HDM-I, all SM fermions couple to one Higgs doublet (conventionally chosen to be Φ_2). This setup can be achieved by assigning a discrete $\mathbb{Z}_2$ symmetry under which $\Phi_1 \to -\Phi_1$. For the 2HDM-II, the up-type quarks u_i couple to one Higgs doublet (conventionally chosen to be Φ_2) and the down-type quarks d_i and the charged leptons ℓ_i couple to the other (Φ_1). This can also be achieved by assigning a discrete $\mathbb{Z}_2$ symmetry under which $\Phi_1 \to -\Phi_1$ together with $(d_i, \ell_i) \to (-d_i, -\ell_i)$. At the tree level, the Yukawa coupling terms for neutral Higgs bosons are expressed as

$$-\mathcal{L}_Y = \sum_f \frac{m_f}{v} \left(\xi_H^f \bar{f} f H - i\xi_A^f \bar{f} \gamma_5 f A \right). \tag{4}$$

The dimensionless coupling strengths of ξ_H^f and ξ_A^f are listed in Table 1 for the 2HDM-I and 2HDM-II cases, respectively. In the alignment limit, all dimensionless

Table 1. The Yukawa couplings of the heavy neutral Higgs bosons H and A in the general 2HDM.

	2HDM-I	2HDM-II
ξ_H^u	$c_{\beta-\alpha} - s_{\beta-\alpha}/t_\beta$	$c_{\beta-\alpha} - s_{\beta-\alpha}/t_\beta$
ξ_H^d	$c_{\beta-\alpha} - s_{\beta-\alpha}/t_\beta$	$c_{\beta-\alpha} + t_\beta s_{\beta-\alpha}$
ξ_H^ℓ	$c_{\beta-\alpha} - s_{\beta-\alpha}/t_\beta$	$c_{\beta-\alpha} + t_\beta s_{\beta-\alpha}$
ξ_A^u	$1/t_\beta$	$1/t_\beta$
ξ_A^d	$-1/t_\beta$	t_β
ξ_A^ℓ	$-1/t_\beta$	t_β

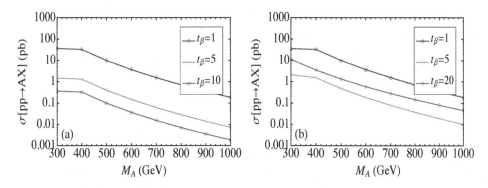

Fig. 1.   (Color online) The inclusive production cross-section $\sigma[pp \to AX]$ for $M_A \in$ (300 GeV, 1 TeV) at the LHC 14 TeV runs. (a) 2HDM-I with inputs of $t_\beta = 1$ (blue), $t_\beta = 5$ (green) and $t_\beta = 10$ (red). (b) 2HDM-II with inputs of $t_\beta = 1$ (blue), $t_\beta = 5$ (green) and $t_\beta = 20$ (red).

Yukawa couplings of ξ_H^f and ξ_A^f are inversely proportional to t_β for the 2HDM-I, while they depend on the t_β inputs in the manner of $\xi_{A/H}^u \propto 1/t_\beta$ and $\xi_{A/H}^d \propto t_\beta$ for the 2HDM-II. With the low-t_β inputs, the heavy neutral Higgs bosons A and H always couple strongly to the top quarks.

Besides of Yukawa couplings, there are also other couplings relevant to the decays of the heavy neutral Higgs bosons A and H. From the 2HDM kinematic terms $|D\Phi_i|^2$, one has Higgs-gauge couplings of $G(HVV)$, $G(AhZ/AHZ)$ and $G(AH^\pm W^\mp)$. The $G(HVV)$ and $G(AhZ)$ couplings are vanishing in the $c_{\beta-\alpha} = 0$ limit. From the general 2HDM potential, one also has the triple Higgs couplings such as $G(Hhh)$, $G(HAA)$ and $G(HH^+H^-)$. The existences of these couplings lead to exotic heavy Higgs boson search strategies. Though we focus on the alignment limit of $c_{\beta-\alpha} = 0$, it is noted that the current global fits to the 125 GeV CP-even Higgs boson h in the 2HDM generally allow the parameter choices of $c_{\beta-\alpha} \sim \mathcal{O}(0.1)$ for 2HDM-I and $c_{\beta-\alpha} \sim \mathcal{O}(0.01)$ for 2HDM-II, respectively. It was shown in the previous discussions[13,44] that some of the heavy Higgs boson decay modes of $A \to hZ$ and $H \to hh$ can become the leading ones, especially for $M_{A/H} \lesssim 2m_t$. The relevant LHC searches can be performed by the boosted Higgs searches plus the opposite-sign-same-flavor (OSSF) dileptons, and via the $b\bar{b} + \gamma\gamma$ final states.

2.2. *The productions of A and H*

Since the global fits to the general 2HDM point to the alignment limit of $c_{\beta-\alpha} = 0$, the production channels of the heavy neutral Higgs bosons A/H at the LHC are most likely due to the gluon fusion and the heavy quark associated processes.[47,48] The VBF and vector boson associated processes are highly suppressed for the heavy CP-even Higgs boson H, and they are absent for the CP-odd Higgs boson A.

At the leading order, the parton-level production cross-section of $\hat{\sigma}(gg \to A)$ is related to the gluonic partial decay width as follows:

$$\hat{\sigma}(gg \to A) = \frac{\pi^2}{8M_A}\Gamma[A \to gg]\delta(\hat{s} - M_A^2)\,, \tag{5a}$$

$$\Gamma[A \to gg] = \frac{G_F \alpha_s^2 M_A^3}{64\sqrt{2}\pi^3}\Big|\sum_q \xi_A^q A_{1/2}^A(\tau_q)\Big|^2\,, \tag{5b}$$

with $\tau_q \equiv M_A^2/(4m_q^2)$ and ξ_A^q being the Yukawa couplings given in Table 1. Here, $A_{1/2}^A(\tau)$ is the fermionic loop factor for the pseudoscalar. In the heavy quark mass limit of $m_q \gg M_A$, this loop factor reaches the asymptotic value of $A_{1/2}^A(\tau) \to 2$, while it vanishes in the chiral limit of $m_q \ll M_A$. In practice, we evaluate the production cross-sections for these processes by SuSHi.[49] The inclusive production cross sections of $\sigma[pp \to AX]$ are shown in Fig. 1 for the LHC runs at 14 TeV, where the CP-odd Higgs boson is considered in the mass range of $M_A \in (300 \text{ GeV}, 1 \text{ TeV})$.

For the decay modes of $A/H \to t\bar{t}$, the previous studies of the gluon fusion to $t\bar{t}$ final states via the spin-0 resonances have suggested strong interference effects with the QCD backgrounds. Therefore, we are left with the heavy-quark associated productions as the possible channels to search for at the LHC. For the 2HDM-I case, all dimensionless Yukawa couplings of the SM fermions to the A/H are universally proportional to $1/t_\beta$. For the 2HDM-II case, the dimensionless Yukawa couplings scale as $\xi_{A/H}^u \propto 1/t_\beta$ and $\xi_{A/H}^d \propto t_\beta$, respectively. Therefore, one expects the production cross-sections of $\sigma[pp \to t\bar{t} + A/H] \sim 1/t_\beta^2$ to become significant with the low-t_β inputs.

2.3. The decays of A and H

All possible decay modes of heavy neutral Higgs boson are listed in Tables 2 and 3 for A and H, respectively. The presence or absence of these decay modes in the alignment limit are also marked. In practice, we evaluate their partial decay widths by using 2HDMC.[50] Some exotic decay modes involving another heavy states, such as $A \to HZ$ and $H \to H^\pm W^\mp$, are always turned off for later discussions. This is reasonable when one assumes the masses of all heavy Higgs bosons are close to each other. The loop-induced decay branching ratios of $\text{Br}[A/H \to \gamma\gamma/\gamma Z]$ are typically smaller than 10^{-5}, which can be neglected. For the 2HDM-I case, the decay branching ratios of $\text{Br}[A/H \to t\bar{t}]$ are always dominant to be $\sim \mathcal{O}(1)$ since all dimensionless Higgs Yukawa couplings scale as $\sim 1/t_\beta$. For the 2HDM-II case, the $\text{Br}[A/H \to t\bar{t}]$ can be suppressed to $\sim \mathcal{O}(0.1)$ with the large-t_β inputs, where the partial decay widths of $\Gamma[A/H \to b\bar{b}]$ and $\Gamma[A/H \to \tau^+\tau^-]$ become dominant. Combining the production cross-sections evaluated by MADGRAPH 5,[51] we demonstrated the cross-sections of $\sigma[pp \to t\bar{t} + A/H] \times \text{Br}[A/H \to t\bar{t}]$ within the mass range of $M_{A/H} \in (350 \text{ GeV}, 1200 \text{ GeV})$ at the LHC 14 TeV runs in Fig. 2. As it turns out, the decay branching ratios of $\text{Br}[A/H \to t\bar{t}]$ tend to unity for both

Table 2. The classification of the CP-odd Higgs boson A decay modes in the general 2HDM. A checkmark ($—$) indicates that the decay mode is present (absent) in the $c_{\beta-\alpha} = 0$ alignment limit.

A decays	Final states	Alignment limit
SM fermions	$A \to (\tau^+\tau^-, \mu^+\mu^-)$	✓
	$A \to (t\bar{t}, b\bar{b})$	✓
Exotics	$A \to hZ$	$—$
	$A \to HZ$	✓
	$A \to H^{\pm}W^{\mp}$	✓
Loops	$A \to (gg, \gamma\gamma, \gamma Z)$	✓

Table 3. The classification of the CP-even Higgs boson H decay modes in the general 2HDM. A checkmark ($—$) indicates that the decay mode is present (absent) in the $c_{\beta-\alpha} = 0$ alignment limit.

H decays	Final states	Alignment limit
SM fermions	$H \to (\tau^+\tau^-, \mu^+\mu^-)$	✓
	$H \to (t\bar{t}, b\bar{b})$	✓
Gauge bosons	$H \to (WW, ZZ)$	$—$
Exotics	$H \to AZ$	✓
	$H \to H^{\pm}W^{\mp}$	✓
	$H \to hh$	$—$
	$H \to AA$	$—$
	$H \to H^+H^-$	✓
Loops	$H \to (gg, \gamma\gamma, \gamma Z)$	✓

2HDM-I and 2HDM-II with the small-t_β inputs of $\sim \mathcal{O}(1)$. For this reason, we combine the cross-sections of $\sigma[pp \to t\bar{t} + A/H] \times \mathrm{Br}[A/H \to t\bar{t}]$ for both 2HDM-I and 2HDM-II into one plot.

When we relax the alignment limit with the $c_{\beta-\alpha}$ inputs subject to the global fit constraints, the possible decay modes of A in our discussions include $A \to (\bar{f}f, gg, hZ)$. Below, we take the alignment parameters of

$$\text{2HDM-I: } c_{\beta-\alpha} = 0.2, \quad \text{2HDM-II: } c_{\beta-\alpha} = -0.02, \tag{6}$$

for the analysis of the $A \to hZ$ decay mode. Other loop-induced decay widths of $\Gamma[A \to \gamma\gamma]$ and $\Gamma[A \to Z\gamma]$ are typically negligible. In Fig. 3, we display the decay branching ratios of the CP-odd Higgs boson A in the mass range of $M_A \in$ (300 GeV, 1 TeV) for the 2HDM-I and 2HDM-II cases, respectively.

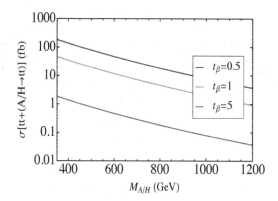

Fig. 2. (Color online) $\sigma[pp \to t\bar{t}A/H] \times \mathrm{BR}[A/H \to t\bar{t}]$ (for both 2HDM-I and 2HDM-II cases) with $M_{A/H} \in (350\ \mathrm{GeV},\ 1200\ \mathrm{GeV})$ at the LHC 14 TeV runs.

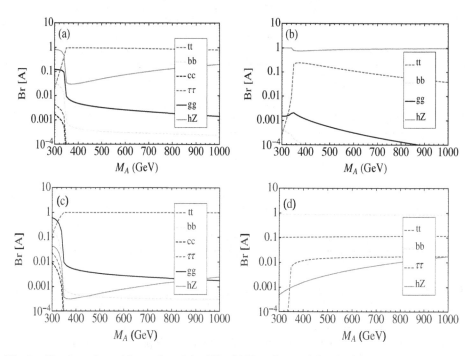

Fig. 3.   The decay branching ratios of the CP-odd Higgs boson A for the (a) 2HDM-I with $t_\beta = 1$, (b) 2HDM-I with $t_\beta = 10$, (c) 2HDM-II with $t_\beta = 1$ and (d) 2HDM-II with $t_\beta = 20$.

Figure 4 shows the $\sigma[pp \to AX] \times \mathrm{BR}[A \to hZ]$ for various cases at the LHC 14 TeV runs. This is done by combining the inclusive production cross-sections of $\sigma[pp \to AX]$ displayed in Fig. 1 and the decay branching ratios of $\mathrm{BR}[A \to hZ]$ displayed in Fig. 3. In descending order, the curves correspond to input parameters of $t_\beta = 1, 5, 10$ for the 2HDM-I signal predictions. This is largely due to the production

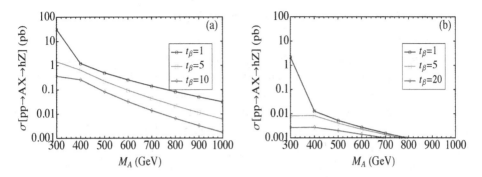

Fig. 4. The $\sigma[pp \to AX] \times \text{BR}[A \to hZ]$ for $M_A \in (300 \text{ GeV}, 1 \text{ TeV})$ at the LHC 14 TeV runs. (a) $M_h = 125$ GeV for 2HDM-I. (b) $M_h = 125$ GeV for 2HDM-II.

cross-section dependence on the t_β inputs, as shown in Fig. 1(a). Meanwhile, the corresponding decay branching ratio of $\text{BR}[A \to hZ]$ for the 2HDM-I case varies moderately in the range of $\mathcal{O}(0.1) - \mathcal{O}(1)$, as shown in Fig. 3. Therefore, the search for the CP-odd Higgs boson via the $A \to hZ$ channel is possible for the 2HDM-I cases at the LHC 14 TeV runs, with the integrated luminosities accumulated up to $\mathcal{O}(10^3)$ fb^{-1}. In comparison, the signal predictions of $\sigma[pp \to AX] \times \text{BR}[A \to hZ]$ for the 2HDM-II case are highly suppressed to $\mathcal{O}(10^{-2}) - \mathcal{O}(10^{-3})$ pb with $M_A \gtrsim 2m_t$. This is obvious as seen from the more dominant decay modes of $A \to \bar{t}t$ for the small $t_\beta = 1$ input and $A \to (\bar{b}b, \tau^+\tau^-)$ for the large $t_\beta = 20$ input, respectively. Thus, the search channel of $A \to hZ$ at the LHC 14 TeV run is of minor interest for the 2HDM-II case.

3. The Searches for the Heavy Neutral Higgs Bosons via the $t\bar{t}$ Decay

In this section, we analyze the LHC 14 TeV searches for the heavy neutral Higgs bosons A and H via the $t\bar{t} + A/H$ production, with the sequential decay modes of $A/H \to t\bar{t}$. We always tag the top jets t_h by using the HEPTOPTAGGER method. For the $t\bar{t} + A/H$ production channel, we shall look for events including a top jet t_h plus SSDL. The corresponding SM background processes should include the final states with SSDL plus multiple jets, where a jet may be mis-tagged as the boosted t_h. Thus, the corresponding SM backgrounds include $t\bar{t}$, $t\bar{t}b\bar{b}$, $(W^\pm Z, ZZ)$ plus jets,[52] and $(t\bar{t}W^\pm, t\bar{t}Z)$.[53] The cross-sections of other SM background processes including $(W^\pm W^\pm, t\bar{t} + W^\pm, t\bar{t} + Z)$ plus jets are less than 1 pb. As we shall show later, the dominant SM background processes after the preselections of t_h plus SSDL are $t\bar{t}$ and $W^\pm Z$ plus jets. Therefore, we neglect all other SM background processes for the $t\bar{t} + (A/H \to t\bar{t})$ signal channel. After the reconstruction of the boosted t_h, we shall select the kinematic variables for the signal events and carry out the TMVA analysis to optimize the signal significance.

3.1. The MC simulations and the top jet tagging

For event generations of the signal processes, we use Universal FeynRules Output[54] simplified models with A or H being the only BSM particles. The relevant coupling terms to be implemented are the Yukawa couplings of $Ab\bar{b}/Hb\bar{b}$, and the Yukawa couplings of $At\bar{t}/Ht\bar{t}$. We generate events for both signal and SM background processes at the parton level by MADGRAPH 5,[51] with the subsequent parton shower and hadronization performed by PYTHIA.[55] Afterwards, DELPHES[56] is used for the fast detector simulations. In our simulations of both signal and background processes, we include up to two extra jets with the MLM matching in order to avoid the double counting. Our fast detector simulations follow the setup of the ATLAS detector. The Delphes output will be used for the jet substructure analysis by FASTJET.[57]

In what follows, we briefly describe the reconstruction of physical objects by the HEPTOPTAGGER method. The energy flow observables from the Delphes output are used for the jet substructure analysis by FASTJET.[57] In each event, we cluster the top jets by using the Cambridge–Aachen (CA) algorithm[59,60] with certain jet cone size R_{CA}. By setting the reconstructed top mass range of $m_t^{rec} \in (140 \text{ GeV}, 210 \text{ GeV})$, the HEPTOPTAGGER algorithm finds a candidate boosted top jet which contains three subjets with their total transverse momenta greater than 200 GeV. The rates of tagging one t_h can be ~ 30–60% with certain choices of R_{CA}. It is also likely to tag a second boosted t_h at the rates of ~ 10–20%. For such cases, we always choose the one with the largest p_T as the t_h. Generally speaking, the tagging rates of top jets vary with different choices of the jet cone sizes R_{CA}. The boost factors of top jets are enhanced with the heavier resonances of $M_{A/H}$. For each signal processes of $pp \to t\bar{t} + A/H$, we scan the jet cone sizes $R_{CA} \in (1.0, 3.0)$ at the step of 0.1 for reconstructing the top jet t_h in the HEPTOP-TAGGER. In addition, the effects due to the underlying events can be eliminated by the filtering procedure[36] in the HEPTOPTAGGER. The remaining particles will be clustered into narrow jets by using the anti-k_t algorithm with a jet cone size of $R_{narrow} = 0.4$, which are required to satisfy $p_T \geq 20$ GeV and $|\eta| < 4.5$.

3.2. The $t\bar{t} + A/H$ search results

For the $t\bar{t} + (A/H \to t\bar{t})$ signal channel, one has four top quarks in the final states. After one boosted top quark t_h be reconstructed through its hadronic decay mode, we select events containing SSDL $\ell_1^\pm \ell_2^\pm$ from the semi-leptonic decays of two other top quarks. It turns out a significant suppression to the SM background can be achieved by selecting the events containing t_h plus SSDL. An example of the pre-selection efficiencies of events for the $M_{A/H} = 500$ GeV case is tabulated in Table 4. The suppression rates of SM background events from the $t\bar{t}$ and $t\bar{t}b\bar{b}$ can be as significant as $\sim 10^{-5}$ when imposing the SSDL selection criterion. Obviously, the $W^\pm Z$

Table 4. The preselection efficiencies of the $M_A = 500$ GeV (with $R_{CA} = 2.1$) signal and background processes at the 14 TeV LHC. We assume the nominal cross-section for the signal process to be $\sigma[pp \to t\bar{t} + A/H] \times \text{BR}[A/H \to t\bar{t}] = 50$ fb.

	Signal	$t\bar{t}$	$t\bar{t}b\bar{b}$	$W^{\pm}Z$	ZZ	$S/\sqrt{B}$
Total cross-section (fb)	50	8.0×10^5	3×10^4	5.0×10^4	1.5×10^4	—
Preselection of t_h (fb)	28	3.1×10^5	1.4×10^4	7.7×10^3	1.9×10^3	—
Preselection of t_h + SSDL (fb)	0.48	0.56	0.11	3.92	0.17	3.4

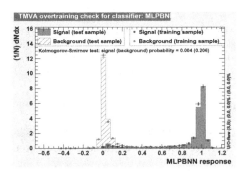 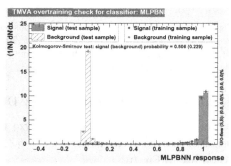

Fig. 5. (Color online) Lower: the normalized distributions of MLP neural network response for signal and background for the $t\bar{t} + (A/H \to t\bar{t})$ channel, left: $M_{A/H} = 500$ GeV, right: $M_{A/H} = 1000$ GeV.

background becomes the most dominant one after the preselections. Meanwhile, one has $\sigma(t\bar{t})_{\text{select}} \approx 0.1\sigma(W^{\pm}Z)_{\text{select}}$ after the preselections.

We perform the multi-variable analysis after the preselections, which is achieved by using the MLP neural network analysis in the ROOT TMVA package. The list of kinematic variables for the later analysis include (p_T, η, ϕ) of $(t_h, \ell_1^{\pm}, \ell_2^{\pm})$, $\slashed{E}_T$, number of (b-jets, non-b jets), $p_T(b_0, j_0)$, $\sum_j p_T(j)$ and $\sum_b p_T(b)$. Here, j_0 and b_0 denote the leading non-b-jet and the leading b-jet ordered in their transverse momenta, respectively. The discriminations between signal and background events are shown in Fig. 5 for the $M_{A/H} = 500$ GeV and $M_{A/H} = 1000$ GeV samples, which are based on the TMVA output displayed in the graphical user interface (GUI). After obtaining the cut efficiencies, we convert the results to the signal cross-sections within the 5σ discovery limits, which read $\sigma[t\bar{t} + (A/H \to t\bar{t})] \sim 2$–$5$ fb for the mass ranges of $M_{A/H} \in (350, 1200)$ GeV. The results are demonstrated on the left panel of Fig. 6. By looking for the t_h plus the SSDL signals, our analysis shows that the HL LHC searches are likely to reach the heavy neutral Higgs boson masses up to $\mathcal{O}(1)$ TeV in the low-t_β regions for the general CP-conserving 2HDM. The model-independent signal cross-sections for the 5σ reaches are further projected to the $(M_{A/H}, t_\beta)$ plane, as shown on the right panel of Fig. 6.

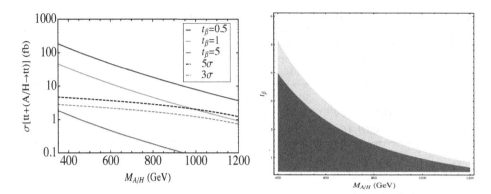

Fig. 6. (Color online) The signal predictions together with the signal reaches (dashed lines) of $t\bar{t} + (A/H \to t\bar{t})$ at the HL-LHC runs. Left: The mode-independent cross-section reaches at the HL LHC runs. Right: The 5σ (blue) and 3σ (green) signal reaches projected on the $(M_{A/H}, t_\beta)$ plane at the HL LHC runs.

4. The Searches for the $A \to hZ$ Decay

In this section, we proceed to analyze the LHC searches for the CP-odd Higgs boson A via the decay mode of $A \to hZ$, where we relax the alignment parameter according to Eq. (6).

4.1. *SM backgrounds and signal benchmark*

The final states to be searched for are the same as the ones in the SM Higgs boson searches via the hZ-associated production channel. Therefore, the dominant irreducible SM backgrounds relevant to our analysis include $\bar{b}b\ell^+\ell^-$, $\bar{t}t$, $ZZ \to \bar{b}b\ell^+\ell^-$ and the $h_{\rm SM}Z \to \bar{b}b\ell^+\ell^-$. In our analysis below, we take the b-tagging efficiency of 70%, and the mistagging rates are taken as

$$\epsilon_{c \to b} \approx 0.2\,, \quad \epsilon_{j \to b} \approx 0.01\,, \tag{7}$$

with j representing the light jets that neither originate from a b quark or a c quark.[58]

4.2. *Jet substructure methods*

Here, we describe the jet substructure analysis and the application to the signals we are interested in. The tracks, neutral hadrons and photons that enter the jet reconstruction should satisfy $p_T > 0.1$ GeV and $|\eta| < 5.0$. The leptons from the events should be isolated, so that they will not be used to cluster the fat jets. The fat jets are reconstructed by using the CA jet algorithm with particular jet cone size R to be specified below and requiring $p_T > 30$ GeV. Afterwards, we adopt the procedures described in the mass-drop tagger[36] for the purpose of identifying a boosted Higgs boson:

(i) Split the fat jet j into two subjets $j_{1,2}$ with masses $m_{1,2}$ and $m_1 > m_2$.

(ii) Require a significant mass drop of $m_1 < \mu m_j$ with $\mu = 0.667$ and also a sufficiently symmetric splitting of $\min(p_{T,1}^2, p_{T,2}^2)\Delta R_{12}^2/m_j^2 > y_{\text{cut}}$ (ΔR_{12}^2 is the angular distance between j_1 and j_2 on the $\eta - \phi$ plane) with $y_{\text{cut}} = 0.09$.

(iii) If the above criteria are not satisfied, define $j \equiv j_1$ and go back to the first step for decomposition.

These steps are followed by the filtering stage using the reclustering radius of $R_{\text{filt}} = \min(0.35, R_{12}/2)$ and selecting the three hardest subjets to suppress the pileup effects.

4.3. *Event selection*

The cut flow we impose to the events is the following:

(i) Cut 1: We select events with the OSSF dileptons $(\ell^+\ell^-)$ in order to reconstruct the final-state Z boson. The OSSF dileptons are required to satisfy the following selection cuts:

$$|\eta_\ell| < 2.5\,, \quad p_T(\ell_1) \geq 20 \text{ GeV}\,, \quad p_T(\ell_2) \geq 10 \text{ GeV}\,, \tag{8}$$

where $\ell_{1,2}$ represent two leading leptons ordered by their transverse momenta.

(ii) Cut 2: The invariant mass of the selected OSSF dileptons should be around the mass window of the Z boson $|m_{\ell\ell} - m_Z| \leq 15$ GeV.

(iii) Cut 3: At least one filtered fat jet is required, which should also contain two leading subjets that pass the b tagging and satisfy $p_T > 20$ GeV and $|\eta| < 2.5$.

(iv) Cut 4: Such a filtered fat jet will be then identified as the SM-like Higgs jet. We impose the cuts to the filtered Higgs jets in the mass window of $M_h(\text{tagged}) \in$ (100 GeV, 150 GeV).

(v) Cut 5: We also impose the cuts on the $p_{T,h}(\text{tagged})$. The SM-like Higgs bosons decaying from the heavier CP-odd Higgs boson A would generally be more boosted. In practice, we vary the $p_{T,h}(\text{tagged})_{\text{cut}} \in$ (50 GeV, 500 GeV) and look for the most optimal cuts on $p_{T,h}(\text{tagged})$ by counting the corresponding cut efficiencies of S/B.

(vi) Cut 6: Combining the filtered Higgs jets and the tagged OSSF dileptons, the invariant mass of the tagged Higgs boson and the OSSF leptons should reconstruct the mass window of the CP-odd Higgs boson A: $|M_{h,\ell^+\ell^-} - M_A| \leq$ 100 GeV.

4.4. *Implications to the LHC searches for A in the general 2HDM*

Here, we present the results after the jet substructure analysis and imposing the kinematic cuts stated previously. As a specific example of the analysis stated above, the distributions of the $M_{h,\ell\ell}$ after Cut 1 through Cut 5 for both signal process and the relevant SM background processes are displayed in Fig. 7. A nominal production cross-section of $\sigma[pp \to AX] \times \text{BR}[A \to hZ] = 500$ fb for the signal process is chosen

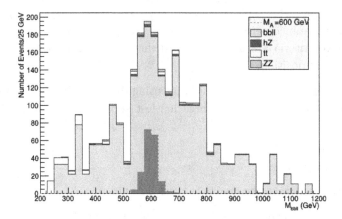

Fig. 7. The $M_{h,ll}$ distributions of the $pp \to AX \to hZ$ signal process (for the $M_A = 600$ GeV case) and all SM background processes after the kinematic cuts. A nominal cross-section of $\sigma[pp \to AX] \times \mathrm{BR}[A \to hZ] = 500$ fb is assumed for the signal. The plot is for the LHC 14 TeV run with integrated luminosity of $\int \mathcal{L} dt = 100$ fb^{-1}.

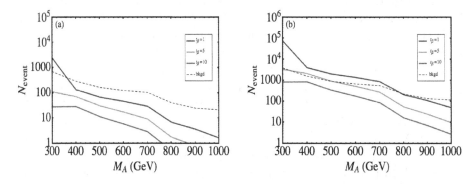

Fig. 8. (Color online) The number of events for the $pp \to AX \to hZ$ signal in the 2HDM-I after the jet substructure analysis. (a) $\int \mathcal{L} dt = 100$ fb^{-1} and (b) $\int \mathcal{L} dt = 3000$ fb^{-1}. We show samples with $t_\beta = 1$ (blue), $t_\beta = 5$ (green) and $t_\beta = 10$ (red) for each plot. The discovery limit (dashed black curve) of $\max\{5\sqrt{B}, 10\}$ is demonstrated for each plot.

for the evaluation. Among all relevant SM background processes, the $\bar{b}b\ell^+\ell^-$ turns out to contribute most after imposing the cuts mentioned above.

In Fig. 8, we display the number of events predicted by the signal process of $pp \to AX \to hZ$ after the cut flows imposed to the 2HDM-I. We demonstrate the predictions at the LHC 14 TeV runs with integrated luminosities of 100 fb^{-1} and high luminosity (HL) runs up to 3000 fb^{-1}. Via the $A \to hZ$ channel, the CP-odd Higgs boson with mass up to ~ 900 GeV is likely to be probed at the HL-LHC runs. The signal reaches on the (M_A, t_β) plane are further displayed in Fig. 9 for the 2HDM-I case. There are significant improvements of the signal reaches when

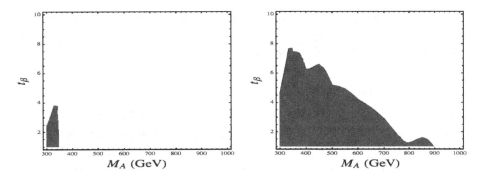

Fig. 9. (Color online) The signal reaches for the $A \to hZ$ on the (M_A, t_β) plane for the 2HDM-I case. Left: $\int \mathcal{L}dt = 100$ fb^{-1}, right: $\int \mathcal{L}dt = 3000$ fb^{-1}. Parameter regions of (M_A, t_β) in blue are within the reach for each case.

increasing the integrated luminosity from 100 fb^{-1} up to the HL-LHC runs up to 3000 fb^{-1}. For the 2HDM-I case, the $\sigma[pp \to AX] \times \mathrm{BR}[A \to hZ]$ decreases with the larger t_β inputs, as is consistent with what is presented in Figs. 4(a) and 4(b).

5. Conclusion and Discussion

In this work, we have carried out an analysis of the LHC searches for the heavy neutral Higgs bosons by reconstructing the boosted top quarks and/or SM-like Higgs bosons in their decays. The decay branching ratios of $A/H \to t\bar{t}$ can be approaching to $\mathcal{O}(1)$ with low-t_β inputs. This is the usual case when setting the alignment limit of $c_{\beta-\alpha} = 0$ and turning off all possible exotic decay modes of heavy neutral Higgs bosons. Correspondingly, the searches for the $A/H \to t\bar{t}$ are of the top priority from the perspective of the production cross-sections. We consider the $t\bar{t} + (A/H \to t\bar{t})$ signal channel in this work, whose interference effects with the QCD background are less severe compared to the gluon fusion channel. In order to suppress the corresponding SM background contributions, we adopt the HEPTOPTAGGER method to reconstruct the boosted top jets t_h. As for the $t\bar{t} + (A/H \to t\bar{t})$ signal channel with multiple top quarks in the final state, we select events containing the boosted t_h plus the SSDL. Much better signal sensitivity is obtained for this production channel by using the MLP neural network analysis. For $M_{A/H} \in (350 \text{ GeV}, 1200 \text{ GeV})$, we find that the production cross-sections of $t\bar{t} + (A/H \to t\bar{t})$ as small as $\sim [2-5]$ fb can be discovered at 5σ CL.

With the deviation from the exact alignment limit of the 2HDM, the possible exotic decay mode such as $A \to hZ$ can become significant, especially for the 2HDM-I case. This decay channel is due to the derivative coupling term AhZ arising from the 2HDM kinematic terms. The technique of the BDRS algorithm of tagging the boosted Higgs jets turns out to be very efficient for suppressing the SM background contributions. The cut flows to capture the kinematical features for the signal processes were applied thereafter. The mass reach can be generally up to ~ 900 GeV for the 2HDM-I with low-t_β inputs at the HL-LHC runs. By projecting the sensitivity regions on the (M_A, t_β) plane for both channel, we find the searches for $t\bar{t}$ and/or $A \to hZ$ modes can become complementary to the conventional search modes motivated by the MSSM, such as $A/H \to b\bar{b}$ and $A/H \to \tau^+\tau^+$.

Acknowledgments

This work is partially supported by the National Science Foundation of China (Grant No. 11575176) and the Fundamental Research Funds for the Central Universities (Grant No. WK2030040069). We would like to thank the conference organizers for the invitation, and the members of the Jockey Club Institute for Advanced Study for their hospitality.

References

1. S. Dimopoulos and H. Georgi, *Nucl. Phys. B* **193**, 150 (1981).
2. Y. Zhang, H. An, X. Ji and R. N. Mohapatra, *Nucl. Phys. B* **802**, 247 (2008), arXiv:0712.4218 [hep-ph].
3. J. Mrazek, A. Pomarol, R. Rattazzi, M. Redi, J. Serra and A. Wulzer, *Nucl. Phys. B* **853**, 1 (2011), arXiv:1105.5403 [hep-ph].
4. TLEP Design Study Working Group Collab. (M. Bicer *et al.*), *J. High Energy Phys.* **1401**, 164 (2014), arXiv:1308.6176 [hep-ex].
5. The CEPC-SPPC Study Group Collaboration, CEPC-SPPC Preliminary Conceptual Design Report, Tech. Rep. IHEP-CEPC-DR-2015-01, 2015.
6. G. C. Branco, P. M. Ferreira, L. Lavoura, M. N. Rebelo, M. Sher and J. P. Silva, *Phys. Rept.* **516**, 1 (2012), doi:10.1016/j.physrep.2012.02.002, arXiv:1106.0034 [hep-ph].
7. N. Craig and S. Thomas, *J. High Energy Phys.* **1211**, 083 (2012), arXiv:1207.4835 [hep-ph].
8. N. Craig, J. A. Evans, R. Gray, C. Kilic, M. Park, S. Somalwar and S. Thomas, *J. High Energy Phys.* **1302**, 033 (2013), arXiv:1210.0559 [hep-ph].
9. B. Coleppa, F. Kling and S. Su, *J. High Energy Phys.* **1401**, 161 (2014), arXiv:1305.0002 [hep-ph].
10. N. Craig, J. Galloway and S. Thomas, arXiv:1305.2424 [hep-ph].
11. B. Coleppa, F. Kling and S. Su, arXiv:1308.6201 [hep-ph].
12. M. Carena, I. Low, N. R. Shah and C. E. M. Wagner, *J. High Energy Phys.* **1404**, 015 (2014), arXiv:1310.2248 [hep-ph].
13. N. Chen, C. Du, Y. Fang and L. C. Lü, *Phys. Rev. D* **89**, 115006 (2014), arXiv:1312.7212 [hep-ph].
14. J. Baglio, O. Eberhardt, U. Nierste and M. Wiebusch, *Phys. Rev. D* **90**, 015008 (2014), arXiv:1403.1264 [hep-ph].
15. B. Coleppa, F. Kling and S. Su, *J. High Energy Phys.* **1409**, 161 (2014), arXiv:1404.1922 [hep-ph].
16. B. Dumont, J. F. Gunion, Y. Jiang and S. Kraml, *Phys. Rev. D* **90**, 035021 (2014), arXiv:1405.3584 [hep-ph].
17. G. C. Dorsch, S. J. Huber, K. Mimasu and J. M. No, *Phys. Rev. Lett.* **113**, 211802 (2014), arXiv:1405.5537 [hep-ph].
18. B. Hespel, D. Lopez-Val and E. Vryonidou, *J. High Energy Phys.* **1409**, 124 (2014), arXiv:1407.0281 [hep-ph].
19. V. Barger, L. L. Everett, C. B. Jackson, A. D. Peterson and G. Shaughnessy, *Phys. Rev. D* **90**, 095006 (2014), arXiv:1408.2525 [hep-ph].

20. D. Fontes, J. C. Romo and J. P. Silva, *J. High Energy Phys.* **1412**, 043 (2014), arXiv:1408.2534 [hep-ph].
21. B. Coleppa, F. Kling and S. Su, *J. High Energy Phys.* **1412**, 148 (2014), arXiv:1408.4119 [hep-ph].
22. B. Grzadkowski, O. M. Ogreid and P. Osland, *J. High Energy Phys.* **1411**, 084 (2014), arXiv:1409.7265 [hep-ph].
23. J. F. Gunion, Y. Jiang and S. Kraml, *Phys. Rev. Lett.* **110**, 051801 (2013), arXiv:1208.1817 [hep-ph].
24. CDF and D0 Collab. (T. Aaltonen *et al.*), *Phys. Rev. D* **86**, 091101 (2012), arXiv:1207.2757 [hep-ex].
25. CMS Collab. (S. Chatrchyan *et al.*), *Phys. Lett. B* **722**, 207 (2013), arXiv:1302.2892 [hep-ex].
26. ALEPH and DELPHI and L3 and OPAL and LEP Working Group for Higgs Boson Searches Collabs. (S. Schael *et al.*), *Eur. Phys. J. C* **47**, 547 (2006), arXiv:hep-ex/0602042.
27. D0 Collab. (V. M. Abazov *et al.*), *Phys. Rev. Lett.* **101**, 071804 (2008), arXiv:0805.2491 [hep-ex].
28. CDF Collab. (T. Aaltonen *et al.*), *Phys. Rev. Lett.* **103**, 201801 (2009), arXiv:0906.1014 [hep-ex].
29. D0 Collab. (V. M. Abazov *et al.*), *Phys. Lett. B* **710**, 569 (2012), arXiv:1112.5431 [hep-ex].
30. CMS Collab. (S. Chatrchyan *et al.*), *Phys. Lett. B* **713**, 68 (2012), arXiv:1202.4083 [hep-ex].
31. ATLAS Collab. (G. Aad *et al.*), *J. High Energy Phys.* **1302**, 095 (2013), arXiv:1211.6956 [hep-ex].
32. ATLAS Collab. (G. Aad *et al.*), *J. High Energy Phys.* **1411**, 056 (2014), arXiv:1409.6064 [hep-ex].
33. D. Dicus, A. Stange and S. Willenbrock, *Phys. Lett. B* **333**, 126 (1994), arXiv:hep-ph/9404359.
34. R. Frederix and F. Maltoni, *J. High Energy Phys.* **0901**, 047 (2009), arXiv:0712.2355 [hep-ph].
35. S. Jung, J. Song and Y. W. Yoon, *Phys. Rev. D* **92**, 055009 (2015), arXiv:1505.00291 [hep-ph].
36. J. M. Butterworth, A. R. Davison, M. Rubin and G. P. Salam, *Phys. Rev. Lett.* **100**, 242001 (2008), arXiv:0802.2470 [hep-ph].
37. J. M. Butterworth, A. R. Davison, M. Rubin and G. P. Salam, arXiv:0810.0409 [hep-ph].
38. D. E. Kaplan, K. Rehermann, M. D. Schwartz and B. Tweedie, *Phys. Rev. Lett.* **101**, 142001 (2008), arXiv:0806.0848 [hep-ph].
39. T. Plehn, G. P. Salam and M. Spannowsky, *Phys. Rev. Lett.* **104**, 111801 (2010), arXiv:0910.5472 [hep-ph].
40. T. Plehn, M. Spannowsky, M. Takeuchi and D. Zerwas, *J. High Energy Phys.* **1010**, 078 (2010), arXiv:1006.2833 [hep-ph].
41. T. Plehn, M. Spannowsky and M. Takeuchi, *Phys. Rev. D* **85**, 034029 (2012), arXiv:1111.5034 [hep-ph].
42. T. Plehn and M. Spannowsky, *J. Phys. G* **39**, 083001 (2012), arXiv:1112.4441 [hep-ph].
43. N. Chen, J. Li and Y. Liu, arXiv:1509.03848 [hep-ph].
44. N. Chen, J. Li, Y. Liu and Z. Liu, *Phys. Rev. D* **91**, 075002 (2015), doi:10.1103/PhysRevD.91.075002, arXiv:1410.4447 [hep-ph].

45. V. Barger, L. L. Everett, H. E. Logan and G. Shaughnessy, *Phys. Rev. D* **88**, 115003 (2013), doi:10.1103/PhysRevD.88.115003, arXiv:1308.0052 [hep-ph].
46. D. Chowdhury and O. Eberhardt, *J. High Energy Phys.* **1511**, 052 (2015), doi:10.1007/JHEP11(2015)052, arXiv:1503.08216 [hep-ph].
47. A. Djouadi, *Phys. Rept.* **457**, 1 (2008), doi:10.1016/j.physrep.2007.10.004, arXiv:hep-ph/0503172.
48. A. Djouadi, *Phys. Rept.* **459**, 1 (2008), doi:10.1016/j.physrep.2007.10.005, arXiv:hep-ph/0503173.
49. R. V. Harlander, S. Liebler and H. Mantler, *Comput. Phys. Commun.* **184**, 1605 (2013), doi:10.1016/j.cpc.2013.02.006, arXiv:1212.3249 [hep-ph].
50. D. Eriksson, J. Rathsman and O. Stal, *Comput. Phys. Commun.* **181**, 189 (2010), doi:10.1016/j.cpc.2009.09.011, arXiv:0902.0851 [hep-ph].
51. J. Alwall *et al.*, *J. High Energy Phys.* **1407**, 079 (2014), arXiv:1405.0301 [hep-ph].
52. J. M. Campbell, R. K. Ellis and C. Williams, *J. High Energy Phys.* **1107**, 018 (2011), arXiv:1105.0020 [hep-ph].
53. Z. Kang, J. Li, T. Li, Y. Liu and G. Z. Ning, *Eur. Phys. J. C* **75**, 574 (2015), arXiv:1404.5207 [hep-ph].
54. N. D. Christensen and C. Duhr, *Comput. Phys. Commun.* **180**, 1614 (2009), arXiv:0806.4194 [hep-ph].
55. T. Sjostrand, S. Mrenna and P. Z. Skands, *J. High Energy Phys.* **0605**, 026 (2006), arXiv:hep-ph/0603175.
56. DELPHES 3 Collab. (J. de Favereau *et al.*), *J. High Energy Phys.* **1402**, 057 (2014), arXiv:1307.6346 [hep-ex].
57. M. Cacciari, G. P. Salam and G. Soyez, *Eur. Phys. J. C* **72**, 1896 (2012), arXiv:1111.6097 [hep-ph].
58. ATLAS Collab., ATLAS-CONF-2012-097.
59. Y. L. Dokshitzer, G. D. Leder, S. Moretti and B. R. Webber, *J. High Energy Phys.* **9708**, 001 (1997), arXiv:hep-ph/9707323.
60. M. Wobisch and T. Wengler, in Hamburg 1998/1999, Monte Carlo generators for HERA physics, pp. 270–279, arXiv:hep-ph/9907280.

Probe Higgs Potential via Multi-Higgs Boson Final States at Hadron Colliders

Qi-Shu Yan

School of Physics Sciences, University of Chinese Academy of Sciences,
Beijing 100039, P. R. China
Center for Future High Energy Physics, CAS, Beijing 100039, P. R. China

I summarize some results related with the determination of the Higgs potential, including a hadron-level Monte Carlo study of the sensitivity of Higgs boson pair production via the WW^*WW^* channel with the final state $3\ell 2j + \not{E}$ and the sensitivity study on the triple Higgs boson final state via $4b2\gamma$ in a 100 TeV collider.

Keywords: Multi-Higgs boson production; a 100 TeV collider.

1. Introduction

The last building block of the Standard Model (SM), the Higgs boson, has been discovered by the ATLAS and CMS Collaborations.[1,2] The analysis of the Higgs self-couplings via Higgs pair and multi-Higgs boson production is achievable at the high luminosity LHC runs and future pp collider, say a 100 TeV collider.[3]

The determination of the Higgs potential is a significant undertaking for several reasons. The potential is directly related to the structure of the vacuum, the electroweak phase transition and electroweak baryogenesis, and the fate of our universe as well.

The importance of Higgs pair production has attracted attention for a long time. Theoretical investigations of the Higgs pair production in the SM began with pioneering works,[4–6] where the gluon–gluon fusion[4] and the vector boson fusion[5,6] processes had been considered. It has been found that at hadron colliders, the gluon–gluon fusion production is almost one order of magnitude larger than the weak boson fusion process.

Here, I report two Monte Carlo studies of the feasibility of Higgs pair production via the final state $gg \to hh \to WW^*WW^* \to 3\ell 2j + \not{E}$ (Ref. 7) and triple Higgs production via the final state $4b2\gamma$ (Ref. 8).

In the language of an effective field theory, we can parametrize the Higgs self-interaction Lagrangian as

$$L \supset -\frac{1}{2}m_H^2 H^2 - \lambda_3 \lambda_{SM} v H^3 - \frac{1}{4}\lambda_4 \lambda_{SM} H^4 + \cdots , \tag{1}$$

where higher-dimensional operators denoted by an ellipsis, like operators $H\partial H \cdot \partial H$ (studied in Ref. 9) and H^5, are neglected here. In Eq. (1), $v = 246$ GeV is the Higgs field vacuum expectation value (VEV) and $m_H = 126$ GeV is the Higgs boson mass. In this Lagrangian, we define two free parameters, λ_3 and λ_4, to describe the triple- and quartic-Higgs vertices.

2. Trilinear Higgs Coupling

We consider the production $gg \rightarrow hh \rightarrow WW^*WW^* \rightarrow 3\ell2j + \not{E}$.[7]

To know the right combination is crucial to reconstruct the kinematic features of the signal and can provide important information to suppress background events. We use the minimal invariant mass approach to determine the correct combination and the extended variable m_{T2} was introduced to extract the information of particle mass in pair-production processes at hadron colliders[10,11] when the information of both the mass and the longitudinal components of invisible particles are missing. The reconstructed visible mass of Higgs bosons $m_h(\ell, \ell)$, $m_h(\ell, jj)$, and the variable m_{T2} are displayed in Fig. 1.

We apply two multivariate analyses: one is based on Boosted Decision Trees (BDT), the other on a Multi-Layer Perceptron (MLP) neural network. The results of these analyses are presented in Table 1. We plot the estimated sensitivity to λ_3 at the LHC, assuming a center-of-mass energy of 14 TeV and a 3 ab^{-1} dataset in Fig. 2. Although there is a large number of background events, we are capable to rule out the ranges $\lambda_3 < -1.0$ and $\lambda_3 > 8.0$; while if λ_3 is within the range $-1.0 < \lambda_3 < 8$, it might be challenging to determine the value of λ_3 due to background fluctuations.

We apply this analysis, demonstrated in the last section, to a 100 TeV collider. It is noticed that both the production rate of signal and background with top quarks are enhanced by factors of 40 or more than 100.

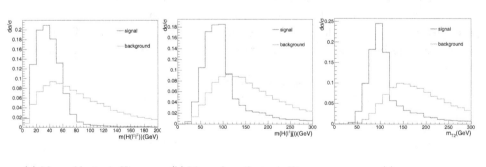

(a) Mass of leptonic Higgs.   (b) Mass of semileptonic Higgs.   (c) m_{T2}.

Fig. 1.   Four crucial reconstructed kinematic observables at hadron level.

Table 1.   Comparison of expected significance among three analysis methods at the LHC running at center-of-mass energy of 14 TeV.

	After preselection	Cut-based method	MLP method $N_{NN} > 0.82$	BDT method $N_{BDT} > 0.41$
No. of signal	13.7	6.2	5.7	3.8
No. of background	913.5	36.8	21.7	6.2
S/B	1.5×10^{-2}	1.7×10^{-1}	2.6×10^{-1}	6.2×10^{-1}
$S/\sqrt{B}$	0.45	1.0	1.2	1.5

Table 2.   Comparison of expected significance among three analysis methods in a 100 TeV collider.

	After preselection	Cut-based method	MLP method $N_{NN} > 0.94$	BDT method $N_{BDT} > 0.22$
No. of signal	416.8	160.0	80.4	104.0
No. of background	14801.8	523.6	107.3	67.1
S/B	2.8×10^{-2}	3.1×10^{-1}	7.5×10^{-1}	1.5
$S/\sqrt{B}$	3.43	7.0	7.8	12.7

In Table 2, we tabulate the optimized results. Similarly to the 14 TeV case, we note that the significance can be improved by a factor of 3.5 or so and the ratio S/B can be improved by two orders of magnitude.

The sensitivity of a 100 TeV collider to the triple coupling λ_3 is provided in Fig. 2. So a 100 TeV collider can exclude all values of λ_3 by simply using the 3 leptons mode considered in this work. Here, our multivariate analysis has been

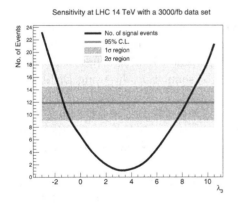

(a) Sensitivity to λ_3 at 14 TeV LHC.   (b) Sensitivity to λ_3 at a 100 TeV collider.

Fig. 2.   The sensitivity to the triple couplings of Higgs boson, λ_3, at the LHC 14 TeV and a 100 TeV collider.

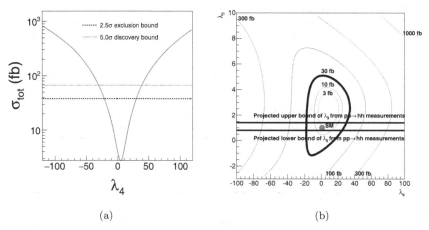

(a)   (b)

Fig. 3.   (a) The fitted cross-section when $\lambda_3 = 1$; (b) The feasibility contours of $\sigma(pp \to hhh)$ in the $\lambda_4 - \lambda_3$ plane.

optimized for the SM, i.e. $\lambda_3 = 1$, all λ_3 out the range $[2.8, 4.5]$ can be discovered. Nonetheless, if we optimize our analysis to different λ_3, we note that even for the minimal cross-section case with $\lambda_3 = 3.6$ or so, the significance can reach up to 5σ.

3. Quartic Higgs Coupling

The cross-section of triple-Higgs production at hadron colliders was calculated in Refs. 12 and 13. Its SM value, via gluon fusion, is $\mathcal{O}(0.01)$ fb at 14 TeV LHC, which is too small to be observed with the current design luminosity. Moreover, the dominant contribution of this process is the top-loop pentagon diagram.[13] A more precise prediction of triple-Higgs production at 100 TeV can be found in Ref. 14, where it is shown that the cross-section can be increased from 3 fb to 5 fb after taking into account the next-to-leading-order (NLO) corrections.

We focused on the feasibility of triple-Higgs production at a future 100 TeV hadron collider via $b\bar{b}b\bar{b}\gamma\gamma$.[8] We include detector simulations by using DELPHES 3.0.[15,16]

We apply two multivariate analysis approaches, (1) Boosted Decision Trees (BDT) and (2) Multi-Layer Perceptron (MLP) neural network, to utilize the correlation of observables in the signal to further suppress backgrounds. The results are presented in Table 3.

To observe the triple-Higgs signal of the SM at a 5σ level, a much larger integrated luminosity is necessary. Table 4 shows the values of $S/\sqrt{S + B}$ at different integrated luminosities. There we scale up the integrated luminosity for both signal and background. From the table, we see that the integrated luminosity should be around 1.8×10^4 ab^{-1} if we want to discover the triple-Higgs production via $b\bar{b}b\bar{b}\gamma\gamma$ mode at a 100 TeV machine.

Table 3. The number of events and the significances of the BDT and MLP neural network method are demonstrated. Here, total integrated luminosity is 30 ab^{-1}.

	Cuts based method	BDT > 0.02	MLP > 0.51
Signal	20	34	49
Background	2.4×10^4	2.8×10^4	9.9×10^4
S/B	8.3×10^{-4}	1.2×10^{-3}	5.0×10^{-4}
$S/\sqrt{S+B}$	0.13	0.20	0.16

Table 4. The values of $S/\sqrt{S+B}$ with BDT > 0.02 at different assumed integrated luminosities.

Integrated luminosity (ab^{-1})	30	300	3000	1.83×10^4
$S/\sqrt{S+B}$	0.2	0.6	2.0	5.0

References

1. ATLAS Collab. (G. Aad *et al.*), *Phys. Lett. B* **716**, 1 (2012), arXiv:1207.7214 [hep-ex].
2. CMS Collab. (S. Chatrchyan *et al.*), *Phys. Lett. B* **716**, 30 (2012), arXiv:1207.7235 [hep-ex].
3. R. Brock *et al.*, arXiv:1401.6081 [hep-ex].
4. O. J. P. Eboli, G. C. Marques, S. F. Novaes and A. A. Natale, *Phys. Lett. B* **197**, 269 (1987).
5. W. Y. Keung, *Mod. Phys. Lett. A* **2**, 765 (1987).
6. D. A. Dicus, K. J. Kallianpur and S. S. D. Willenbrock, *Phys. Lett. B* **200**, 187 (1988).
7. Q. Li, Z. Li, Q. S. Yan and X. Zhao, *Phys. Rev. D* **92**, 014015 (2015), doi:10.1103/PhysRevD.92.014015, arXiv:1503.07611 [hep-ph].
8. C. Y. Chen, Q. S. Yan, X. Zhao, Y. M. Zhong and Z. Zhao, *Phys. Rev. D* **93**, 013007 (2016), doi:10.1103/PhysRevD.93.013007, arXiv:1510.04013 [hep-ph].
9. H. J. He, J. Ren and W. Yao, arXiv:1506.03302 [hep-ph].
10. C. G. Lester and D. J. Summers, *Phys. Lett. B* **463**, 99 (1999), arXiv:hep-ph/9906349.
11. A. Barr, C. Lester and P. Stephens, *J. Phys. G* **29**, 2343 (2003), arXiv:hep-ph/0304226.
12. T. Plehn and M. Rauch, *Phys. Rev. D* **72**, 053008 (2005), arXiv:hep-ph/0507321.
13. T. Binoth, S. Karg, N. Kauer and R. Ruckl, *Phys. Rev. D* **74**, 113008 (2006), arXiv:hep-ph/0608057.
14. F. Maltoni, E. Vryonidou and M. Zaro, *J. High Energy Phys.* **1411**, 079 (2014), arXiv:1408.6542 [hep-ph].
15. S. Ovyn, X. Rouby and V. Lemaitre, arXiv:0903.2225 [hep-ph].
16. DELPHES 3 Collab. (J. de Favereau *et al.*), *J. High Energy Phys.* **1402**, 057 (2014), arXiv:1307.6346 [hep-ex].

Testable SUSY Spectra from GUTs at a 100 TeV *pp* Collider

Stefan Antusch*,†,‡ and Constantin Sluka*,§

*Department of Physics, University of Basel,
Klingelbergstr. 82, CH-4056 Basel, Switzerland
† Max-Planck-Institut für Physik (Werner-Heisenberg-Institut),
Föhringer Ring 6, D-80805 München, Germany
‡ stefan.antusch@unibas.ch
§ constantin.sluka@unibas.ch

Grand Unified Theories (GUTs) are attractive candidates for more fundamental elementary particle theories. They cannot only unify the Standard Model (SM) interactions but also different types of SM fermions, in particular quarks and leptons, in joint representations of the GUT gauge group. We discuss how comparing predictive supersymmetric GUT models with the experimental results for quark and charged lepton masses leads to constraints on the SUSY spectrum. We show an example from a recent analysis where the resulting superpartner masses where found just beyond the reach of LHC Run 1, but fully within the reach of a 100 TeV *pp* collider.

Keywords: Grand Unified Theories; supersymmetry; fermion masses.

1. Introduction

Supersymmetry (SUSY) has various attractive features. Most prominent among them are the properties that SUSY ameliorates considerably the hierarchy problem of the SM, by introducing new particles, superpartners of the SM states, with spin that differs from that of the SM counterparts by half a unit. Furthermore, these new states modify the renormalization group (RG) running of the gauge couplings in a way that simple schemes for Grand Unification of the fundamental interactions become possible with a unification scale high enough to be consistent with bounds on proton decay.

Observations tell us that supersymmetry has to be broken, such that the masses of the additional superpartner particles are (in general) heavier than the electroweak (EW) scale. From a bottom-up perspective, and from the theory point of view, the scale(s) where these masses lie is essentially a free parameter of the respective SUSY extension of the SM. On the other hand, Grand Unified Theories (GUTs) have the potential to constrain these scales, as we recently investigated in Ref. 1, and as we will discuss in this note.

The mass scale(s) of the SUSY particles (= sparticles) is relevant for various reasons. To start with, at the LHC various searches for them have been performed with negative results. As a general rule, the lighter the sparticles the better the solution to the hierarchy problem. Because of this, many people were hoping for an early discovery of SUSY at the LHC, which did not happen (so far). However, the measure of the "severeness" of the hierarchy problem is not possible without some ambiguity.

On the other hand, currently envisioned future 100 TeV pp colliders such as the FCC-hh and the SPPC could probe SUSY particles up to mass scales of $\mathcal{O}(10$ TeV).[2–4] In this note, we like to discuss an example where a GUT scenario predicts a SUSY spectrum which may be fully testable at the FCC-hh or SPPC. We will also go through the general arguments behind this result and discuss how it may generalize to other GUT models.

2. Predictive GUTs for Quark and Lepton Mass Ratios

GUTs are defined as theories which unify the three forces of the SM into a single unified force described by a single gauge symmetry group. As a consequence, also the particles get unified into joint representations of this gauge group. This can lead to predictions for the ratios of quark and lepton masses, respectively their Yukawa couplings, at high energies where the GUT description holds. Which predictions for the ratios are realized depends on the model, however there is only a limited number of options. GUT models which feature such predictions for the quark–lepton Yukawa coupling ratios are much more predictive than models without this property, and are thus of high interest in the theoretical community.

One prominent example is the so-called bottom-τ unification, or top-bottom-τ unification, i.e. the possible prediction that the respective third family Yukawa couplings are equal at the GUT scale. For the second generation, Georgi and Jarlskog postulated the GUT scale ratio $y_\mu/y_s = 3$ for the strange quark Yukawa coupling and the Yukawa coupling of the muon.[5] More recently, driven for example by the changed experimental results for the mass of the strange quark, alternative ratios have been proposed,[6,7] for example in $SU(5)$ GUTs $y_\tau = \pm\frac{3}{2}y_b$, $y_\mu = 6y_s$ or $y_\mu = \frac{9}{2}y_s$, and $y_e = -\frac{1}{2}y_d$. These ratios emerge when the Yukawa couplings are generated by dimension 5 operators, which involve the GUT symmetry breaking field H_{24} in the representation 24 of $SU(5)$. For example, the ratio $y_\tau = -\frac{3}{2}y_b$ can arise as a simple consequence of H_{24} getting a vacuum expectation value of $\langle H_{24}\rangle \propto \mathrm{diag}(2,2,2,-3,-3)$ in order to break the $SU(5)$ gauge symmetry spontaneously to the one of the SM.

To compare the GUT predictions for the quark–lepton Yukawa ratios, which hold at high energies, with the experimental results for quark and lepton masses at low energies, one has to calculate their RG running. At the scale of the SUSY particles, the sparticles have to be integrated out of the theory and the SUSY extension has to be matched to the SM at loop level. It is known that these SUSY threshold corrections[8] can have a large effect on the Yukawa couplings, especially when they are enhanced by a large (or moderate) value of $\tan \beta$.

The crucial point here is that these SUSY threshold corrections depend on the sparticle spectrum, i.e. on the masses of the SUSY particles. All the sets of GUT predictions known to date for the quark–lepton Yukawa ratios for all three families require a certain size of the threshold corrections, i.e. impose specific constraints on the SUSY spectrum. As we are going to show in an example, these requirements, combined with the measured value of the SM-like Higgs mass, can be powerful enough to constrain the sparticle spectrum to a range accessible by future 100 TeV pp colliders.[1]

3. GUTs and the Boundary Conditions for the SUSY Parameters at the GUT Scale

The fact that GUTs also unify the SM particles (and their superpartners) in joint representations of the GUT symmetry group also reduces the number of free SUSY parameters at the GUT scale. In $SU(5)$ GUTs, for example, one is left with only two soft breaking mass matrices at the GUT scale per family, one for the fermions in the five-dimensional matter representation and one for the fermions in the ten-dimensional representation. In $SO(10)$ GUTs, there is only one unified sfermion mass matrix.

In addition, the symmetries of GUT flavor models like[10–13] include various (non-Abelian) "family symmetries," which lead to hierarchical Yukawa matrices and impose (partially) universal soft breaking mass matrices among different generations. The combination of these effects can indeed lead to GUT scale boundary conditions which are very "universal" and can be described by only a few parameters.

Furthermore, universal boundary conditions may also be a result of a specific SUSY breaking mechanism. In the example to be presented below, for simplification, we will assume constrained MSSM (CMSSM) boundary conditions for the soft breaking parameters at the GUT scale, which is a quite strong assumption that will probably often be relaxed in realistic models.

Finally, we like to note that the absence of deviations from the SM in flavor physics processes implies constraints on flavor nonuniversalities in the SUSY spectrum (if the sparticles are not too heavy) and provides an experimental hint that, if SUSY exists at a comparatively low scale, it should be close to flavor-universal. In any case, it will be interesting to investigate in future works how the constraints on the SUSY spectrum get modified when the assumption of exact CMSSM boundary conditions at the GUT scale is relaxed.

4. Example: SUSY Spectrum from GUT Scenarios with $y_\tau = \pm\frac{3}{2}y_b$, $y_\mu = 6y_s$ and $y_e = -\frac{1}{2}y_d$

As an example, we will consider the class of GUT models which features the GUT-scale Yukawa relations $\frac{y_e}{y_d} = -\frac{1}{2}$, $\frac{y_\mu}{y_s} = 6$, and $\frac{y_\tau}{y_b} = -\frac{3}{2}$ (cf. Ref. 6). These GUT relations have been proven promising for GUT flavor model building and can emerge as direct result of CG factors in SU(5) GUTs or as approximate relation after diagonalization of the GUT-scale Yukawa matrices Y_d and Y_e (cf. Refs. 10–13).

For the GUT scale boundary conditions for the soft-breaking parameters, we restrict our analysis to the constrained MSSM, with parameters m_0, $m_{1/2}$, and A_0, with μ determined from requiring the breaking of electroweak symmetry, and set $sgn(\mu) = +1$. We have not included $\tan\beta$ explicitly in the fit, however, we have scanned over various different values of $\tan\beta$ and found that the best fit can be obtained for values of $\tan\beta \approx 30$. We have therefore set $\tan\beta = 30$ for our main analysis. The RG running including the calculation of the SUSY threshold corrections for all families has been performed with the REAP[14] extension SusyTC.[1]

We use the experimental constraints for the running $\overline{\text{MS}}$ Yukawa couplings at the Z-boson mass scale calculated in Ref. 15, and set the uncertainty of the charged lepton Yukawa couplings to 1% to account for the estimated theoretical uncertainty (which exceeds here the experimental uncertainty). When applying the measured Higgs mass $m_H = 125.7 \pm 0.4$ GeV (Ref. 16) as constraint, we use a 1σ interval

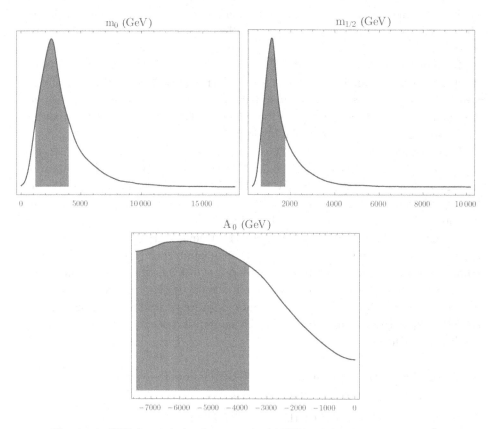

Fig. 1. 1σ HPD intervals for the constrained MSSM soft-breaking parameters.[1]

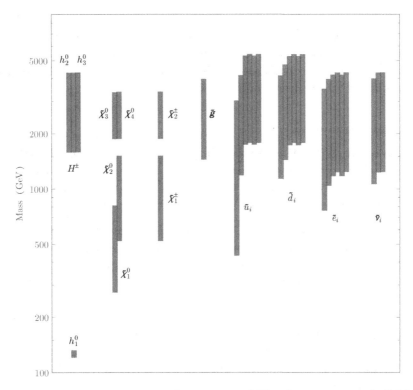

Fig. 2. 1σ HPD intervals for the sparticle spectrum and Higgs boson masses with $SU(5)$ GUT scale boundary conditions $\frac{y_e}{y_d} = -\frac{1}{2}$, $\frac{y_\mu}{y_s} = 6$, and $\frac{y_\tau}{y_b} = -\frac{3}{2}$. The LSP is always $\tilde{\chi}_1^0$ and the NLSP is always a stop.[1]

of ± 3 GeV, including the estimated theoretical uncertainty. For calculating m_H at the two-loop level, we have used the external software package `FeynHiggs` 2.11.2,[17] the current version when our numerical analysis was performed.

The confidence intervals for the masses of the sparticles are obtained as Bayesian "highest posterior density" (HPD) intervals from a Markov chain Monte Carlo sample of two million points, using a Metropolis algorithm. As an additional constraint, we restricted $|A_0| < 7.5$ TeV to make sure to avoid too large vacuum decay rates. It would be desirable to compute the lifetime of the vacuum for each point of the Markov chain, however, this clearly would take too much computation time. We remark that a more accurate inclusion of the lifetime constraint in the MC analysis may somewhat enlarge the predicted ranges for the masses of the sparticles. Our results for the 1σ intervals for the constrained MSSM parameters are shown in Fig. 1. The 1σ HPD results of the sparticle masses are presented in Fig. 2. For all parameter points, the LSP and NLSP are a neutralino and stop, respectively. The interval for the SUSY scale is $Q_{\text{HPD}} = [841, 3092]$ GeV.

We note that the analyzed GUT scale relation $\frac{y_\tau}{y_b} = -\frac{3}{2}$, $\frac{y_\mu}{y_s} = 6$ and $\frac{y_e}{y_d} = -\frac{1}{2}$ is indeed only one of the possible predictions that can arise from GUTs. We have chosen the above set of GUT scale predictions since they are among the ones recently used successfully in GUT model building.[10–13] In the future, it will of course be interesting to also test other combinations of promising GUT relations, and compare the resulting predictions for the SUSY spectra. Further details, comments and discussion of this analysis can be found in Ref. 1.

5. General Arguments

Although we have analyzed here a specific example only, some of the effects that lead to a predicted sparticle spectrum seem rather general, as long as the quark–lepton Yukawa ratios are predicted at the GUT scale together with (close-to) universal soft-breaking parameters:

- The main reason for the predictions/constraints on the SUSY spectrum is the fact that, to our knowledge, all the possible sets of GUT predictions for the quark–lepton Yukawa ratios require a certain amount of SUSY threshold corrections for each generation.[a] In general, to obtain the required size of the threshold corrections, one cannot have a sparticle spectrum which is too "split" (as e.g. in Ref. 18), since otherwise the loop functions (cf. Ref. 1) get too suppressed. More specifically, the required threshold corrections constrain the ratios of trilinear couplings, gaugino masses, μ and sfermion masses. In a CMSSM-like scenario, this implies that the ratios between m_0, $m_{1/2}$ and A_0 are constrained. Furthermore, since the most relevant threshold corrections are the ones which are $\tan\beta$-enhanced, it also implies that $\tan\beta$ cannot be too small.

- With the ratios between m_0, $m_{1/2}$, and A_0 constrained and a moderate to large value of $\tan\beta$, the measured value of the mass m_h of the SM-like Higgs allows to constrain the SUSY scale. We emphasize that this is an important ingredient, since the threshold corrections themselves depend only on the ratios of trilinear couplings, gaugino masses, μ and sfermion masses, and do not constrain the overall scale of the soft-breaking parameters. The combination of the two effects results in a predicted sparticle spectrum from the assumed GUT boundary conditions.

Since the Higgs mass m_h plays an important role, we would like to remark that it would be highly desirable to have a more precise computation of m_h available, especially for the "large stop-mixing" regime. In our analysis, we have used a theoretical uncertainty of ± 3 GeV, which is dominating the 1σ interval for m_h. This theoretical uncertainty should of course, strictly speaking, not be treated on the same footing as a pure statistical uncertainty. Furthermore, there are indications that the theoretical uncertainty in the m_h calculation in the most relevant regions of

---

[a]Note that with a CMSSM-like spectrum the SUSY threshold corrections are very similar for the first two families, and therefore the argument is also valid even if the quark–lepton Yukawa ratios are predicted for two of the families only, i.e. for the third family and either the second or the first family.

parameter space of our analysis, with "large stop-mixing," may be larger (as recently discussed[19]), however there is no full agreement on this. For our example analysis, as mentioned above, we have used the external software `FeynHiggs` 2.11.2[17] for a two-loop calculation of the Higgs mass, the current version when our numerical analysis was performed, and the most commonly assumed estimate ± 3 GeV for the theoretical uncertainty.

6. Summary

We have discussed how certain classes of predictive GUT models are capable of predicting a testable SUSY spectrum at a future 100 TeV *pp* collider such as the FCC-hh or the SPPC. The predictions for the sparticle spectrum can be understood as follows:

When GUT models predict the ratios of quark and charged lepton masses for all three generations at the GUT scale, as a result of the unification of the SM particles in GUT representation, they impose constraints on the amount of SUSY threshold corrections. This in turn implies constraints on the SUSY spectrum. These constraints, combined with the measured value of the SM-like Higgs mass, can be powerful enough to constrain the sparticle spectrum to a compact region.[1]

We have discussed an example where we found (cf. Fig. 2) that the resulting superpartner masses are beyond the reach of LHC Run 1, but fully within the reach of a 100 TeV *pp* collider.

Acknowledgments

This work was supported by the Swiss National Science Foundation. We thank the organizers of the IAS Program on the Future of High Energy Physics in Hong Kong for their hospitality.

References

1. S. Antusch and C. Sluka, arXiv:1512.06727 [hep-ph].
2. T. Cohen, T. Golling, M. Hance, A. Henrichs, K. Howe, J. Loyal, S. Padhi and J. G. Wacker, *J. High Energy Phys.* **1404**, 117 (2014), arXiv:1311.6480 [hep-ph].
3. T. Cohen, R. T. D'Agnolo, M. Hance, H. K. Lou and J. G. Wacker, *J. High Energy Phys.* **1411**, 021 (2014), arXiv:1406.4512 [hep-ph].
4. S. A. R. Ellis and B. Zheng, *Phys. Rev. D* **92**, 075034 (2015), arXiv:1506.02644 [hep-ph].
5. H. Georgi and C. Jarlskog, *Phys. Lett. B* **86**, 297 (1979).
6. S. Antusch and M. Spinrath, *Phys. Rev. D* **79**, 095004 (2009), arXiv:0902.4644 [hep-ph].
7. S. Antusch, S. F. King and M. Spinrath, *Phys. Rev. D* **89**, 055027 (2014), arXiv:1311.0877 [hep-ph].
8. R. Hempfling, *Phys. Rev. D* **49**, 6168 (1994); L. J. Hall, R. Rattazzi and U. Sarid, *Phys. Rev. D* **50**, 7048 (1994), arXiv:hep-ph/9306309; M. Carena, M. Olechowski, S. Pokorski and C. E. M. Wagner, *Nucl. Phys. B* **426**, 269 (1994), arXiv:hep-ph/

9402253; T. Blazek, S. Raby and S. Pokorski, *Phys. Rev. D* **52**, 4151 (1995), arXiv:hep-ph/9504364; S. Antusch and M. Spinrath, *Phys. Rev. D* **78**, 075020 (2008), arXiv:0804.0717 [hep-ph].

9. S. Antusch, C. Gross, V. Maurer and C. Sluka, *Nucl. Phys. B* **866**, 255 (2013), arXiv:1205.1051 [hep-ph].

10. S. Antusch, C. Gross, V. Maurer and C. Sluka, *Nucl. Phys. B* **877**, 772 (2013), arXiv:1305.6612 [hep-ph].

11. S. Antusch, C. Gross, V. Maurer and C. Sluka, *Nucl. Phys. B* **879**, 19 (2014), arXiv:1306.3984 [hep-ph].

12. S. Antusch, I. de Medeiros Varzielas, V. Maurer, C. Sluka and M. Spinrath, *J. High Energy Phys.* **1409**, 141 (2014), arXiv:1405.6962 [hep-ph].

13. J. Gehrlein, J. P. Oppermann, D. Schäfer and M. Spinrath, *Nucl. Phys. B* **890**, 539 (2014), doi:10.1016/j.nuclphysb.2014.11.023, arXiv:1410.2057 [hep-ph].

14. S. Antusch, J. Kersten, M. Lindner and M. Ratz, *Nucl. Phys. B* **674**, 401 (2003), arXiv:hep-ph/0305273.

15. S. Antusch and V. Maurer, *J. High Energy Phys.* **1311**, 115 (2013), arXiv:1306.6879 [hep-ph].

16. Particle Data Group (K. A. Olive *et al.*), *Chin. Phys. C* **38**, 090001 (2014).

17. S. Heinemeyer, W. Hollik and G. Weiglein, *Comput. Phys. Commun.* **124**, 76 (2000), arXiv:hep-ph/9812320; S. Heinemeyer, W. Hollik and G. Weiglein, *Eur. Phys. J. C* **9**, 343 (1999), arXiv:hep-ph/9812472; G. Degrassi, S. Heinemeyer, W. Hollik, P. Slavich and G. Weiglein, *Eur. Phys. J. C* **28**, 133 (2003), arXiv:hep-ph/0212020; M. Frank, T. Hahn, S. Heinemeyer, W. Hollik, H. Rzehak and G. Weiglein, *J. High Energy Phys.* **0702**, 047 (2007), arXiv:hep-ph/0611326; T. Hahn, S. Heinemeyer, W. Hollik, H. Rzehak and G. Weiglein, *Phys. Rev. Lett.* **112**, 141801 (2014), arXiv:1312.4937 [hep-ph].

18. N. Arkani-Hamed and S. Dimopoulos, *J. High Energy Phys.* **0506**, 073 (2005), arXiv:hep-th/0405159; G. F. Giudice and A. Romanino, *Nucl. Phys. B* **699**, 65 (2004), arXiv:hep-ph/0406088.

19. Workshop "Precision SUSY Higgs Mass Calculation Initiative," Heidelberg, 20-22.01.2016, https://sites.google.com/site/kutsmh/home.

Accelerator

A Conceptual Design of Circular Higgs Factory

Yunhai Cai

SLAC National Accelerator Laboratory, Stanford University,
2575 Sand Hill Road, Menlo Park, CA 94025, USA
yunhai@slac.stanford.edu

Similar to a super B-factory, a circular Higgs factory (CHF) will require strong focusing systems near the interaction points and a low-emittance lattice in the arcs to achieve a factory luminosity. At electron beam energy of 125 GeV, beamstrahlung effects during the collision pose an additional challenge to the collider design. In particular, a large momentum acceptance at the 2% level is necessary to retain an adequate beam lifetime. This turns out to be the most challenging aspect in the design of a CHF. In this paper, an example will be provided to illustrate the beam dynamics in a CHF, emphasizing the chromatic optics. Basic optical modules and advanced analysis will be presented. Most importantly, we will show that 2% momentum aperture is achievable.

Keywords: Charged particle optics; electron storage rings; circular colliders.

1. Introduction

Since the discovery of the Higgs particle at large hadron collider (LHC), the recent results for ATLAS and CMS have shown that the discovered particle resembles the Higgs boson in the standard model of elementary particles. Because of this remarkable discovery, it becomes increasingly important to precisely measure the property of the particle that gives the mass to all and to study the nature of the spontaneous symmetry breaking in the standard model.

The relatively low mass of the Higgs boson provides an opportunity to build an e^+ and e^- collider to efficiently and precisely measure its properties. In the production channel of $e^+e^- \rightarrow HZ$, the beam energy required for such a collider is 125 GeV, which is about 20% higher than the energy reached about two decades ago at LEP2. Can we design and build a circular Higgs factory (CHF) within a decade? What are the major challenges in the design? In this paper, we will address these questions.

2. Luminosity

In a collider, aside from its energy, its luminosity is the most important design parameter. For Gaussian beams, we can write the bunch luminosity as

$$\mathcal{L}_b = f_0 \frac{N_b^2}{4\pi\sigma_x\sigma_y} R_h \, , \tag{1}$$

where f_0 is the revolution frequency, N_b the bunch population, $\sigma_{x,y}$ transverse beam sizes, and R_h is a factor of geometrical reduction due to a finite bunch length σ_z and is given by

$$R_h = \sqrt{\frac{2}{\pi}} a e^{a^2} K_0(a^2) \, , \tag{2}$$

$a = \beta_y^*/(\sqrt{2}\sigma_z)$, β_y^* is the vertical beta function at the interaction point (IP), and K_0 is the modified Bessel function. In order to prevent R_h from becoming too small, we shall require $\sigma_z \approx \beta_y^*$. Obviously, for a number of n_b bunches, the total luminosity is $\mathcal{L} = n_b \mathcal{L}_b$.

In general, the beam sizes in the luminosity formula are not static variables. They are subject to the influence of the electromagnetic interaction during the collision. Typically, for flat beams, the vertical beam size will be blown up by the beam–beam force. To take this effect into account, we introduce the beam–beam parameter as[1]

$$\xi_y = \frac{r_e N_b \beta_y^*}{2\pi\gamma\sigma_y(\sigma_x + \sigma_y)} \, , \tag{3}$$

where γ is the Lorentz factor and r_e the classical electron radius. Using this formula for ξ_y, we can rewrite the luminosity as[2]

$$\mathcal{L} = \frac{cI\gamma\xi_y}{2r_e^2 I_A \beta_y^*} R_h \, , \tag{4}$$

where I is the beam current and $I_A = ec/r_e \approx 17,045$ A, the Alferov current. Since ξ_y is limited below 0.1 in most colliders, this formula is often used for estimating an upper bound of the luminosity.

In Table 1, we tabulated a set of consistent parameters for a CHF. In contrast to the B-factories,[3,4] the beam current is severely limited by the power of synchrotron radiation at very high energy. To reach the factory luminosity, we need to have very strong final focusing systems (FFS) and a very low emittance lattice. This combination makes the design of optics much more difficult compared with that of the B-factories.

3. Synchrotron Radiation

When an electron is in circular motion with a bending radius ρ, its energy loss per turn to synchrotron radiation is given by

$$U_0 = \frac{4\pi r_e mc^2 \gamma^4}{3\rho} \, . \tag{5}$$

Table 1. Main parameters of a circular Higgs factory.

Parameter	LEP2	CHF
Beam energy, E_0 (GeV)	104.5	125.0
Circumference, C (km)	26.7	52.7
Beam current, I (mA)	4	13
SR power, P_{SR} (MW)	11	50
Beta function at IP, β_y^* (mm)	50	1
Hourglass factor, R_h	0.98	0.73
Beam–beam parameter, ξ_y	0.07	0.10
Luminosity/IR, $\mathcal{L}$ (10^{34} cm^{-2} s^{-1})	0.0125	2.55

This loss has to be compensated by an RF system. The required RF power per ring is

$$P_{SR} = U_0 I / e. \tag{6}$$

For the beam energy of 125 GeV with a bending radius of $\rho = 6.1$ km in arcs, we have $U_0 = 3.56$ GeV. Adding additional bends in the interaction region, it increases to $U_0 = 3.85$ GeV, which means that electron loses about 3.1% of its energy every turn. The loss has to be compensated by the RF cavities. Here, we have used a RF system with $f_{RF} = 650$ MHz and $V_{RF} = 8.45$ GV. The voltage also provides the longitudinal focusing to the beam so that its length is not too long comparing to the vertical beta function at the IP. Assuming P_{SR} has to be less than 50 MW, the beam current is limited to 13.0 mA in the ring. Applying the expression of P_{SR} to the luminosity formula, we obtain

$$\mathcal{L} = \frac{3c\xi_y \rho P_{SR}}{8\pi r_e^3 \gamma^3 \beta_y^* P_A} R_h, \tag{7}$$

where $P_A = mc^2 I_A / e \approx 8.7$ GW. This scaling property of luminosity in $e^+ e^-$ colliders at extremely high energy was given by Richter.[5]

For a CHF with beam energy larger than 125 GeV, its beam current will be severely capped by the electrical power consumed by the RF system and therefore a smaller β_y^* seems the only available option to reach the required factory luminosity.

4. Beamstrahlung

Another important aspect of very high energy colliding beams is the emission of photons during collision. In general, this phenomenon is well known and is called beamstrahlung. Recently, Telnov[6] found that the most limiting effects to a CHF is an event when a high-energy photon is emitted by an electron in the beamstrahlung process. The electron energy loss can be so large that it falls outside of the momentum aperture η in the colliding ring. For a typical CHF, it was suggested that the

Table 2. Additional parameters selected to mitigate the beamstrahlung effects so that beamstrahlung beam lifetime is longer than 30 min.

Parameter	LEP2	CHF
Beam energy, E_0 (GeV)	104.5	125.0
Circumference, C (km)	26.7	52.65
Horizontal emittance, ϵ_x (nm)	48	4.5
Vertical emittance, ϵ_y (nm)	0.25	0.0045
Momentum acceptance, η (%)	1.0	2.0
Bunch length, σ_z (mm)	16.1	1.85
Momentum compaction, α_p (10^{-5})	18.5	2.5

following condition:

$$\frac{N_b}{\sigma_x \sigma_z} < \frac{0.1 \eta \alpha}{3 \gamma r_e^2}, \tag{8}$$

has to be satisfied to achieve 30 min of beam lifetime. Here, $\alpha \approx 1/137$ is the fine structure constant. If we introduce aspect ratios of beta functions at the IP and emittances in the ring, namely $\kappa_\beta = \beta_y^*/\beta_x^*$ and $\kappa_e = \epsilon_y/\epsilon_x$, this criteria can be rewritten as

$$\frac{N_b}{\sqrt{\epsilon_x}} < \frac{0.1 \eta \alpha \sigma_z}{3 \gamma r_e^2} \sqrt{\frac{\beta_y^*}{\kappa_\beta}}. \tag{9}$$

On the other hand, to achieve the beam–beam parameter ξ_y, we need

$$\frac{N_b}{\epsilon_x} = \frac{2\pi \gamma \xi_y}{r_e} \sqrt{\frac{\kappa_e}{\kappa_\beta}}. \tag{10}$$

Combining this equation with Eq. (9), we have

$$\epsilon_x < \frac{\beta_y^*}{\kappa_e} \left(\frac{0.1 \eta \alpha \sigma_z}{6 \pi \gamma^2 \xi_y r_e} \right)^2. \tag{11}$$

Since the quantities like ξ_y, β_y^* and σ_z are largely determined by the required luminosity and γ by the particle to be studied, this inequality specifies a low-emittance lattice that is required to achieve 30 min of beam lifetime. Normally, the natural emittance scales as γ^2. Here, it requires a scaling of γ^{-4}, indicating another difficulty in designing a factory with much higher energy beyond 125 GeV.

As shown in Table 2, we need to design a lattice with much smaller emittance than the one in LEP2 to mitigate the beamstrahlung effect. In particular, to satisfy the condition in Eq. (11), the emittance has to be smaller than 7 nm. Typically, a low emittance lattice requires smaller dispersion and stronger focusing. Both will lead to an increase in the strength of the sextupole, therefore dramatically reducing the dynamic aperture of the storage ring.

In the choice of the main design parameters, we want a factor of 100 increase in luminosity from LEP2. Due to the limit of the electric power, the increase of luminosity is largely achieved by a combination of very small beta functions at the IP and a low emittance lattice. In summary, the lattice of a CHF has the following main challenges:

- Low emittance lattice at high energy,
- High packing factor of magnets,
- Strong final focusing,
- Large momentum acceptance,
- Short bunches.

A high packing factor is required to reduce synchrotron radiation in the bending magnets and not increase the circumference of the ring. In this design, the dipoles in the arcs occupy 73% of the space in the ring. We will proceed to this specific design to show how to meet these challenges.

5. Arc

For a simple electron ring, the horizontal emittance is given by

$$\epsilon_x = C_q \gamma^2 \theta^3 F_c \,, \tag{12}$$

where C_q is a constant,

$$C_q = \frac{55}{32\sqrt{3}} \frac{\hbar}{mc} \,, \tag{13}$$

and $\theta = 2\pi/N_c$ is the bending angle per cell, and N_c the number of cell. F_c depends only on the structure of the cell. For FODO cells with equal phase advances, $\mu_x = \mu_y = \mu$, we have,

$$F_c^{\text{FODO}} = \frac{1 - \frac{3}{4}\sin^2\frac{\mu}{2} + \frac{1}{60}\sin^4\frac{\mu}{2}}{4\sin^2\frac{\mu}{2}\sin\mu} \,, \tag{14}$$

which is plotted as a function of μ in Fig. 1. For a 60° cell, we have $F_c^{\text{FODO}}(\pi/3) = 781/480\sqrt{3} \approx 0.94$.

Clearly, as seen in Eq. (12), the most effective way to reduce the emittance is to make the bending angle in a cell small. This implies that we use more cells. To reach 4 nm emittance, we used $N_c = 1176$.

In the arcs, we choose FODO cells because of their high packing factor and use many cells to reach the required emittance. The 60° phase advance is selected due to its property of resonance cancellation that we will explain later. The optics of the cell is illustrated in Fig. 2. Every six cells makes a unit transformation of betatron oscillation. In our design, each arc consists of 24 units and ends with dispersion suppressors. Similar to LEP2, we have eight arcs and eight straight sections to complete a ring with parameters shown in Table 2.

In this study, we set two families of sextupoles to make the linear chromaticity zero in the ring. For the third-order resonances, the contribution of sextupoles to all driving terms along the storage ring are computed[7] using the Lie method and

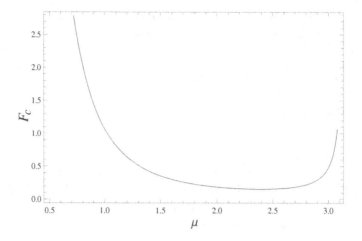

Fig. 1.   The emittance scaling parameter F_c as a function of phase advance μ in FODO cells.

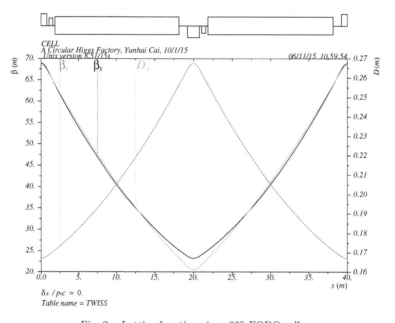

Fig. 2.   Lattice functions in a 60° FODO cell.

plotted in Fig. 3. As one can see from the figure, they are all canceled out within one betatron unit (made with six cells), as predicted by theorem.[8]

For the fourth-order resonances, we find similar cancellations[7] as shown in Fig. 4 except for one resonance: $2\nu_x - 2\nu_y = 0$. Since this resonance overlaps the same line as the linear coupling resonance in the betatron tune space, we can ignore it because the ring cannot operate near the linear resonance anyway.

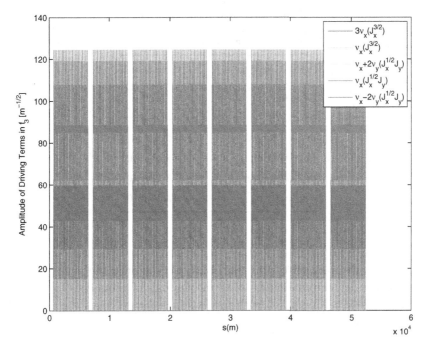

Fig. 3. All third-order resonances driven by sextupoles.

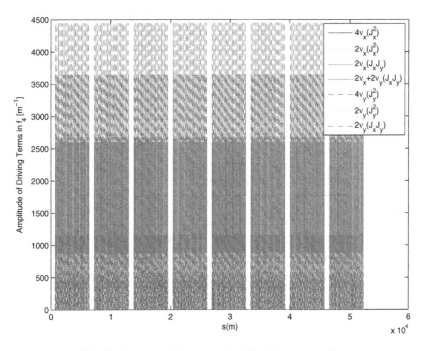

Fig. 4. Fourth-order resonances driven by sextupoles.

Table 3. The nonlinear chromaticities and tune shifts due to betatron amplitudes in the lattice that consists of arcs and simple straight sections.

Derivatives of tunes	Values
$\partial \nu_{x,y}/\partial \delta$	$0, 0$
$\partial^2 \nu_{x,y}/\partial \delta^2$	$-52, +102$
$\partial^3 \nu_{x,y}/\partial \delta^3$	$+1152, +197$
$\partial \nu_x/\partial J_x$ (m^{-1})	-8.43×10^4
$\partial \nu_{x,y}/\partial J_{y,x}$ (m^{-1})	-3.11×10^5
$\partial \nu_y/\partial J_y$ (m^{-1})	-5.34×10^4

It is also worth noting that there are three more terms of geometric aberrations in f_4. To quantify their effects on the beam, we compute the tune shifts along with the high-order chromaticities using the normal form analysis[9] and tabulate the result in Table 3. Compared with the existing storage rings, these tune shifts are too large at least by an order of magnitude.

6. Final Focusing System

Note that the beam lifetime condition in Eq. (11) does not depend on κ_β. Therefore, according to Eq. (10), κ_β (or β_x^*) can be used to adjust the bunch population N_b or equivalently, the number of bunches n_b when the total current is limited by the electrical power. Here, we would like to choose a large β_x^*, leading to a smaller n_b. Our choice of the parameters in the interaction region are tabulated in Table 4.

It is always challenging to design a FFS in a circular collider. In the CHF, it becomes even more because of a smaller β_y^* (1 mm) and a longer distance L^* (2 m) which is the distance between the IP and the first focusing quadrupole.

Table 4. Other parameters determined by a specific design of final focusing system.

Parameter	LEP2	CHF
Beam energy, E_0 (GeV)	104.5	125.0
Circumference, C (km)	26.7	52.65
β_x^* (mm)	1500	100
β_y^* (mm)	50	1
Bunch population, N_b (10^{10})	57.5	7.77
Number of bunches, n_b	4	184

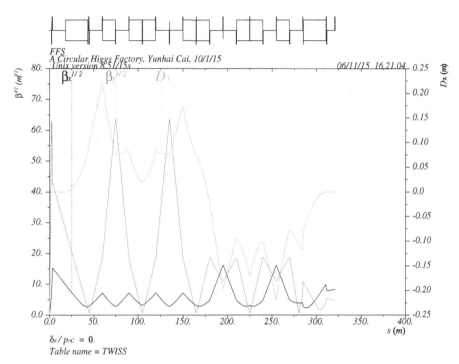

Fig. 5. Lattice functions in a final focusing system with local chromatic compensation section.

Here, we adopt an optics similar to the design of a linear collider. The optics of the FFS is shown in Fig. 5. The FFS starts with a final transformer (FT), continues with a chromatic correction in the vertical (CCY) and then the horizontal plane (CCX), and ends with a matching section. The FFS has two secondary imaging points and fits in a 321 m section.

The FT contains four quadrupoles, including the final focusing doublet. The betatron phase advances are 180° in both planes. At the end of the FT, we have the first imaging point where the beta functions remain very small.

The CCY consists of four 90° FODO cells and makes a unit of betatron transformer. The module starts at the middle of the defocusing quadrupole to enhance the peak of the vertical beta function at the positions of a pair of sextupoles separated by "-I" transformation. Five dipoles with an equal bending angle provide dispersions at the locations of the sextupoles. At the end of the CCY, we have the second imaging point at which the lattice functions are identical to those at the first one.

Similarly, we construct the CCX, but starting at the middle of the focusing quadrupole. There are five dipoles that generate the dispersion with negative bending angles. The amplitude of the angles are chosen to be the same as those in the CCY so that there is no net bending from the FFS. At the end of the CCX, we have a section matching to the optics of the dispersion suppressor.

The nonlinear chromatic effects can be characterized by the high order derivatives of the lattice functions. These derivatives can be computed[10] using the technique of the differential algebra.[11] For the FFS, we start with the initial condition: $\beta_x = 0.1$ m and $\beta_y = 1$ mm and calculate the lattice functions and their derivatives element-by-element down to the end.

The first-order chromaticity is compensated by two pairs of sextupoles in the CCY and CCX, respectively in the horizontal and vertical planes; the second-order ones by slight changes of betatron phases between the final doublet and the sextupole pairs; and finally, the third-order ones by the two sextupoles at the two secondary imaging points where the beta functions are at the minimum. The results of the chromatic compensation is summarized in Table 5. Clearly, the nonlinear

Table 5.   The chromatic tune shifts from the FFS after the correction.

Derivatives of tunes	Values
$\partial \nu_{x,y}/\partial \delta$	$-0.43, -0.34$
$\partial^2 \nu_{x,y}/\partial \delta^2$	$+127, +49$
$\partial^3 \nu_{x,y}/\partial \delta^3$	$-954, -852$

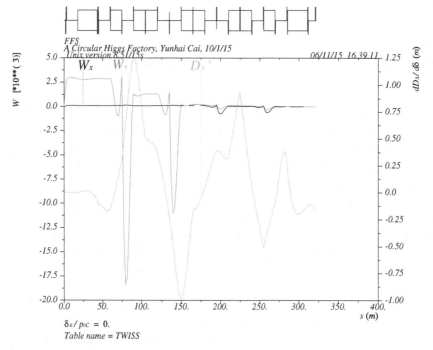

Fig. 6. W function and the second-order dispersion in the final focusing system with the local chromatic compensation.

chromatic effects in the FFS have been reduced to similar values as in the arcs shown in Table 3. In addition, we plot the W functions and the second-order dispersion in Fig. 6. It is worth noting that there is a small amount of second-order dispersion leaking out of the FFS.

7. Collider

Replacing two interaction regions with two simple straights in the arc lattice, we build a collider lattice shown in Fig. 7. The main parameters are summarized in Table 6. Since the lattice has a twofold symmetry, the half of the betatron tunes are slightly above the half integer, which enhances the dynamic focusing from the beam–beam interaction. As a result, the beam–beam parameter becomes larger as demonstrated in the B-factories.[3,4]

Since the strongest quadrupoles and sextupoles are positioned at the highest beta functions in the FFS, naturally the IR contains many high-order aberrations. We compute the third-order and fourth-order driving terms in the collider. The cancellation of the resonances at third-order remains intact. But the fourth-order resonance driving terms become much larger as shown in Fig. 8. Clearly, the aberrations in the IR are dominant in the entire ring.

To quantify their nonlinear effects in the collider, we compute the tune shifts, the high-order chromaticities and geometric and chromatic tune shifts using the

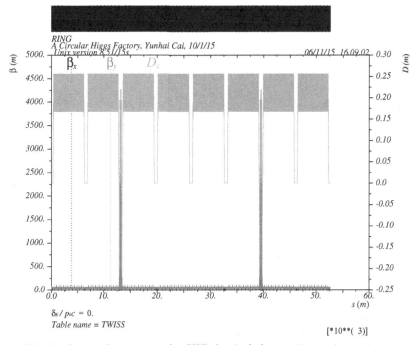

Fig. 7.   Lattice functions in the CHF that includes two interaction regions.

Table 6.   Main parameters of the circular Higgs factory.

Parameter	Value
Energy, E_0 (GeV)	125.0
Circumference, $\mathcal{C}$ (km)	52.7
Tune, ν_x, ν_y, ν_z	225.04, 227.14, 0.165
Natural emittance, ϵ_x (nm)	4.5
Bunch length, σ_z (mm)	1.85
Energy spread, σ_δ	1.44×10^{-3}
Momentum compaction	1.25×10^{-5}
Damping time, τ_x, τ_y, τ_z (ms)	11.4, 11.4, 5.7
Energy loss per turn, U_0 (GeV)	3.85
RF voltage, V_{RF} (GV)	8.45
RF frequency, f_{RF} (MHz)	650.0
Harmonic number	114,144

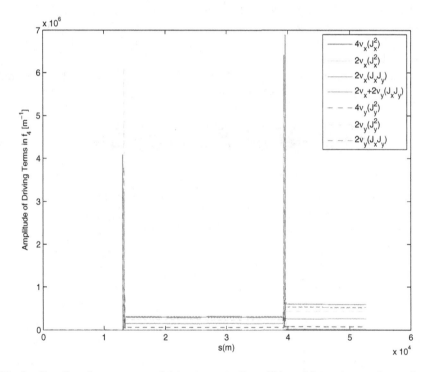

Fig. 8.   Fourth-order resonances driving terms in the collider with two interaction regions.

normal form analysis and tabulate the results in Table 7. The table shows that the geometric–chromatic tune shifts are at the same level at $\delta = 0.01$ than the geometric tune shifts. This result is achieved by adding a few octopoles and decapoles at the beta peaks in the interaction regions.

Table 7. The nonlinear chromaticities and tune shifts due to betatron amplitudes in the collider that contains two interaction regions.

Derivatives of tunes	Values
$\partial \nu_{x,y}/\partial \delta$	$0, 0$
$\partial^2 \nu_{x,y}/\partial \delta^2$	$-167, +790$
$\partial^3 \nu_{x,y}/\partial \delta^3$	$+27978, -19146$
$\partial \nu_x/\partial J_x \ (m^{-1})$	-8.18×10^4
$\partial \nu_{x,y}/\partial J_{y,x} \ (m^{-1})$	-4.03×10^5
$\partial \nu_y/\partial J_y \ (m^{-1})$	$+6.09 \times 10^4$
$\partial^2 \nu_x/\partial \delta \partial J_x \ (m^{-1})$	-2.23×10^6
$\partial^2 \nu_{x,y}/\partial \delta \partial J_{y,x} \ (m^{-1})$	-8.95×10^7
$\partial^2 \nu_y/\partial \delta \partial J_y \ (m^{-1})$	-1.49×10^7

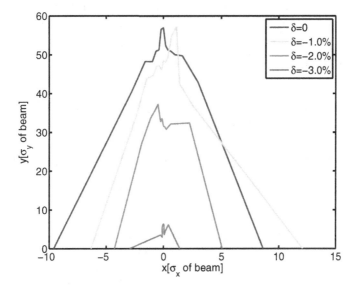

Fig. 9. Dynamic aperture of the collider with various momentum deviations.

Finally, we evaluate the dynamic aperture of the collider by tracking the particles with various momentums. The tracking is carried out with synchrotron oscillation and the radiation damping. The orbit and optics errors due to the saw-tooth energy profile are corrected by tapering the settings for all magnetic elements. As shown in Fig. 9, though the degradation of the off-momentum aperture is large, there is sufficient momentum acceptance to retain the particles in the long tail distribution of energy due to beamstrahlung. Most importantly, there is a large enough dynamic aperture in the vertical plane to accommodate the large tail generated by the beam–beam interaction.

8. Conclusion

We have analyzed the impact on lattice design due to beamstrahlung in a CHF. In particular, we found that a minimum emittance is necessary to retain an adequate beam lifetime. As a result, we developed a systematic procedure that can be applied to the lattice design.

Furthermore, we have developed a method to effectively compensate nonlinear chromaticity in the FFS. In particular, we have demonstrated that the chromatic aberration can be reduced down to the required level in the arcs.

We have achieved 2% momentum acceptance in a lattice with an ultra-low beta interaction region. Six families of sextupoles are used in the chromatic correction. Octupoles and decapoles in the FFS are helpful to correct high order chromatic-geometric aberrations.

As shown in our paper, a CHF requires not only a FFS with an ultra-low beta at the interaction point but also a very low-emittance lattice at very high energy. Such optics in a collider with a consistent set of accelerator parameters and especially with a large momentum aperture has been demonstrated in design.

Acknowledgments

This work is supported by the Department of Energy under Contract No. DE-AC02-76SF00515. I would like to thank Alex Chao, Yuri Nosochkov, Katsunobu Oide, Richard Talman, Uli Wienands, and Frank Zimmermann for many stimulating discussions. Special thanks to Professor Henry Tye for his hospitality during my visit in the Institute for Advanced Study in the Hong Kong University of Science and Technology.

References

1. F. Amman and D. Ritson, Design of electron–positron colliding beam rings, in *1961 Int. Conf. on High Energy Accelerators*, National Laboratory, Brookhaven, New York, 1961, p. 471.
2. J. T. Seeman, Beam–beam interaction: luminosity, tails and noise, SLAC-PUB-3182, July 1983.
3. PEP-II: An Asymmetric B Factory, Conceptual Design Report, SLAC-418, June 1993.
4. KEKB *B*-factory Design Report, KEK-Report-95-7 (1995).
5. B. Richter, *Nucl. Instrum. Methods A* **136**, 47 (1976).
6. V. I. Telnov, *Phys. Rev. Lett.* **110**, 114801 (2013).
7. Y. Cai, *Nucl. Instrum. Methods A* **645**, 168 (2011).
8. K. L. Brown and R. V. Servranckx, *Nucl. Instrum. Methods A* **258**, 480 (1987).
9. E. Forest, M. Berz and J. Irwin, *Part. Accel.* **24**, 91 (1989).
10. Y. Cai, *Nucl. Instrum. Methods A* **797**, 172 (2015).
11. M. Berz, *Part. Accel.* **24**, 109 (1989).

Pretzel Scheme for CEPC

Huiping Geng

Institute of High Energy Physics,
19B Yuquan Road, Shijingshan, Beijing 100049, P. R. China
genghp@ihep.ac.cn

CEPC was proposed as an electron and positron collider ring with a circumference of 50–100 km to study the Higgs boson. Since the proposal was made, the lattice design for CEPC has been carried out and a preliminary conceptual design report has been written at the end of 2014. In this paper, we will describe the principles of pretzel scheme design, which is one of most important issues in CEPC lattice design. Then, we will show the modification of the lattice based on the lattice design shown in the Pre-CDR. The latest pretzel orbit design result will also be shown. The issues remained to be solved in the present design will be discussed and a brief summary will be given at the end.

Keywords: Electron positron collider; lattice design; pretzel orbit.

1. Introduction

After the discovery of Higgs-like boson at CERN,[1,2] many proposals have been raised to build a Higgs factory to explicitly study the properties of the particle. One of the most attractive proposals is the Circular Electron and Positron Collider (CEPC) project in China.[3]

CEPC is a ring with a circumference of 50–70 km, which will be used as electron and positron collider at phase-I and will be upgraded to a Super proton–proton Collider (SppC) at phase-II. The designed beam energy for CEPC is 120 GeV, the main constraints in the design is the synchrotron radiation power, which should be limited to 50 MW, the target luminosity is on the order of 10^{34} cm^{-2} s^{-1}.

As beam energy is high, CEPC favors a lattice with more arcs which will enable RF cavities to compensate the energy loss in the straight section, thus can reduce energy variation from synchrotron radiation. SppC needs long straight sections for collimators etc. To compromise between CEPC and SppC, the ring is decided to have eight arcs and eight straight sections, RF cavities will be distributed in each straight section.

The lattice design for CEPC has been carried out and a preliminary conceptual design report has been written at the end of 2014.[4] Many work has been done since the publication of the Pre-CDR. In this paper, we will describe the principles of pretzel scheme design, which is one of most important issues in CEPC lattice design. Then, we will show the modification of the lattice based on the lattice design shown in the Pre-CDR. The latest pretzel orbit design result will also be shown. The issues remained to be solved in the present design will be discussed and a brief summary will be given at the end.

2. Principles of Pretzel Scheme

In single ring collider, the pretzel orbit is used to avoid the beam collision at positions except the IP.

For ideal pretzel orbit, the following relationship should be fulfilled: $\phi = N \cdot 2\pi$, where ϕ is the phase advance between the adjacent collision points, N is an integer. This relation guarantees that if the beam is properly separated at the first parasitic collision point, then it can be automatically properly separated at other parasitic collision points.

For our lattice, it is comprised of $60°/60°$ FODO cells, every 6 cells have a phase advance of 2π, so the distance between the adjacent parasitic points L_{pc} can be written as $L_{pc} = N \cdot 6 \cdot 47.2 = N \cdot 283.2$ m. For 50 bunches, there are 100 collision points in total, thus the ring circumference C must be $C = 100 \cdot L_{pc} = 28{,}320 \cdot N$ m.

As the circumference of the CEPC ring is about 50 km, the integer number N has to be 2, which means the ring circumference has to be 56,640 m and there will be one collision point every 4π phase advance.

3. Modification of the Main Ring Lattice

In our Pre-CDR, the ring circumference is 54,752 m. To make the pretzel scheme works for 50 bunches, two options (assuming the phase advance per cell keeps constant) can be done to modify the ring lattice. First, the cell length can be changed. This will result in the change of the circumference and emittance etc. Second, the number of cells, or the length of the straight section can be changed. This will only change the circumference while keeping the emittance unchanged.

In the following, we take the easy way, i.e. we change the circumference of the ring to make the lattice works for the pretzel orbit of 50 bunches. We keep the arc length and short straight section length constant, while making the long straight section length increased from 1132.8 m to 1604.8 m, or from 20 FODO cells to 34 FODO cells. After making this change, the circumference of the ring becomes 56,640 m, which is suitable for the pretzel orbit of 50 bunches as shown in Sec. 2.

4. Pretzel Scheme Design

To avoid big coupling between horizontal and vertical plane, we use horizontal separation scheme to generate the pretzel orbit here. Also, in order to avoid beam

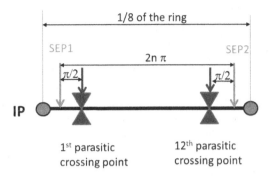

Fig. 1. A schematic drawing of the positions of the electrostatic separators for 1/8th of the ring. SEP1 and SEP2 in the drawing means the first and second electrostatic separators.

instability and high order mode in the RF cavities, we require there is no off-center orbit in any RF sections. Thus, we use one pair of electrostatic separators for each arc.

For each pair, the position of the first electrostatic separator is chosen such that it is $\pi/2$ phase advance before the first parasitic crossing point, and the position of the second electrostatic separator is chosen such that it is $\pi/2$ phase advance after the last parasitic crossing point in this arc. A schematic drawing is shown in Fig. 1.

The separation distance between the two beams is about $10\sigma_x$, which is a empirical number, to allow for a reasonable beam lifetime. The final orbit of the beam is shown in Fig. 2.

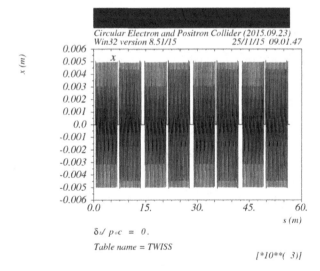

Fig. 2. The pretzel orbit in the ring for one beam. The separation distance between the two beams is about $10\sigma_x$ in the horizontal plane.

4.1. *Issues with pretzel orbit*

When there is an off-center orbit, the beam will experience extra fields in magnets. To be specific, in quadrupole magnets, the beam will see an extra dipole field when it is off-centered. The dipole strength can be estimated with a simple formula: $\Delta B = K_1 \cdot B\rho \cdot \Delta x$, where K_1 is the normalized quadrupole strength, $B\rho$ is the magnetic rigidity of the beam, and Δx is the orbit of the beam. With a simple calculation, we can see that the extra dipole field seen by the off-center beam has a strength that is comparable to the strength of the main bending magnets.

In sextupole magnets, the beam will experience extra dipole and quadrupole fields. The field strength can be estimated similarly. These extra fields (dipole field in quadrupole, and both dipole and quadrupole fields in sextupole) will break the periodicity and achromatic condition of the lattice, and this effect has to be corrected.

The distortion of pretzel orbit effects on beta functions and dispersion function can be corrected by making quadrupoles individually adjustable, which can be done by adding shunts on each quadrupoles. Then a new lattice period can be found to be agree with the pretzel orbit period, namely 6 FODO cells, or phase advance of 2π. A new lattice can be found accordingly.

5. Summary

In this paper, we have described how the pretzel orbit of 50 bunches has been designed. The distortion of lattice due to pretzel orbit has also been explained. We have also shown how the pretzel orbit distortion effect on the lattice can be corrected. The work to achieve a reasonable dynamic aperture result is still ongoing.

Acknowledgment

This work was supported by National Natural Science Foundation of China, under contract No. 11405188.

References

1. A. Cho, *Science* **337**, 6091 (2012).
2. CMS Collab., *Phys. Lett. B* **716**, 1 (2012).
3. Q. Qin, Accelerators for a Higgs Factory: Linear versus Circular (HF2012), Fermi National Laboratory, November 14–16, 2012.
4. The CEPC-SPPC Study Group, CEPC–SPPC Preliminary Conceptual Design Report, Vol. 2, Accelerator (2015).

Beam–Beam and Electron Cloud Effects in CEPC/FCC-ee

Kazuhito Ohmi

KEK, 1-1 Oho, Tsukuba, Ibaraki 305-0801, Japan

ohmi@post.kek.jp

We discuss beam dynamics issues in CEPC/FCC-ee, especially focusing on the beam–beam and electron cloud effects. Beamstrahlung is strong in extreme high energy collision such as Higgs and top factory. Beam–beam simulations considering beamstrahlung are now ready. Several points of beam–beam effects for FCC-ee are presented. Electron cloud effects are serious for high current positron machine, especially in Z factory that many bunches are stored. Analytical estimate for threshold of electron density and electron build-up for CEPC are presented.

Keywords: Electron–positron collider; beam–beam effect; electron cloud effect.

1. Introduction

Beam–beam and electron cloud effects are most serious issues of beam dynamics in electron–positron circular colliders. Beam–beam tune shift limit is traditional problem of e^+e^- circular colliders. The tune shift limit is $\xi \sim 0.03$–0.1 depending on the radiation damping time. Luminosity depends on total current ($N_e f_{\rm rep}$), vertical beta function at the interaction point (β_y^*) and the beam–beam tune shift,

$$L = \frac{\gamma}{r_e} \frac{N_e f_{\rm rep} \xi}{\beta^*} \, . \tag{1}$$

In extreme high energy colliders as Higgs and top factories, beamstrahlung during beam–beam collision becomes serious.

Electron cloud in positron ring limits the performance of e^+e^- circular colliders. Electrons produced by photo-emission for synchrotron radiation cause coupled bunch and single bunch instabilities of positron beam. The two issues are discussed for CEPC and TLEP's in this document.

2. Study of Beam–Beam Effects

Four topics concerning to the beam–beam interactions are discussed. First, tune operating point is important to realize high luminosity performance in electron–positron colliders. Second subject, beamstrahlung is serious issue for extreme high energy lepton colliders. It is key to choose which circular or linear collider. Beam–beam simulation to predict luminosity performance is third subject. Finally, coherent instability due to beam–beam interaction is investigated.

2.1. *Golden tune*

Most of e^+e^- colliders are operated at tune where horizontal tune ν_x is close to slightly upper of half integer and vertical ν_y is around 0.55–0.6. CESR is operated (0.52, 0.58), KEKB, PEP-II and BEPC-II are operated closer to half integer (0.505–0.51, 0.55). The luminosity gain compare with other operating point is quite high, 20%–50%. There are reasons why high luminosity has been delivered in this operating point.

2.1.1. *Tune shift/spread is suppressed near half integer tune*

The nominal expression of the beam–beam tune shift is given by

$$\xi_{\pm,x(y)} = \frac{N_\mp r_e}{2\pi\gamma_\pm} \frac{\beta_{x(y),\pm}}{\sigma_{x(y),\mp}(\sigma_{x,\mp} + \sigma_{y,\mp})}, \tag{2}$$

where $\gamma_\pm$, $N_\pm$, $\sigma_{x(y),\pm}$ and $\beta_{x(y),\pm}$ are the relativistic factor, bunch populations, the horizontal (vertical) size and beta function of positron/electron beam, respectively. The tune shift is suppressed near half integer tune. Dynamic beta squeezes horizontal beta strongly. The dynamic beta gives both merit and demerit in the beam–beam performance. The merit is simply the increase of luminosity due to squeezes of σ_x, while the demerit is beta variation at other than collision point, which induces beam loss at a narrow aperture area.

2.1.2. *Horizontal force is independent of vertical amplitude*

Beam–beam force of a flat Gaussian distribution is expressed by so-called Basetti–Erskine formula using complex error function. Aspect ratio of transverse beam size σ_x/σ_y is more than 100 for e^+e^- colliders. The ratio of CEPC/FCC-ee is around 500. The horizontal beam–beam force is independent of the vertical amplitude, while the vertical beam–beam force is dependent of the horizontal amplitude. This fact means that the horizontal betatron motion can be treated independently of the vertical motion.

2.1.3. *Horizontal motion is integrable near half integer for the beam–beam force*

Figure 1 shows phase space plot for horizontal betatron motion. Clear lines for every initial conditions are seen up to 0.52, while chaotic behavior is seen in $\nu_x = 0.55$. We can say the horizontal motion is integrable $\nu_x \leq 0.52$.

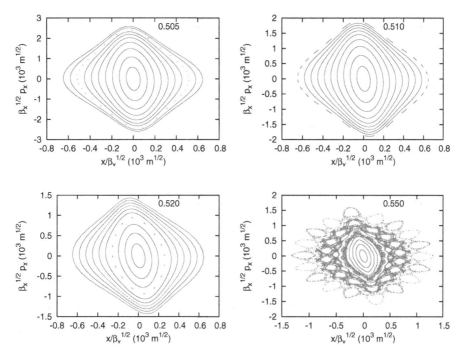

Fig. 1.   Phase space plot for particles with betatron amplitude $x = 0.5 - 5\sigma_x$ under the beam–beam force. The horizontal tunes are 0.505, 0.510, 0.52 and 0.55 in plots top to bottom.

2.1.4. *Vertical motion is integrable near half integer, but x–y coupling should be avoided*

The vertical beam–beam force depends on the horizontal amplitude. It is easy to conjecture chaotic vertical motion under the chaotic horizontal motion: that is vertical motion is studies for $\nu_x < 0.52$. Figure 2 shows phase space plot for a particle with $(x_0, y_0) = (3\sigma_x, 3\sigma_y)$ under the beam–beam force. Top two plots and bottom right plots with $(\nu_x, \nu_y) = (0.505, 0.54)$, $(0.51, 0.54)$ and $(0.52, 0.54)$ show good integrability; that is, the motion seems regular and clear boundary. While bottom left plot $(0.51, 0.56)$ indicates chaotic behavior. It is not preferable that both tunes are close to half integer, because of x–y coupling. Top right and bottom right plots show that x–y motion coupled in the phase space motion. Motion with $(x_0, y_0) = (3\sigma_x, 2\sigma_y)$ is plotted blue dots in bottom right plot. The phase space trajectory with $(3\sigma_x, 3\sigma_y)$ is located close to that with $(3\sigma_x, 2\sigma_y)$, that is, a perturbation for example the radiation excitation can make transition easily from trajectory with $(3, 2)$ to that of $(3, 3)$. Emittance growth related to the mechanism is seen in strong–strong simulation later, in Fig. 4. Synchrotron oscillation enhances emittance growth, because the phase space structure is mixed due to change of tune shift depending on the synchrotron amplitude z.

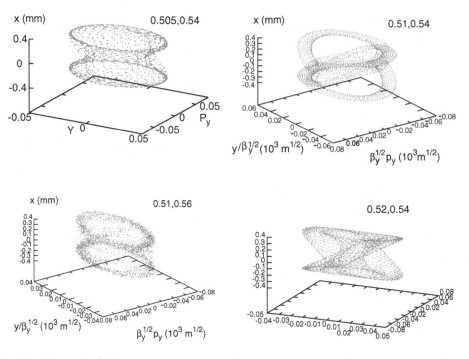

Fig. 2. (Color online) Phase space plot for a particle with betatron amplitude, $y = 3\sigma_y$, $x = 3\sigma_x$ under the beam–beam force.

2.2. *Beamstrahlung*

Beamstrahlung is serious issue for extreme high energy e^+e^- colliders. Bending radius during beam–beam collision is very small compared with that of bending magnets. For CEPC, bending radius in magnets is 6000 m, while that during the beam–beam interaction is 23 m. Synchrotron radiation due to the beam–beam bending is called beamstrahlung. The photon emission is stochastic; the emission process is stochastic in the timing and energy, and the number of photon per crossing is less than 1, 0.2 in CEPC. The stochastic emission induces energy spread in the beam. The characteristic energy of photon is 0.16 GeV, 0.13% of the beam energy. Energy distribution of the photon contains a higher energy tail than Gaussian. Particles with large energy losses are lost due to the limitation of energy acceptance of the ring. The photon emission process during beam–beam collision has been taken into account to the beam–beam simulations, Lifetrack and BBWS/BBSS.

2.3. *Simulation of luminosity performance and beam lifetime*

Two types of beam–beam simulations, weak–strong and strong–strong, have been done to evaluate luminosity performance. The weak–strong simulation is used to evaluate beam lifetime, because the calculation time is fast and long term tracking is required for the lifetime evaluation.

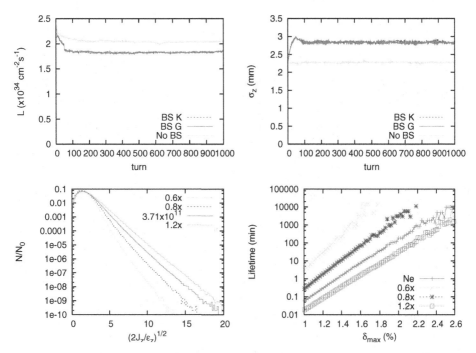

Fig. 3. Top two plots depict luminosity and beam sizes for CEPC. Bottom two plots depict equilibrium longitudinal distribution and beam lifetime. Weak–strong simulation code BBWS is used.

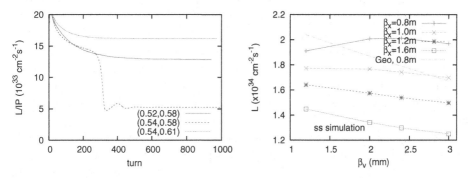

Fig. 4. Results of a strong–strong simulation using BBSS code. Left plot depicts luminosity evolution at three operating point in CEPC. Right plot depicts simulated luminosity as function of (β_x^*, β_y^*).

Figure 3 shows luminosity and bunch length evolutions, equilibrium longitudinal distribution and beam lifetime estimated by the equilibrium distribution.

The strong–strong simulation is capable to know self-consistent beam distributions of two beams. The simulation is used for luminosity evaluation as shown in Fig. 4. The left plot shows luminosity evolution for CEPC at some operating

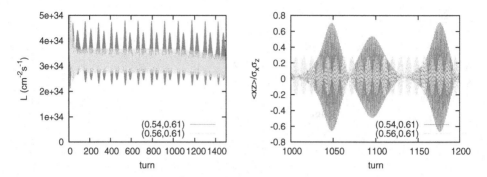

Fig. 5. Luminosity and $\langle xz \rangle$ evolutions given by a strong–strong simulation using BBSS code.

points. The luminosity drops suddenly at 300 turns for tune $(0.54, 0.58)$ due to x–y coupling mentioned above. The behavior is not seen in weak–strong simulation. Deviation from Gaussian distribution is essential for the emittance growth of both beam seen in the simulation. The right plot depicts simulated luminosity as function of (β_x^*, β_y^*). Dynamic beta contributes a flat luminosity dependence on $\beta_y^* = 2$–3 mm at the design $\beta_x = 0.8$ m.

2.4. Coherent beam–beam instability

Colliding two beams are attracted at the collision point and oscillate with betatron tune in the arc. Two modes for two beams, which have the same tune (ν_0) without interaction, are degenerated by the beam–beam interaction. They are σ and π modes, in which two beams oscillate in-phase and out-phase each other. The tunes are $\nu_\sigma = \nu_0$, $\nu_\pi \approx \nu_0 + \xi$.

When one beam consists of two macro-particles located at head and tail part, 4 modes exist. Two are above σ and π modes, in which two macro-particles in each beam move in-phase. Other two modes are head–tail modes, in which two macro-particles in each beam move out-phase. Two head–tail modes are classified so that the head–tail phases of two beams are in-phase or out-phase.

Strong–strong simulation, which is capable to study the coherent instability, was performed for CEPC and TLEP-H, t. Figure 5 shows luminosity and $\langle xz \rangle$ evolutions for TLEP-H. The simulation was performed at two operating points, $(\nu_x, \nu_y) = (0.54, 0.61)$ and $(0.58, 0.61)$. The correlation $\langle xz \rangle$ of two beams oscillates in-phase each other like / \ or \ /. The coherent motion is not seen in TLEP-t and CEPC. The reason is conjectured that a faster damping time in TLEP-t and zero crossing angle collision in CEPC.

3. Study of Electron Cloud Effects

Many positron rings which stores multi-bunches have suffered from electron cloud effects. Photo-electrons are created by the synchrotron radiation of the positron

Table 1. Parameters for electron cloud instability.

Parameter		CEPC	TLEP-Z	TLEP-W	TLEP-H	TLEP-t
Energy	E (GeV)	120	45.5	80	120	175
Bunch population	N_b (10^{10})	37.1	3.3	6	8	17
Number of bunch		50	90300	5162	770	78
Beam line density	λ_{beam} (10^{10} m^{-1})	0.74	3.0	0.3	0.06	0.0013
Beam size	σ_x/σ_y (μm)	583/32	95/10	164/10	247/11	360/16
Bunch length	σ_z (mm)	2.6	5	3	2.4	2.5
Averaged vert. beta	β_y (m)	50	100	100	100	100
Synchrotron tune	ν_z	0.216	0.015	0.037	0.056	0.075
Electron frequency	$\omega_e/2\pi$ (GHz)	137	127	171	174	171
Electron osc. period	$\omega_e\sigma_z/c$	7.5	13	11	8.7	9.0
Threshold density	$\rho_{e,\text{th}}$ (10^{10} m^{-3})	104	0.8	3.4	7.7	15

beam. The electrons are absorbed in the chamber wall, but they are amplified by multi-pactoring or trapped in magnets.

We discuss single bunch instability caused by electron cloud. The threshold density is estimated density for CEPC and TLEP, and then how dense electron cloud is formed.

3.1. *Threshold of electron density for single bunch instability*

Electrons oscillate in electric field near the positron beam, especially vertical oscillation is fast and is more serious for the beam instability. The frequency with the amplitude $\sim \sigma_y$ is expressed by

$$\omega_e = \sqrt{\frac{\lambda_p r_e c^2}{\sigma_y(\sigma_x + \sigma_y)}}, \tag{3}$$

where λ_p is the line density of positron inside a bunch, $\lambda_p = N_p/(\sqrt{2\pi}\sigma_z)$. The electron oscillates in the beam passage with a period of $\omega_e\sigma_z/c$. The oscillation induces instability inner bunch. The threshold is given by[9]

$$\rho_{e,\text{th}} = \frac{2\gamma\nu_z\omega_e\sigma_z/c}{\sqrt{3}KQr_e\beta_y L}, \tag{4}$$

where $K = \omega_e\sigma_z/c$ and $Q = \min(\omega_e\sigma_z/c, 7)$. Table 1 shows beam and electrons cloud parameters for TLEP and CEPC.

3.2. *Electron build-up*

Electron density built-up is evaluated by simulations using a code PEI. Figure 6 shows electron cloud build-up in CEPC. Secondary emission rates $\delta_2(E_e)$ for left

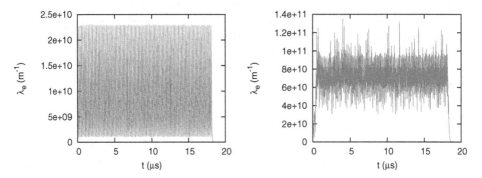

Fig. 6. Electron cloud build-up in CEPC. Left and right plots are given for $\delta_{2,\max} = \delta_2$ (300 eV) = 1.8 and 2.2, respectively.

and right plots are set to $\delta_{2,\max} = \delta_2$ (300 eV) = 1.8 and 2.2, respectively. Bunch separation is assumed as 50 m. Partial two ring scheme is considered in CEPC. 50 bunches are injected in ≈ 3000 m section in the scheme. Photo-electrons are produced with the line density 10^{10} m^{-1} for every bunch passage. In the simulation for $\delta_{2,\max} = 1.8$, 50% more electrons are produced by reflection of the radiation. Including multipactoring effect, electrons with the line density 2×10^{10} m^{-1} is created every bunch passage. The electrons are absorbed into chamber wall, and remains about 10% at next bunch arrival. The electron density at the bunch arrival determines the instability, because the electrons are created after the passage of the beam. Assuming the vacuum chamber cross-section, 0.01 m^2, the density is 2×10^{11} m^{-3}, that is factor 5 lower than the threshold $\rho_{e,\text{th}} = 10^{12}$ m^{-3}. For higher secondary rate $\delta_{2,\max} = 2.2$, electron line density is one order higher than that for $\delta_{2,\max} = 1.8$. The density is close to the threshold. The secondary rate is around 2.2 for bear aluminum. It is known that the rate decreases due to conditioning. Efforts to reduce the secondary rate is necessary, though the rate value is not very serious.

The same studies are performed for TLEP's. The electron cloud effects are more serious for more number of bunches. Z factory should be most serious.

4. Summary

Beam–beam and electron cloud effects are discussed in CPEC and TLEP's. Simulation tools for the effects are being ready. Studies of the effects have to be continued to fix the design parameters of CEPC and TLEP's.

Acknowledgment

The author thanks Drs. W. Chou, K. Oide, D. Shatilov, Y. Zhang, D. Zhou and F. Zimmermann for fruitful discussions.

References

1. K. Yokoya, Scaling of high-energy e^+e^- ring colliders, *KEK Accelerator Seminar*, 15 March 2012.
2. K. Yokoya, *Nucl. Instrum. Methods A* **251**, 1 (1986).
3. A. Bogomyagkov, E. Levichev and D. Shatilov, *Phys. Rev. ST-AB* **17**, 041004 (2014).
4. V. Telnov, *Phys. Rev. Lett.* **110**, 114801 (2013).
5. F. Zimmermann, M. Benedikt, D. Schulte and J. Wenninger, Challenges for highest energy circular colliders, in *Proceedings of IPAC14, MOXAA01,1* (2014).
6. Y. Zhang, K. Ohmi and D. Shatilov, Beam–beam simulation study for CEPC, in *Proceedings of IPAC14, THPRI003*, p. 3763 (2014).
7. K. Ohmi *et al.*, *Phys. Rev. Lett.* **92**, 214801 (2004).
8. K. Ohmi, Beam–beam effects in CEPC and TLEP, in *Int. Proc. 55th ICFA Advanced Beam Dynamics Workshop on High Luminosity Circular e^+e^- Colliders — Higgs Factory*, Beijing, 9–12 October 2014, `http://jacow.web.psi.ch/conf/hf2014ihep/prepress/FRT3B1.PDF`.
9. K. Ohmi, F. Zimmermann and E. Perevedentsev, *Phys. Rev. E* **65**, 016502 (2001).

Beam–Beam Effects of Single Ring and Partial Double Ring Scheme in CEPC

Yuan Zhang

Institute of High Energy Physics,
19B Yuquanlu, Beijing 100049, P. R. China
zhangy@ihep.ac.cn

After the Higgs discovery, it is believed that a circular e^+e^- collider could serve as a Higgs factory. The high energy physics community in China launched a study of a 50–100 km ring collider. A preliminary conceptual design report (Pre-CDR) has been published in early 2015. This report is based on a 54-km ring design. Some progress on beam–beam effect study after Pre-CDR is shown in the paper. We estimate the beamstrahlung lifetime using a pure strong–strong code as a comparison with the result obtained using a quasi-strong–strong method. The effect of parasitic crossing in the pretzel scheme is also estimated for the very first time. The feasibility of the main parameters for partial double ring scheme are evaluated from the point view of beam–beam interaction.

Keywords: CEPC; beam–beam.

1. Introduction

During the preparation of the preliminary conceptual design report (Pre-CDR) of circular electron–positron collider (CEPC), Shatilov, Ohmi and the author have worked on the beam–beam effect for the optimization of the machine main parameters.[1,2] The focus was on the luminosity and lifetime considering the beamstrahlung effect with the ideal ring at that time. The baseline design was the pretzel scheme with one single ring, and there were about 100 parasitic crossing points in the arc. Since the lattice with pretzel scheme was not finished yet, we only evaluated the parasitic beam–beam effect roughly using the weak–strong LIFETRAC code.[3]

The priority of Z-factory luminosity with the same machine was not very high in the Pre-CDR. It was just assumed that we will build the Higgs factory and will keep the option for a Z-factory. It is different now. The luminosity of a Z-factory would be as high as 1×10^{34} cm^{-2} s^{-1}. In the appendix of the Pre-CDR, the potential of the CEPC to operate at lower energy and higher luminosity as a Super-Z-factory using one ring has been discussed.[4] It is known however that the pretzel scheme is very complicated, and there is no detailed design for the time being. As a matter of fact, the crab-waist collision scheme[5] with double ring,

Table 1.   Design parameters of CEPC.

	Pre-CDR	Partial double ring
Beam energy	120 GeV	120 GeV
Circumference	53.6 km	53.6 km
Luminosity	1.8×10^{34} cm^{-2} s^{-1}	3.0×10^{34} cm^{-2} s^{-1}
SR power/beam	50 MW	50 MW
Number of IP	2	2
Number of bunch	50	50
Momentum compaction factor	4.15×10^{-5}	3×10^{-5}
Energy acceptance	0.02	0.02
Bunch population	3.79×10^{11}	3.79×10^{11}
Horizontal emittance	6.79×10^{-9} mrad	3.34×10^{-9} mrad
Emittance coupling	0.003	0.003
Bunch length (SR)	2.26 mm	3.3 mm
Beam–beam parameter (x/y)	0.104/0.074	0.04/0.11

which is the base line design of FCC-ee,[6] is easier and more flexible. It is much easier to achieve high luminosity in a wider energy region with crab-waist scheme in such a machine, since the emittance is smaller at lower energy which is a must to implement crab-waist. Koratzinos[7] proposes a "bowtie" scheme to implement a crab-waist scheme with partial double ring, that is to say that complicated pretzel scheme can be avoided. Wang proposes new machine parameters for CEPC with the new scheme. Here, we show some beam–beam results to check the reasonability of these parameters.

In previous work, when estimating the lifetime limited by momentum acceptance and transverse dynamic aperture, the assumption is made that the aperture boundary is rectangular. The boundary is more like an ellipse than a rectangle. Here, we compare the lifetime between the two cases. We use the weak–strong codes, LIFETRAC and BBWS to calculate the lifetime. The two codes are both weak–strong codes, but different quasi-strong–strong method is used.[1] Here, we use the code IBB,[8] which now considers the beamstrahlung effect. IBB is a strong–strong code, it is reasonable for us to simulate the lifetime in the Higgs factory.

The main parameters of Pre-CDR and the new partial double ring scheme are listed in Table 1.

2. Simulation Results

2.1. *Benchmark*

With momentum acceptance 0.02, the beamstrahlung lifetimes estimated by LIFETRAC and BBWS are 85 min and 250 min, respectively. The difference between the two codes comes from the different quasi-strong–strong model. The cause has been discussed in Refs. 1 and 9. The result of IBB is 336 min for 0.021. The difference is comparable and more detailed benchmark should be done in the future.

2.2. *Elliptical versus rectangular dynamic aperture boundary*

For the same rectangular dynamic aperture, the lifetime given by LIFETRAC is about 20 min for $40\sigma_y$, and by BBWS is about 200 min. But it is only 3 min for IBB. It is not clear that if this difference comes from the model.

It should be noted that for the pure strong–strong simulation, we have to enlarge the field solution region to enclose the halo particle. It is also a must to enlarge the grid number to ensure the accuracy for the core particle with the PIC method. This will result in unreasonably long simulation time. The better method is to calculate the field felt acting on the halo particle with the synchro-beam mapping.[10] In fact, we have noted that the lifetime is sensitive to the grid region if the region area is too small. The preliminary result show that the lifetime limited by elliptical boundary reduces by a factor of 1.5 compared to the rectangle.

2.3. *Parasitic beam–beam effect*

In the present design of the pretzel scheme, the beam is separated in horizontal direction with $10\sigma_x$. There is one parasitic collision point every 4π phase advance. There are 100 parasitic crossing points in total. The tune shift coming from one parasitic point is about -0.0007, and the total tune shift is about -0.07. We use LIFETRAC to simulate the parasitic beam–beam effect. All PCs are in phase. The result shows that the lifetime would reduces by three orders of magnitude. It could not be accepted due to the short lifetime. One solution is to enlarge the separation.

2.4. *Check of the new machine parameters*

The strong–strong code IBB is used to simulate the beam–beam performance of new parameters. The luminosity/IP could achieve 2.7×10^{34} cm^{-2} s^{-1}. The beam-strahlung lifetime is about 100 min with momentum acceptance 2%. It seems reasonable from the point view of beam–beam interaction.

3. Summary

The beam–beam code IBB now considers the beamstrahlung effect. It is used to evaluate the beam–beam performance of CEPC. The new machine parameters with partial double ring scheme are reasonable for beam–beam. The parasitic beam–beam effect in the pretzel scheme is simulated with LIFETRAC, which tells us, we may need larger separation.

References

1. K. Ohmi, D. Shatilov, D. Zhou and Y. Zhang, Beam–beam simulation study for CEPC, in *Proc. 5th Int. Particle Accelerator Conf.*, Dresden, Germany, 15–20 June 2014, pp. 3763–3765.
2. K. Ohmi and F. Zimmermann, FCC-ee/CEPC beam–beam simulations with beam-strahlung, in *Proc. 5th Int. Particle Accelerator Conf.*, Dresden, Germany, 15–20 June 2014, pp. 3766–3769.
3. D. Shatilov, *Part. Accel.* **52**, 65 (1996).
4. A. Apyan *et al.*, CEPC-SppC Preliminary Conceptual Design Report, Volume II: Accelerator (2015), http://cepc.ihep.ac.cn/preCDR/volume.html.
5. P. Raimondi, *Proc. 2nd SuperB Workshop*, Frascati, Italy, 16–18 March 2006.
6. J. Wenninger, M. Benedikt, K. Oide and F. Zimmermann, Future circular collider study: Lepton collider parameters (2016), FCC-ACC-SPC-0003, http://cern.ch/fcc.
7. M. Koratzinos, Mitigating performance limitations of single beam-pipe circular e^+e^- colliders, in *Proc. 6th Int. Particle Accelerator Conf.*, Richmond, VA, USA, May 2015, pp. 2160–2168.
8. Y. Zhang, K. Ohmi and L. Chen, *Phys. Rev. Spec. Top. Accel. Beams* **8**, 074402 (2005).
9. K. Ohmi and D. Shatilov, Status of benchmark of LIFETRAC and BBWS, FCC-ee meeting, CERN (2015).
10. K. Hirata, H. Moshammer and F. Ruggiero, *Part. Accel.* **40**, 205 (1993).

CEPC Partial Double Ring Scheme and Crab-Waist Parameters

Dou Wang,* Jie Gao, Feng Su, Yuan Zhang, Jiyuan Zhai, Yiwei Wang,
Sha Bai, Huiping Geng, Tianjian Bian, Xiaohao Cui, Na Wang, Zhe Duan,
Yuanyuan Guo and Qing Qin

Accelerator Center, Institute of High Energy Physics (IHEP),
19B Yuquan Road, Beijing 100049, China
**wangd93@ihep.ac.cn*

In order to avoid the pretzel orbit, CEPC is proposed to use partial double ring scheme in CDR. In this paper, a general method of how to make an consistent machine parameter design of CEPC with crab-waist by using analytical expression of maximum beam–beam tune shift and beamstrahlung beam lifetime started from given IP vertical beta, beam power and other technical limitations were developed. FFS with crab sextupoles will be developed and the arc lattice will be redesigned to acheive the lower emittance for crab-waist scheme.

Keywords: CEPC; partial double ring; parameter choice; crab-waist; lattice design.

1. Introduction

CEPC is a ring with a circumference of 54 km to house an electron–positron collider in phase-I and be upgraded to a super proton–proton collider (SPPC) in phase-II. The designed beam energy for CEPC is 120 GeV, aims for Higgs study. The main constraint in the design is the synchrotron radiation power, which should be limited to 50 MW per beam, in order to control the total AC power of the whole machine. The target luminosity for CEPC is $\sim 2 \times 10^{34}$ cm^{-2} s^{-1}.[1]

In Pre-CDR, CEPC is a single ring machine.[1-3] All 50 bunches are equally spaced, and the collisions are head-on. This design requires a pretzel orbit in order to avoid parasitic collisions in the arcs. From the experience of LEP and CESR, the pretzel orbit is difficult to operate and control, and is also difficult for injection. After Pre-CDR, we developed a new idea called partial double ring scheme[4,5] showed in Fig. 1. Therefore, a pretzel orbit is not needed. With partial double ring scheme, we can consider crab-waist on CEPC. Figure 2 shows the principle of crab

---

*Corresponding author.

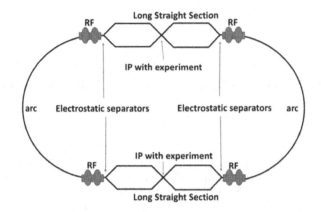

Fig. 1.   Layout of CEPC partial double ring scheme.

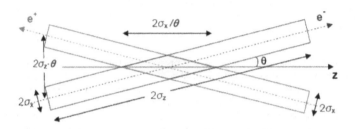

Fig. 2.   Sketch of collision with crab-waist.

waist scheme. The most important advantage of crab-waist is that the beam–beam limit can be increased greatly. In this paper, we try to apply crab-waist scheme on CEPC either to increase the luminosity or to reduce the beam power.

2. CEPC Parameter Design

2.1. Machine constraints/given parameters

- Energy $E_0 = 120$ GeV.
- Circumference $C_0 = 54$ km.
- $N_{\rm IP} = 2$.
- Beam power $P_0 = 50(30)$ MW/beam.
- $\beta_y^* = 1.2$ mm.
- $r = \sigma_y/\sigma_x$.
- Emittance coupling factor κ_ε.
- Bending radius $\rho = 6.1$ km.
- Piwinski angle $\Phi = 2$.
- ξ_y enhancement by crab-waist $F_l \sim 1.5$ (2.6).
- Energy acceptance (DA) $\sim 2\%$.
- Phase advance per cell (FODO) $\sim 60°$ or $90°$.

- Ratio of nature energy spread and the one due to beamstrahlung $A = \delta_0/\delta_{\mathrm{BS}}$ ($A \geq 3$).
- RF HOM power per cavity $P_{\mathrm{HOM}} \lesssim 1$ kW.

2.2. *Parameter calculations*

Step 1: *Beam–beam limit*

In our method, the energy of the ring E_0, the bending radius of the main dipole magnets ρ, the synchrotron radiation power P_0 (machine technical constraint), the aspect ratio R and the IP number N_{IP} are the known quantity. From these input parameters, one gets first

$$U_0 = 88.5 \times 10^3 \frac{E_0^4 \, (\mathrm{GeV})}{\rho \, (\mathrm{m})}, \tag{1}$$

$$I_b = \frac{eP_0}{U_0}, \tag{2}$$

$$\delta_0 = \gamma \sqrt{\frac{C_q}{J_\varepsilon \rho}} \tag{3}$$

and the maximum beam–beam tune shift is[6]

$$\xi_y = \frac{2845}{2\pi} \sqrt{\frac{U_0}{2\gamma E_0 N_{\mathrm{IP}}}} \times F_l, \tag{4}$$

where F_l is the beam–beam limit (ξ_y) enhancement factor by crab-waist scheme and so far, we assume it is 1.5 for Higgs and 2.6 for Z.

Step 2: *Luminosity*

The luminosity of circular collider is expressed by

$$L_0 \, [\mathrm{cm}^{-2}\,\mathrm{s}^{-1}] = 2.17 \times 10^{34} \, (1 + r)\xi_y \frac{eE_0 \, (\mathrm{GeV})N_b N_e}{T_0 \, (\mathrm{s}) \, \beta_y^* \, (\mathrm{cm})}. \tag{5}$$

According to Eqs. (4) and (5) can be expressed by another way

$$L_0 \, [\mathrm{cm}^{-2}\,\mathrm{s}^{-1}] = 0.7 \times 10^{34} \, (1 + r)\frac{1}{\beta_y \, [\mathrm{cm}]} \sqrt{\frac{E_0 \, [\mathrm{GeV}] \, I_b \, [\mathrm{mA}] \, P_0 \, [\mathrm{MW}]}{\gamma N_{\mathrm{IP}}}}. \tag{6}$$

Step 3: *Transverse beam size*

Telnov pointed out that at energy-frontier e^+e^- storage ring colliders, beamstrahlung determines the beam lifetime through the emission of single photons in the tail of the beamstrahlung spectra. If we want to achieve a reasonable beamstrahlung-driven beam lifetime of at least 30 min, we need to confine the relation of the bunch population and the beam size as Eq. (7)[7,8]

$$\frac{N_e}{\sigma_x \sigma_z} \leq 0.1 \eta \frac{\alpha}{3\gamma r_e^2}. \tag{7}$$

Recalling the definition of the vertical beam–beam tune shift, for the flat beam, we get

$$\frac{N_e}{\sigma_x \sigma_y \sqrt{1 + \Phi^2}} = \frac{2\pi\gamma}{r_e \beta_y} \xi_y. \tag{8}$$

Combining Eq. (7) with Eq. (8), one has

$$\frac{N_e^2}{\sigma_x^2 \sigma_y \sigma_z} = \frac{0.2\pi\eta\alpha\xi_y\sqrt{1 + \Phi^2}}{3r_e^3\beta_y}. \tag{9}$$

In order to control the extra energy spread by beamstrahlung to a certain degree, we introduce a constraint in this paper as

$$\delta_{\text{BS}} = \frac{\delta_0}{A} \quad (A \geq 3). \tag{10}$$

From Eq. (10) and the definition of beamstrahlung energy spread, one finds

$$\frac{N_e^2}{\sigma_x \sigma_y \sigma_z} = \frac{3\delta_0}{2.6r_e^3 \gamma r A}. \tag{11}$$

So, according to Eqs. (9) and (11), we get

$$\sigma_x = \frac{5.77\delta_0 \beta_y}{\pi\eta\alpha\xi_y\sqrt{1 + \Phi^2}\gamma r A}. \tag{12}$$

With certain given coupling factor κ_ε (0.003 for example) and the aspect ratio r, one can get the vertical beam size/emittance and horizontal emittance/beta:

$$\sigma_y = r\sigma_x, \quad \varepsilon_y = \frac{\sigma_y^2}{\beta_y},$$

$$\varepsilon_x = \frac{\varepsilon_y}{\kappa_\varepsilon}, \quad \beta_x^* = \frac{\sigma_x^2}{\varepsilon_x}. \tag{13}$$

Assuming we use 60° FODO cell for the arc, then we can get the bending angle per FODO cell knowing the horizontal emittance

$$\varepsilon_x = \frac{C_q\gamma^2\varphi^3 \left(1 - \frac{3}{4}\sin^2\left(\frac{\mu}{2}\right) + \frac{1}{60}\sin^4\left(\frac{\mu}{2}\right)\right)}{8J_x \sin^3\left(\frac{\mu}{2}\right)\cos\left(\frac{\mu}{2}\right)} \to \varphi. \tag{14}$$

Furthermore, one can make an estimation of the momentum compaction factor

$$\alpha_p = \left(\frac{\varphi}{2}\right)^2 \left(\frac{1}{\sin^2\frac{\mu}{2}} - \frac{1}{12}\right). \tag{15}$$

Step 4: *Crossing angle*

From Eq. (8), one gets

$$N_e = \frac{2\pi\gamma\xi_y\sqrt{1+\Phi^2}}{r_e\beta_y}\sigma_x\sigma_y\,.$$ (16)

Then, having the bunch population Eq. (16), it is easy to get the bunch number

$$N_b = \frac{I_bT_0}{eN_e}$$ (17)

and also from Eq. (7), one gets

$$\sigma_z = \frac{3\gamma r_e^2 N_e}{0.1\eta\alpha\sigma_x}\,.$$ (18)

Definition of Piwinski angle:

$$\Phi = \frac{\sigma_z}{\sigma_x}tg\theta_h\,.$$ (19)

Combining Eqs. (18) and (19), one finds out the value of half crossing angle for the partial double ring scheme

$$\theta_h = \text{Arc tg}\left(\frac{0.1\eta\alpha\sigma_x^2\Phi}{3\gamma r_e^2 N_e}\right)\,.$$ (20)

Step 5: *Hour glass effect*

By the principle of crab-waist, the overlap area of colliding bunches is much shorter than the bunch length. Here, we define a new parameter, effective bunch length as

$$\sigma_{z\text{eff}} = \frac{\sigma_x}{\sin\theta_h}\,.$$ (21)

Then, the hour glass effects can be evaluate by

$$F_h = \frac{\beta_y}{\sqrt{\pi}\sigma_{z\text{eff}}}\exp\left(\frac{\beta_y^2}{2\sigma_{z\text{eff}}^2}\right)K_0\left(\frac{\beta_y^2}{2\sigma_{z\text{eff}}^2}\right)\,.$$ (22)

Finally, the real luminosity can be expressed by the product of peak luminosity L_0 and the hour glass factor F_h

$$L = L_0F_h\,.$$ (23)

Step 6: *RF parameters*

First, considering the synchrotron radiation energy loss have to be compensated by the RF cavities, one finds

$$U_0 = eV_{\text{RF}}\sin\phi_s\,.$$ (24)

The nature bunch length is expressed by

$$\sigma_{z0} = \sigma_z \times \frac{A}{1+A} = \sqrt{-\frac{2\pi E_0 \alpha_p}{f_{RF} T_0 e V_{RF} \cos \phi_s}} \bar{R} \delta_0 \,. \tag{25}$$

From Eqs. (24) and (25), we can get the RF voltage V_{RF} and RF accelerating phase ϕ's. Then, we can get the energy acceptance for RF system

$$\eta_{RF} = \sqrt{\frac{2U_0}{\pi \alpha_p f_{RF} T_0 E_0}} \left(\sqrt{q^2 - 1} - \arccos \left(\frac{1}{q} \right) \right) \quad \left(q = \frac{e V_{RF}}{U_0} \right). \tag{26}$$

Step 7: Beam lifetime and RF HOM power

At last, we need to check the beam lifetime due to radiative Bhabha scattering and the beam lifetime due to beamstrahlung effect[9,10]

$$\tau_{Bhabha} = \frac{I_b}{e L N_{IP} \sigma_{ee} f_0} (\sigma_{ee} = 1.52 \times 10^{-25} \text{ cm}^2) \,, \tag{27}$$

$$\tau_{BS} = \frac{2\pi R}{c} \frac{\sqrt{6\pi} r_e \gamma e^{1.475u}}{0.057 \alpha^2 \eta \sigma_z} \,. \tag{28}$$

In addition, we need to check the HOM power per cavity

$$P_{HOM} = k(\sigma_z) e N_e \cdot 2 I_b \leq 1 \text{ kW} \,, \tag{29}$$

where the HOM loss factor is

$$k(\sigma_z) = \frac{1.8}{\sqrt{\sigma_z / 0.00265}} V/pC \,. \tag{30}$$

2.3. *Primary parameter choice*

Using the method above, we get a set of new designs for CEPC with crab-waist scheme. For the column of high luminosity, we keep the same beam power as Pre-CDR (50 MW/beam) and get almost 50% increment of luminosity. Otherwise, in the column of low power, we can decrease the beam power from 50–30 MW with same luminosity as Pre-CDR.

3. FFS Design and Crab Sextupole Parameter

The lattice design of FFS ($\beta_x = 0.25$ m, $\beta_y = 0.00136$ m) for CEPC partial double ring is shown in Fig. 3. The L^* is 1.5 m and the strength of first quadrupole (twin aperture) is 200 T/m. The critical energy of the whole system is under 190 keV.

The crab sextupole should be placed on both sides of the IP in phase with the IP in the horizontal plane and at $\pi/2$ in the vertical one. As Oide said, the second FFS sextupoles of the CCS-Y section can work as the crab sextupoles, if their strengths and phases to the IP are properly chosen.

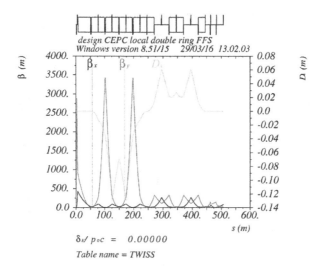

Fig. 3.   FFS optics for CEPC partial double ring.

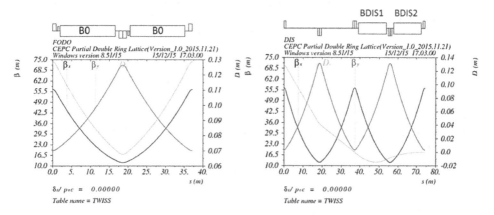

Fig. 4.   Low emittance arc with 90°/60° phase advance (left: FODO, right: dispersion suppressor).

The crab sextupole strength should satisfy the following condition depending on the crossing angle and the beta functions at the IP and the sextupole locations:

$$KL = \frac{1}{2\theta} \frac{1}{\beta_y^* \beta_y} \sqrt{\frac{\beta_x^*}{\beta_x}} = 1.27 \text{ m}^{-2} , \tag{31}$$

$$K_2 = 4.2 \text{ m}^{-3} .$$

4. Low Emittance Arc

We tried to get smaller emittance as the parameter Table 1. By reducing the FODO length from 47 m to 37 m and increasing the phase of FODO cell, we got 2.3 nm emittance. Two kinds of arc optics are shown in Figs. 4 and 5.

Table 1. CEPC crab-waist parameters.

	Pre-CDR	H-high lumi.	H-low power	W	Z
Number of IPs	2	2	2	2	2
Energy (GeV)	120	120	120	80	45.5
Circumference (km)	54	54	54	54	54
SR loss/turn (GeV)	3.1	2.96	2.96	0.59	0.062
Half crossing angle (mrad)	0	15	15	15	15.
Piwinski angle	0	2.5	2.6	5	7.6
Ne/bunch (10^{11})	3.79	2.85	2.67	0.74	0.46
Bunch number	50	67	44	400	1100
Beam current (mA)	16.6	16.9	10.5	26.2	45.4
SR power/beam (MW)	51.7	50	31.2	15.6	2.8
Bending radius (km)	6.1	6.2	6.2	6.1	6.1
Momentum compaction (10^{-5})	3.4	2.5	2.2	2.4	3.5
$\beta_{IPx/y}$ (m)	0.8/0.0012	0.25/0.00136	0.268/0.00124	0.1/0.001	0.1/0.001
Emittance x/y (nm)	6.12/0.018	2.45/0.0074	2.06/0.0062	1.02/0.003	0.62/0.0028
Transverse σ_{IP} (um)	69.97/0.15	24.8/0.1	23.5/0.088	10.1/0.056	7.9/0.053
ξ_x/IP	0.118	0.03	0.032	0.008	0.006
ξ_y/IP	0.083	0.11	0.11	0.074	0.073
V_{RF} (GV)	6.87	3.62	3.53	0.81	0.12
f_{RF} (MHz)	650	650	650	650	650
Nature σ_z (mm)	2.14	3.1	3.0	3.25	3.9
Total σ_z (mm)	2.65	4.1	4.0	3.35	4.0
HOM power/cavity (kW)	3.6	2.2	1.3	0.99	0.99
Energy spread (%)	0.13	0.13	0.13	0.09	0.05
Energy acceptance (%)	2	2	2		
Energy acceptance by RF (%)	6	2.2	2.1	1.7	1.1
n_γ	0.23	0.47	0.47	0.3	0.24
Life time due to beamstrahlung cal (min)	47	36	32		
F (hour glass)	0.68	0.82	0.81	0.92	0.95
L_{max}/IP (10^{34}cm^{-2}s^{-1})	2.04	2.96	2.01	3.09	3.09

5. Dynamic Aperture

With FFS, partial double ring, low emittance arc (90°/90°) and the bypasses of *pp* detectors together, we got a satisfying dynamic aperture for on momentum particle shown in Fig. 6. So far, we have just used two groups of sextupoles in the arc. The further optimization of DA bandwidth is undergoing.

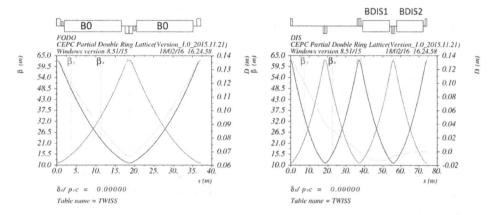

Fig. 5.   Low emittance arc with 90°/90° phase advance (left: FODO, right: dispersion suppressor).

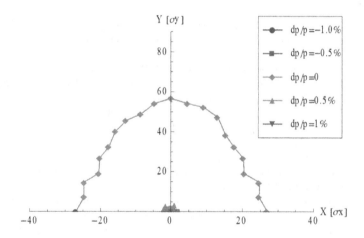

Fig. 6.   Dynamic aperture for CEPC partial double ring.

6. Conclusion

In this paper, a general method of how to make an consistent machine parameter design of CEPC with crab-waist by using analytical expression of maximum beam–beam tune shift and beamstrahlung beam lifetime started from given IP vertical beta, beam power and other technical limitations was developed. Based on this method, a set of optimized parameter designs for 54 km CEPC with partial double ring scheme were proposed. Crossing angle was fixed at 30 mrad both for Higgs W and Z. Thanks to the beam–beam limit enhancement effect of crab-waist, we can either get higher luminosity with same beam power as Pre-CDR, or reduce the beam power by 40% keeping the same luminosity in Pre-CDR. Both proposals should improve the performance of CEPC. For "H-high lumi" mode, the HOM power per cavity cannot be reduced under 1 kW. The minimum value is about

2 kW. In addition, the optics of FFS has been designed and the strength of crab sextupole has been estimated, and so far this kind of sextupole is available.

Based on partial double ring scheme, we get a set of Z-parameter with 3.1×10^{34} cm^{-2} s^{-1} luminosity using 1100 bunches.

Lower emittance arc with 2.3 nm emittance has been designed both for $90°/60°$ case and $90°/90°$ case. The dynamic aperture of the whole ring is good enough for on momentum particle while the off momentum DA is still needed to be optimized.

Acknowledgments

This work was supported by the National Foundation of Natural Sciences (11575218 and 11505198) and the innovation foundation of IHEP with the contract number of Y4545240Y2, CAS, 2015.

References

1. The CEPC-SPPC Study Group, CEPC-SPPC Preliminary Conceptual Design Report, Volume II-Acceleraor, IHEP-CEPC-DR-2015-01, IHEP-AC-2015-01, March 2015.
2. D. Wang *et al.*, *Chin. Phys. C* **37**, 097003 (2013).
3. M. Xiao, J. Gao, D. Wang, F. Su, Y. Wang, S. Bai and T. Bian, *Chin. Phys. C* **40**, 087001 (2016), arXiv:1512.07348 [physics.acc-ph].
4. J. Gao, Ultra-low beta and crossing angle scheme in CEPC lattice design for high luminosity and low power, private note, IHEP-AC-LC-Note2013-012 (in Chinese).
5. M. Koratzinos, private communication (2014).
6. J. Gao, *Nucl. Instrum. Methods A* **533**, 270 (2004).
7. V. Telnov, Restriction on the energy and luminosity of e^+e^- storage rings due to beamstrahlung, arXiv:1203.6563v.
8. V. Telnov, Limitation on the luminosity of e^+e^- storage rings due to beamstrahlung, HF2012, 15 November 2012, arXiv:1307.3915 [physics.acc-ph].
9. V. Telnov, Issues with current designs for e^+e^- and $\gamma\gamma$ colliders, in *Int. Conf. Structure and the Interactions of the Photon including the 20th Int. Workshop on Photon-Photon Collisions and the Int. Workshop on High Energy Photon Linear Colliders*, Paris, *PoS* Photon 2013, 070 (2013), https://inspirehep.net/record/1298149/files/Photon%202013_070.pdf
10. J. Zhai, private discussion (2015).

CEPC Partial Double Ring Lattice Design and SPPC Lattice Design

Feng Su,* Jie Gao, Dou Wang, Yiwei Wang, Jingyu Tang, Tianjian Bian,
Sha Bai, Huiping Geng, Yuan Zhang, Yuanyuan Guo, Yuemei Peng,
Ye Zou, Yukai Chen and Ming Xiao

*Key Laboratory of Particle Acceleration Physics and Technology,
Institute of High Energy Physics, Chinese Academy of Sciences,
Beijing 100049, P. R. China*
sufeng@ihep.ac.cn

In this paper, we introduce the layout and lattice design of Circular-Electron-Positron-Collider (CEPC) partial double ring scheme and the lattice design of Super-Proton-Proton-Collider (SPPC). The baseline design of CEPC is a single beam-pipe electron positron collider, which has to adopt pretzel orbit scheme and it is not suitable to serve as a high luminosity Z factory. If we choose partial double ring scheme, we can get a higher luminosity with lower power and be suitable to serve as a high luminosity Z factory. In this paper, we discuss the details of CEPC partial double ring lattice design and show the dynamic aperture study and optimization. We also show the first version of SPPC lattice although it needs lots of work to do and to be optimized.

Keywords: CEPC partial double ring; lattice design; dynamic aperture; SPPC lattice.

1. Introduction

With the discovery of the Higgs boson at the LHC, the world high-energy physics community is investigating the feasibility of a Higgs factory as a complement to the LHC for studying the Higgs and pushing the high-energy frontier. CERN physicists are busy planning the LHC upgrade program, including HL-LHC and HE-LHC. They also plan a more inspiring program called FCC, including FCC-ee and FCC-hh. Both the HE-LHC and the FCC-hh are proton–proton (pp) colliders aiming to explore the high energy frontier and expecting to find new physics.[1,2] Chinese accelerator physicists also plan to design an ambitious machine called Circular Electron Positron Collider (CEPC)–Super Proton–Proton Collider (SPPC). The CEPC-SPPC program contains two stages. The first stage is an electron–positron collider with center-of-mass energy 240 GeV to study the Higgs properties carefully. The second stage is a pp collider at center-of-mass energy of more than 70 TeV.[3]

- **Advantages**
- ➢ No pretzel
- ➢ More bunches

- **Challenges**
- ➢ Crossing angle & crab waist design.
- ➢ Electron cloud issues.
- ➢ Bunch train operation introduces an uneven load to the RF system.

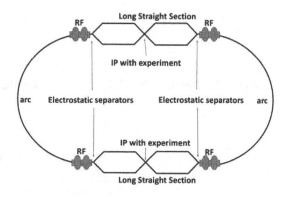

Fig. 1.   Advantages and challenges of CEPC partial double ring scheme.

In March 2015, we finished and published the Preliminary Conceptual Design Report (Pre-CDR) of CEPC–SPPC. In this paper, we choose the single ring scheme for CEPC. The synchrotron radiation power of CEPC is as high as 50 MW and its goal is to deliver a peak luminosity greater than 1×10^{34} cm^{-2}s^{-1} per IP.[4,5] The e^+e^- beams are in the same pipe, which has to adopt pretzel orbit and it is not suitable to serve as a high luminosity Z factory. To solve the problems above, we raised a partial double ring scheme.[6,7] We can get a higher luminosity with lower power and be suitable to serve as a high luminosity Z factory though there are many challenges we should overcome. Figure 1 shows the advantages comparing with the single beam-pipe scheme and the challenges we should overcome. In the first part of this paper, we will discuss and show the lattice design of CEPC partial double ring and in the second part, we will show the preliminary think and lattice design of SPPC.

2. CEPC Partial Double Ring Lattice Design

2.1. *CEPC partial double ring layout*

We choose double ring scheme for e^+e^- at IP1 and IP3. The total length of this part is about $3 \sim 4$ Km. The arcs of both side of IP1 and IP3 are kept the same as the Pre-CDR single ring scheme designed by Huiping Geng. The other straight sections' length keeps the same with the single ring scheme. Figure 2 is the whole layout of CEPC partial double ring part and arc part. Figure 3 is the detail of partial double ring part of CEPC.

The full crossing angle for CEPC partial double ring scheme is 30 mrad. We assume the final ficus system (FFS) length is about 500 m, then the largest distance at the end of FFS is about 7.5 m and between the two separated pipes is about 15 m. At the start of the double ring, we need to use electrostatic separator to separate the electron and positron beams. We choose the parameter of electrostatic

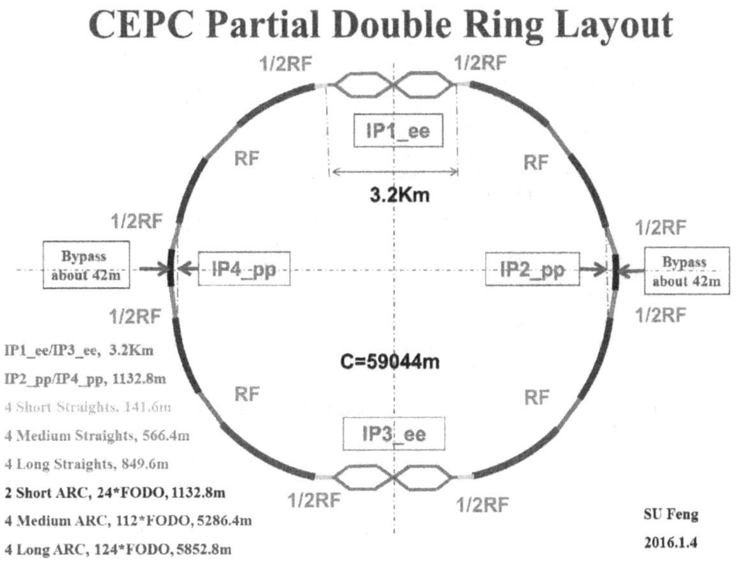

CEPC Partial Double Ring Layout

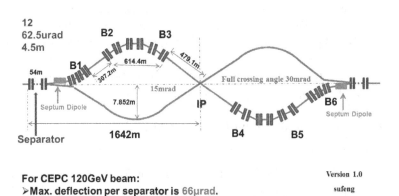

Fig. 2.   CEPC partial double ring layout (whole ring).

CEPC Partial Double Ring Layout

Fig. 3.   CEPC partial double ring layout (double ring part).

separator according to the experience on LEP.[8] Figure 4 is the parameter list chose by LEP. The maximum operating field strength is 2 MV/m. The length of electrostatic separator is 4.5 m. For the beam energy $E_0 = 120$ GeV, the maximum deflection per separator is about 66 μrad. We choose 12 electrostatic separators work together to obtain a deflection 0.75 mrad, each separator deflect 62.5 μrad. After those separators, we use a pair of septum dipoles to obtain 4.25 mrad and a

```
| Separator length                          4.5 m      |
| Inner diameter of separator tank          540 mm     |
| Electrode length                          4.0 m      |
| Electrode width                           260 mm     |
| Nominal gap                               110 mm     |
| Maximum operating field strength          20 kV/cm   |
| Maximum operating voltage                 ± 110 kV   |
| Max. deflection per separator at 55 GeV   145 μrad   |
| Conditioning voltage on the test bench    ± 200 kV   |
| Conditioning voltage after installation   ± 160 kV   |
| Maximum voltage for vernier adjustment     ±  35 kV   |
| Range of vernier adjustment at 55 GeV     76 μm      |
| Horizontal good field region (1% limit)   ± 80 mm    |
| Maximum tilt per electrode                 ±   5 mrad  |
| Pumping speed of sputter ion pumps        800 l/s    |
| Pumping speed of sublimation pumps        1300 l/s   |
| Nominal vacuum pressure in the low-beta              |
|   insertions                              2.7·10⁻⁸ Pa|
| Number of separators per collision point  4          |
| Total number of separators                32         |
| Total number of high voltage circuits     32         |
```

Fig. 4.   Parameter for electrostatic separator of LEP.

group of dipole (B1) to acquire the other 10 mrad and suppress the dispersion to zero. But the simulation of separators in MAD has a problem when we use SURVEY command, so we use dipoles stand for the separators, whose length and deflect angle are both same as the separators. The scheme is shown in Fig. 3. Then we use a group of dipoles to bend the beam, which is schemed in Fig. 3 as B2. B3 and B4 are the symmetrical elements according to B2 and B1. Each group (for example B2) has 16 dipoles in 8 FODO cell and 8 half-strength dipoles in 2 dispersion suppressor section, and each dipole bend the beam by 1.5 mrad and half-strength dipole bend the beam by 0.75 mrad. This scheme can keep the dispersion 0 at the two side of this bend part. At beginning, we use straight FODO instead of final ficus system optics. The total length of this scheme is about 3.4 km.

2.2. CEPC partial double ring lattice without FFS

2.2.1. Orbit and optics without FFS

We design the lattice of CEPC partial double ring by MAD. Following is the optics and orbit. Figure 5 shows the beta function and the dispersion of CEPC partial double ring part without FFS. Figure 6 shows the orbit of CEPC partial double ring part without FFS.

2.2.2. Dynamic aperture without FFS

Figure 7 is the dynamic aperture of CEPC partial double ring lattice without FFS. The on-momentum dynamic aperture is about $33\sigma_x$ in horizontal and $700\sigma_y$ in vertical. But the off-momentum particles dynamic aperture is only about $13\sigma_x$ in horizontal and $300\sigma_y$ in vertical for $dp/p = \pm1\%$ and $13\sigma_x$ in horizontal and $200\sigma_y$ in vertical for $dp/p = \pm2\%$.

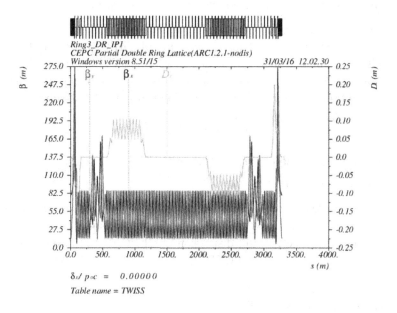

Fig. 5.   Beta function and dispersion of CEPC partial double ring part.

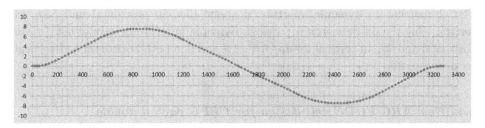

Fig. 6.   Orbit of CEPC partial double ring part.

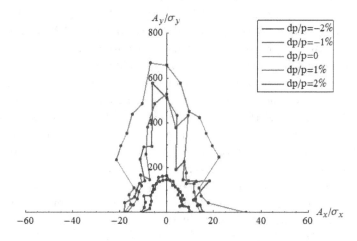

Fig. 7.   Dynamic aperture of CEPC partial double ring without FFS.

2.2.3. *Comparison of dynamic aperture with single ring lattice*

In this section, we compare the dynamic aperture of CEPC partial double ring lattice (without FFS) with the dynamic aperture of CEPC single ring lattice (the so-called CEPC main ring lattice designed by Huiping Geng 2014.9). The results are shown in Figs. 8 and 9. We can see the decrease of dynamic aperture obviously. This is reasonable and indicates that we should optimize the lattice design of partial double ring to acquire a larger dynamic aperture.

2.3. *CEPC partial double ring lattice with FFS*

2.3.1. *Orbit and optics with FFS*

In this section, we insert the FFS lattice designed by Dou Wang into the partial double ring lattice. Figure 10 is the beta function and dispersion of FFS lattice. Figure 11 is the beta function and dispersion of CEPC partial double ring lattice with FFS. Figure 12 is the orbit of CEPC partial double ring lattice with FFS.

2.3.2. *Dynamic aperture with FFS*

Figure 13 is the dynamic aperture of CEPC partial double ring lattice with FFS. The on-momentum dynamic aperture is about $13\sigma_x$ in horizontal and $55\sigma_y$ in vertical. But the off-momentum particles dynamic aperture is only about $1\sigma_x$ in horizontal and $4\sigma_y$ in vertical for $dp/p = \pm 1\%$ and 0 in both horizontal and vertical for $dp/p = \pm 2\%$.

2.4. *New ARC FODO cell design for CEPC partial double ring lattice*

The above results are based on the phase advance of ARC FODO cell as 60°. According to the new parameter of CEPC partial double ring, we need to design a smaller emittance lattice. So we design a 90° phase advance FODO cell and divided the sextupoles into 12 groups. The sextupoles arrangement is shown in Fig. 14. We choose noninterleaved arrangement, the phase between the two neighboring focusing sextupoles is π, between the two neighboring defocusing sextupoles is π and the phase between the two neighboring focusing and defocusing sextupoles is $\pi/4$. The other parts, like the straight sections, the bypass sections and partial double ring part's phase advance also change to 90°. Figure 15 shows the dynamic aperture of this new design without FFS. We can see that the dynamic aperture of on-momentum particles become larger but the off-momentum particles' aperture is still very small. Figure 16 shows the dynamic aperture of this new design with FFS. We can see that the dynamic aperture of on-momentum particles become larger both in horizontal and vertical but the off-momentum particles' aperture is still 0. All of this is still under studying and optimization.

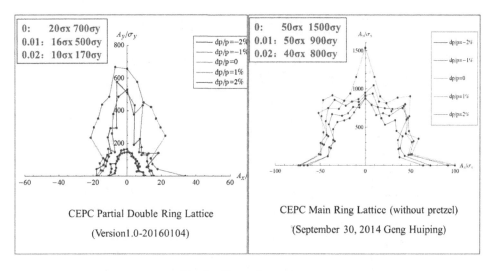

Fig. 8.   Dynamic aperture.

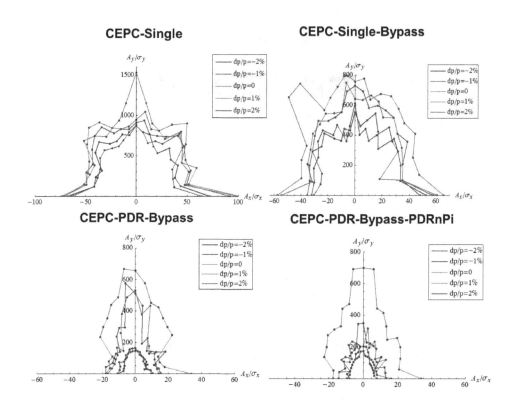

Fig. 9.   Dynamic aperture.

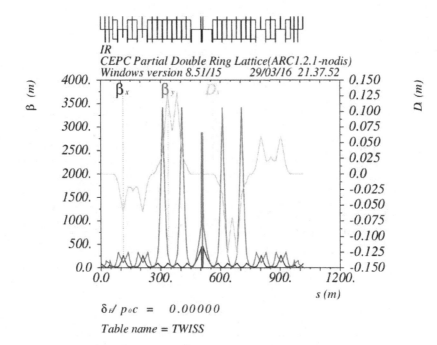

Fig. 10.   FFS optics.

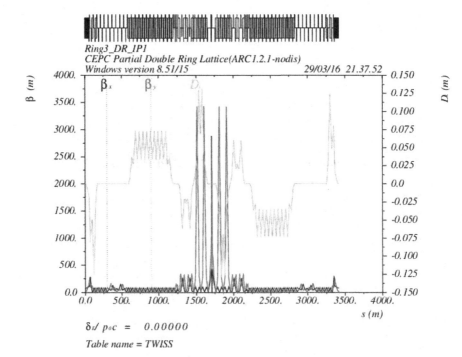

Fig. 11.   Optics of CEPC partial double ring (with FFS).

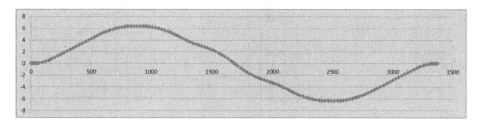

Fig. 12.   Orbit of CEPC partial double ring (with FFS).

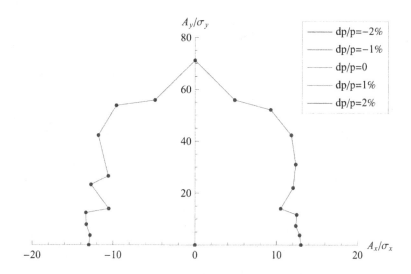

Fig. 13.   Dynamic aperture of CEPC partial double ring with FFS.

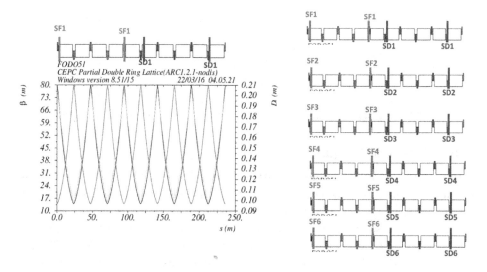

Fig. 14.   90° phase advance FODO cell and sextupoles families.

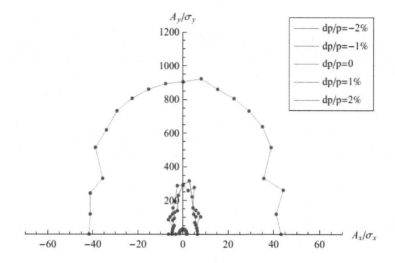

Fig. 15.   Dynamic aperture of CEPC partial double ring without FFS (New FODO).

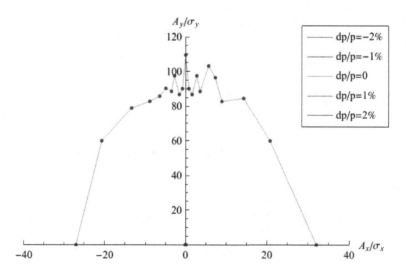

Fig. 16.   Dynamic aperture of CEPC partial double ring with FFS (New FODO).

3. SPPC Lattice Design

In this section, we will introduce the lattice design of SPPC. The SPPC design is just starting, and first we developed a systematic method of how to make an appropriate parameter choice for a circular pp collider by using an analytical expression of beam–beam tune shift, starting from the required luminosity goal, beam energy, physical constraints at the interaction point (IP) and some technical limitations.[9,10] Then we start the lattice design according to the parameter list and have the first version SPPC lattice.

3.1. *SPPC parameter choose*

The energy design goal of the SPPC is about 70–100 TeV, using the same tunnel as the CEPC, which is about 54 km in circumference. A larger circumference for the SPPC, like 100 km, is also being considered. It is planned to use superconducting magnets of about 20 T.[10] We obtain a set of parameters for the 54.7 km SPPC. In this set of parameters, the full crossing angle θ_c keeps the separation of 12 RMS beam sizes for the parasitic crossings. The luminosity reduction factor due to the crossing angle is larger than 0.9 and the ratio of β^* and σ_z is about 15. We also give a set of parameters for the larger circumference SPPC, considering both 78 km and 100 km. Table 1 is the parameter list for the SPPC. As a comparison, we put the parameters for LHC, HL-LHC, HE-LHC and FCC-hh together in Table 1.[10,11] The first plan for SPPC uses the same tunnel as the CEPC. The circumference is 54.7 km, which is determined by CEPC. We choose the dipole field as 20 T and get a center-of-mass energy of 70 TeV. If we want to explore the higher energy, we should make the circumference larger. To explore a center-of-mass energy of 100 TeV while keeping the dipole field at 20 T, the circumference should be at 78 km at least. With this condition, there is hardly any space to upgrade, so a 100 km SPPC is much better because the dipole field is then only 14.7 T. If the dipole field is kept at 20 T in a 100 km SPPC, we can get a center-of-mass energy as high as 136 TeV.

3.2. *SPPC lattice consideration*

According to the Pre-CDR, in the future, SPPC will be running at the same time with CEPC, and they will be in the same tunnel. So, the layout of SPPC should considered the CEPC layout. Figure 17 shows the layout of SPPC according the layout of CEPC partial double ring.

3.3. *SPPC lattice design*

3.3.1. *ARC and FODO cell*

In this section, we introduce the preliminary lattice design of SPPC. There are eight arcs and eight long straight sections. We use FODO in the ARC, and Fig. 18 shows the parameters of FODO cell in ARC. Each cell has eight dipoles whose length is 14.8 m and strength is 20 T. The total cell length is 144.4 m, maximum beta function is 244.8 m, minimum beta function is 42.6 m and phase advance is 90° in both horizontal and vertical. The quadrupole gradient and dipole parameter are reasonable according to the Pre-CDR choose and the aperture of quadrupole is also reasonable for both injection and collision energy. Figures 19 and 20 is the optics of FODO cell and ARC.

Table 1. Parameter lists for LHC HL-LHC HE-LHC FCC-hh and SPPC.

	LHC	HL-LHC	HE-LHC	FCC-hh	SPPC (Pre-CDR)	SPPC 54.7 Km	SPPC 100 Km	SPPC 100 Km	SPPC 78 Km	Unit
	Value									
Main parameters and geometrical aspects										
Beam energy [E_0]	7	7	16.5	50	35.6	35.0	50.0	68.0	50.0	TeV
Circumference [C_0]	26.7	26.7	26.7	100 (83)	54.7	54.7	100	100	78	km
Dipole field [B]	8.33	8.33	20	16 (20)	20	19.69	14.73	20.03	19.49	T
Dipole curvature radius [ρ]	2801	2801	2250	10416 (8333.3)	5928	5922.6	11315.9	11315.9	8549.8	m
Bunch filling factor [f_2]	0.78	0.78	0.63	0.79	0.79	0.79	0.8	0.8	0.79	
Arc filling factor [f_1]	0.79	0.79	0.79	0.79	0.79	0.79	0.79	0.79	0.79	
Total dipole length [L_{Dipole}]	17599	17599	14062	65412 (52333)	37246	37213	71100	71100	53720	m
Arc length [L_{ARC}]	22476	22476	22476	83200 (66200)	47146	47105	90000	90000	68000	m
Straight section length [L_{ss}]	4224	4224	4224	16800	7554	7595	10000	10000	10000	m
Physics performance and beam parameters										
Peak luminosity per IP [L]	1.0E+34	5.0E+34	5.0E+34	5.0E+34	1.1E+35	1.2E+35	1.52E+35	1.02E+36	1.52E+35	cm^{-2} s^{-1}
Beta function at collision [β^*]	0.55	0.15 (min)	0.35	1.1	0.75	0.85	0.97	0.24	1.06	m
Max B–B tune shift per IP [ξ_y]	0.0033	0.0075	0.005	0.005	0.006	0.0065	0.0067	0.008	0.0073	
Number of IPs contribute to ΔQ	3	2	2	2	2	2	2	2	2	
Max total B–B tune shift	0.01	0.015	0.01	0.01	0.012	0.013	0.0134	0.016	0.0146	
Circulating beam current [I_b]	0.5805	1.12	0.478	0.5	1.0	1.024	1.024	1.024	1.024	A
Bunch separation [Δt]	25/5	25/5	25/5	25/5	25	25	25	25	25	ns
Number of bunches [n_b]	2808	2808	2808	10600 (8900)	5835	5835	10667	10667	8320	
Bunch population [N_p]	1.15	2.2	1	1.0/0.2	2.0	2.0	2.0	2.0	2.0	10^{11}
Normalized RMS transverse emittance [ϵ]	3.75	2.5	1.38	2.2/0.44	4.10	3.72	3.65	3.05	3.36	μm
RMS IP spot size [σ^*]	16.7	7.1	5.2	6.8	9.0	8.85	7.85	3.04	7.86	μm
Beta at the 1st parasitic encounter [$\beta 1$]	26.12	93.9	40.53	13.88	19.5	18.70	16.51	64.1	15.36	m
RMS spot size at the 1st parasitic encounter [$\sigma 1$]	114.6	177.4	62.3	23.9	45.9	43.2	33.6	51.9	31.14	μm
RMS bunch length [σ_z]	75.5	75.5	75.5	80 (75.5)	75.5	65	65	15.8	70.6	mm
Full crossing angle [θ_c]	285	590	185	74	146	138	108	166	99	μrad
Reduction factor according to cross angle [Fca]	0.8391	0.314	0.608	0.910	0.8514	0.9257	0.9248	0.9283	0.9248	
Reduction factor according to hour glass effect [Fh]	0.9954	0.9491	0.9889	0.9987	0.9975	0.9989	0.9989	0.9989	0.9989	
Energy loss per turn [U_0]	0.0067	0.0067	0.201	4.6 (5.86)	2.10	1.97	4.30	14.7	5.69	MeV
Critical photon energy [Ec]	0.044	0.044	0.575	4.3 (5.5)	2.73	2.60	3.97	9.96	5.25	KeV
SR power per ring [P_0]	0.0038	0.0073	0.0962	2.4 (2.9)	2.1	2.0	4.4	15.1	5.82	MW
Transverse damping time [τ_x]	25.8	25.8	2.0	1.08 (0.64)	1.71	1.80	2.15	0.86	1.27	h
Longitudinal damping time [τ_e]	12.9	12.9	1.0	0.54 (0.32)	0.85	0.90	1.08	0.43	0.635	h

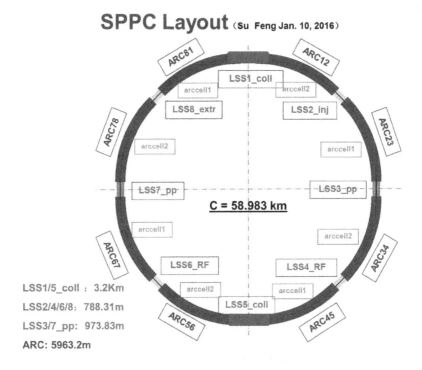

Fig. 17. SPPC lattice layout.

ARC CELL

	LQ	DQS	LS	DSB	LB	DBB
SPPC	4m	1m	0.5m	1m	14.8m	1m
FCC-hh	6.3137m	1m	0.5m	2.184m	14.3m	1.36m

B max [T]	G max [T/m]	k1	k2
19.61	582.156	4.9899E-3	0

Betax: 244.878/42.57
Betay: 42.569/244.869   (ϵ_n =4.1um)

E (Collision: 35TeV) $\epsilon = \dfrac{\epsilon_n}{\gamma}$
  (Injection: 2.1TeV)

ϵ (Collision: 1.099×10^{-10}m=0.1099nm)
(Injection: 1.83×10^{-9}m=1.83nm)

Pre-CDR:

Dipole: L=15m   B=20T

Quadrupole:

    D = 45 mm B_{pole}=16 T

    G=711.1T/m   K1=6.097×10^{-3}

σ (Collision: 1.66×10^{-4}m=166um)
(Injection: 6.76×10^{-4}m=676um)

R=20*σ_{inj}=13.52mm

D=27.04

Fig. 18. SPPC FODO cell parameter choice.

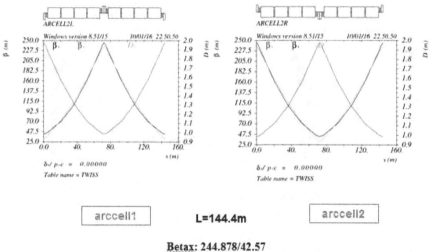

Fig. 19.   SPPC FODO cell optics.

ARC （ARCDSPL,36 CELL, ARCDSPR）

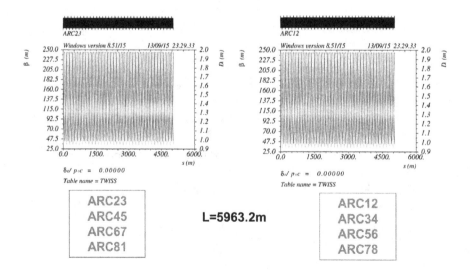

Fig. 20.   SPPC ARC optics.

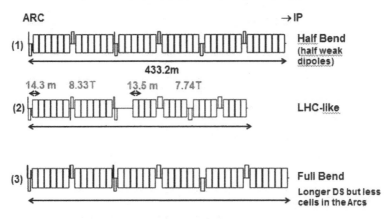

Fig. 21.   Dispersion suppressor section for FCC-hh.

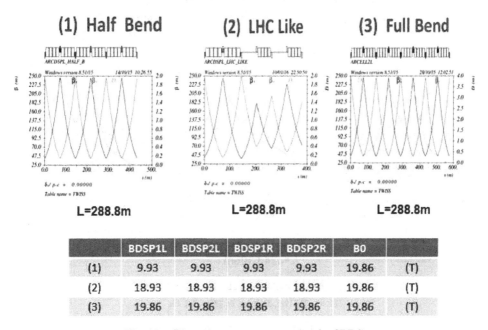

	BDSP1L	BDSP2L	BDSP1R	BDSP2R	B0	
(1)	9.93	9.93	9.93	9.93	19.86	(T)
(2)	18.93	18.93	18.93	18.93	19.86	(T)
(3)	19.86	19.86	19.86	19.86	19.86	(T)

Fig. 22.   Dispersion suppressor section for SPPC.

3.3.2. Dispersion suppressor section

For 90° phase advance FODO cell, the dispersion suppressor section has three schemes, called full-bend scheme, half-bend scheme and missing-dipole scheme. Figure 21 shows this three schemes for FCC-hh and Fig. 22 shows these for SPPC and in our design, we choose the missing-dipole scheme as the space can be used for collimation in the future.

LSS3_pp/LSS7_pp

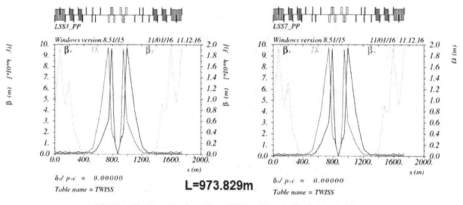

ARCDSPL, ARC to STR, 21.5*STRCELL, STR to ARC, ARCDSPR

382.4m, 71.719m, 973.829m, 66.789m, 382.4m

Fig. 23.   Long straight section for low β pp collision.

LSS1/5_coll

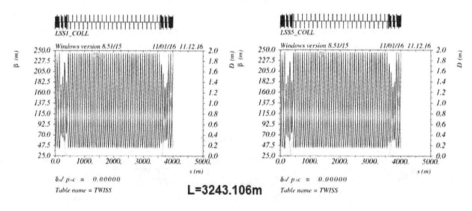

ARCDSPL, ARC to STR, 21.5*STRCELL, STR to ARC, ARCDSPR

382.4m, 71.719m, 3104.6m, 66.789m, 382.4m

Fig. 24.   Long straight section for collimation.

3.3.3. *Long straight section and interaction region*

There are eight long straight sections in SPPC lattice which are named as *LSS1_coll, LSS2_inj, LSS3_pp, LSS4_RF, LSS5_coll, LSS6_RF, LSS7_pp* and *LSS8_extr*. Long straight sections 3 and 7 are for low β pp collision, long straight sections 1 and 5 are for collimation using the long space as 3.4 km, long straight sections 4 and 6 are for RF system and long straight sections 2 and 8 are for injection and extraction. Figures 23–26 show the optics of these long straight sections.

LSS4/6_rf

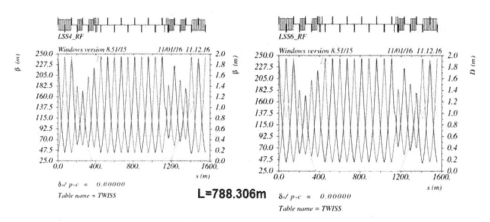

-ARCDSPR, ARC to STR, 4.5*STRCELL, STR to ARC, -ARCDSPL

382.4m, 71.719m, 649.8m, 66.787m, 382.4m

Fig. 25.   Long straight section for RF system.

LSS2_inj/LSS8_extr

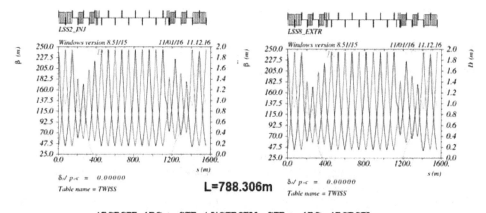

-ARCDSPR, ARC to STR, 4.5*STRCELL, STR to ARC, -ARCDSPL

382.4m, 71.719m, 649.8m, 66.787m, 382.4m

Fig. 26.   Long straight section for injection and extraction.

Figure 27 shows the quadrupole strength of *LSS3_pp* and *LSS7_pp*, and the gradient and aperture are reasonable according to the Pre-CDR parameter choice of quadrupoles.

R	K1(m^-2)	G (T/M)	L(M)	βmax	L	K1(m^-2)	G (T/M)	L(M)	βmax
K1.QT.1R	4.9751e-03	580.428	6	3543.69	K1.QT.1L	-4.9751e-03	-580.428	6	3543.69
K1.QT.A2R	-5.2595e-03	-613.668	9	9601.686	K1.QT.A2L	5.2595e-03	613.668	9	9601.686
K1.QT.B2R	-5.2595e-03	-613.668	9	9601.686	K1.QT.B2L	5.2595e-03	613.668	9	9601.686
K1.QT.3R	5.3434e-03	623.369	8	9731.53	K1.QT.3L	-5.3434e-03	-623.369	8	9731.53
K1.QM.4R	-2.2804E-04	-266.04	4	3798.29	K1.QM.4L	2.2804E-04	266.04	4	3798.29
K1.QM.5R	8.8592E-04	103.36	4	1506.53	K1.QM.5L	-8.8592E-04	-103.36	4	1506.53
K1.QM.6R	-1.2144E-03	-141.68	4	587.87	K1.QM.6L	1.2144E-03	141.68	4	587.87
K1.QM.7R	1.0640E-04	124.133	4	531.25	K1.QM.7L	-1.0640E-04	-124.133	4	531.25
K1.QM.8R	-4.2431E-03	-495.028	4	162.20	K1.QM.8L	4.2431E-03	495.028	4	162.20

Fig. 27.   Quadrupole gradient and aperture in LSS3 and LSS7.

4. Summary

In this paper, we discussed the details of CEPC partial double ring lattice design and showed the dynamic aperture study and optimization. The first version of CEPC partial double ring lattice has been done and the dynamic aperture need to be optimized. Now the DA of CEPC with PDR and Bypass (at IP2/4) and without FFS is better than before, but the DA with FFS is not good enough. All the above results and figures indicate that we have lots of work to do to optimize the design. We also showed the first version of SPPC lattice although it needs lots of work to do and to be optimized, especially the $LSS3\_pp$ low β pp optics and the dynamic aperture.

Acknowledgments

This work was supported by the National Natural Science Foundation of China (NSFC 11575218).

References

1. F. Zimmermann, HE-LHC & VHE-LHC accelerator overview (injector chain and main parameter choices), Report of the Joint Snowmass-EuCARD/AccNet-HiLumi LHC Meeting, Switzerland, 2013.
2. F. Zimmermann et al., FCC-ee overview, in Proc. HF2014, eds. N. Zhao and V. R. Schaa, Beijing, China, 2014, pp. 6–15.
3. The CEPC-SPPC Study Group, CEPC-SPPC: Pre-CDR, Volume II — Accelerator (March, 2015, IHEP-CEPC-DR-2015-01), pp. 28–35.
4. D. Wang et al., Chin. Phys. C **37**, 97003 (2013).
5. M. Xiao et al., Chin. Phys. C **40**, 08701 (2016), arXiv:1512.07348 [physics.acc-ph].
6. J. Gao, Private note about ultra-low beta and crossing angle scheme in CEPC lattice design for high luminosity and low power, IHEP-AC-LC-Note2013-012, June 16th, 2013.
7. M. Koratzinos and F. Zimmermann, Mitigating performance limitations of single beam-pipe circular e^+e^- colliders, IPAC2015, Shanghai, China, May 3–8, 2015.

8. W. Kalbreier *et al.*, Layout, design and construction of the electrostatic separation system of the LEP e^+e^- collider, CERN, Geneva, Switzerland.

9. J. Gao, *Mod. Phys. Lett. A* **30**, 1530006 (2015).

10. F. Su *et al.*, *Chin. Phys. C* **40**, 017001 (2016).

11. LHC Design Report, Vol. 1, CERN-2004-003.

R&D Steps of a 12-T Common Coil
Dipole Magnet for SPPC Pre-study

Chengtao Wang, Kai Zhang and Qingjin Xu*

Institute of High Energy Physics, Chinese Academy of Sciences,
Beijing 100049, P. R. China
*xuqj@ihep.ac.cn

IHEP (the Institute of High Energy Physics, Beijing, China) has started the R&D of high field accelerator magnet technology from 2014 for recently proposed CEPC-SppC (Circular Electron Positron Collider, Super proton–proton Collider) project. The conceptual design study of a 20-T dipole magnet is ongoing with the common coil configuration, and a 12-T model magnet will be fabricated in the next two years. A 3-step R&D process has been proposed to realize this 12-T common-coil model magnet: first, a 12-T subscale magnet will be fabricated with Nb_3Sn and NbTi superconductors to investigate the fabrication process and characteristics of Nb_3Sn coils, then a 12-T subscale magnet will be fabricated with only Nb_3Sn superconductors to test the stress management method and quench protection method of Nb_3Sn coils; the final step is fabricating the 12-T common-coil dipole magnet with HTS (YBCO) and Nb_3Sn superconductors to test the field optimization method of the HTS and Nb_3Sn coils. The characteristics of these R&D steps will be introduced in the paper.

Keywords: Accelerators; generation of magnetic fields; superconducting magnets.

1. Introduction

After the discovery of Higgs bosons in 2012,[1] a two-stage particle collider is proposed by the Institute of High Energy Physics (IHEP, Beijing, China) to carry out high precision study on them. CEPC is a 240–250 GeV Circular Electron Positron Collider with 54 km circumference, which will be upgraded to a 70 TeV or higher proton–proton collider (SppC) in the same tunnel to study the new physics beyond the Standard Model.[2,3] To reach 70 TeV center-of-mass energy, SppC needs thousands of 20-T level dipole magnets to bend proton beams. The highest field strength of superconducting magnets used in present accelerators is 8.3 T (operated at 1.8 K at LHC).[4] So R&D of the high field superconducting magnets is the most challenging technology for SppC project. There are totally four types of coil configurations which can provide dipole magnetic field for accelerators, i.e. cos-theta type,[5] common-coil type,[6] block type[7] and canted cos-theta type.[8] The conceptual design study of a 20-T two-in-one dipole magnet is ongoing with common-coil configuration,[9,10] and a 12-T model magnet will be fabricated firstly in the next two years. Details of the R&D process will be discussed in this paper.

2. A Subscale Magnet Fabricated with Nb₃Sn and NbTi Superconductors

A 12-T subscale magnet will be firstly fabricated with Nb_3Sn and $NbTi$ superconductors. The electromagnetic design study of this 12-T subscale magnet has been carried out. The main parameters of this subscale magnet are listed in Table 1. Figure 1 shows the cross-section of the magnet. The red section in the figure represents the iron yoke with the outer diameter of 600 mm. The blue section represents the superconducting coils. Detail of the cross-section for one quadrant is shown in Fig. 2. The whole coil width is 52 mm, and the height is 36 mm. Two types of cables are used to fabricate the superconducting coils, named IHEPW1 and IHEPW2. The cables have the same width and thickness. The parameters of the cables are shown in Table 2. The cable IHEPW1 is made up of IHEPWCJC Nb_3Sn strands and the cable IHEPW2 is made up of IHEPWNJC $NbTi$ strands. Table 3 shows the parameters of these strands. Totally 8 such double-pancake coils are needed (4 Nb_3Sn + 4 $NbTi$) to reach the 12-T peak field in coil at 82% load line at 4.2 K. For each layer of coil, there are 20 turns of cables. The bending radius of the coil is 60 mm. Table 4 shows the required amount of superconductors per meter of such magnet: the required length of the IHEPWCJC strands (Nb_3Sn) and IHEPWNJC strands ($NbTi$) is 4.5 km for each. The maximum coil field with 100% load line

Table 1. Main parameters of 12-T magnets.

Parameter	Unit	First subscale magnet	Second subscale magnet	Third dipole magnet
Number of aperture	—	NONE	NONE	2
Aperture diameter	mm	—	—	15
Inter-aperture spacing	mm	—	—	156
Operating current	A	3970	6250	280 (HTS)/7900 (Nb₃Sn)
Operating temperature	K	4.2	4.2	4.2
Coil peak field	T	12.01	12.41	12.17
Margin along the loadline	%	18.4	20	19
Stored energy	MJ/m	0.27	0.44	0.81
Inductance	mH/m	19.9	12.9	12.1 (HTS)/16.2 (Nb₃Sn)
Outer diameter of iron	mm	600	600	600
Lorentz force Fx/Fy (per aperture)	MN/m	2.81/0.15	3.35/0.26	3.99/0.67
Peak stress in coil	MPa	81.4	102.9	137.8
Minimum bending radius	mm	60	60	60

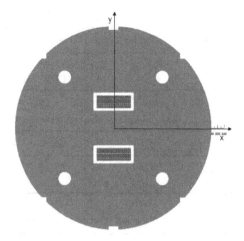

Fig. 1.   (Color online) The cross-section of a 12-T subscale magnet.

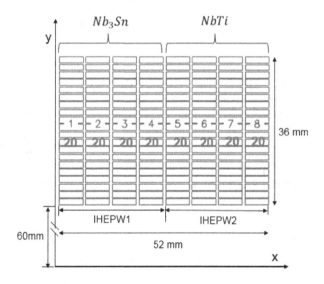

Fig. 2.   Detail of the cross-section for the first coil quadrant.

ratio at 4.2 K is 14.6 T, corresponding to an operating current of 5000 A, located at the center of the Nb3Sn coil, as shown in Fig. 3. The peak field of the NbTi coil is around 6.7 T. Or we can get a coil field of 12 T with an operating margin of 18.4% at 4.2 K corresponding to an operating current of 3970 A. The peak field of the NbTi coil is around 5.6 T.

Different from conventional solenoid superconducting magnets which coils are winded directly with strands, the accelerator magnets use Rutherford-type cables to form the coils.[11] There is almost no practical experience in fabricating this kind of cables in China. Since 2014, a few companies have worked on it in collaboration

Table 2.   Main parameters of the cables.

Cable	Width	Thickness-I	Thickness-o	Ns	Strand	Filament	Insulation
IHEPW1	5.8	1.5	1.5	14	IHEPWCJC	Nb_3Sn	0.15
IHEPW2	5.8	1.5	1.5	14	IHEPWNJC	NbTi	0.15

Table 3.   Main parameters of the strands.

Strand	Diam.	cu/sc	RRR	Tref	Bref	Jc@BrTr	dJc/dB
IHEPWCJC	0.82	1	100	4.2	12	2400	400
IHEPWNJC	0.82	1	100	4.2	7.2	1500	550

Table 4.   The required amount of superconducting per meter of such magnet.

Cable	Cable turns (1/4)	Cable turns (all)	Strand length (km)
IHEPW1	80	320	4.5
IHEPW2	80	320	4.5

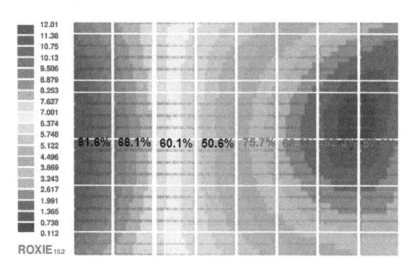

Fig. 3.   Field distribution in coil (in first quadrant) with the operating current of 3970 A.

with IHEP. Recently, they have successfully wound some Rutherford-type cables with NbTi strands, which having a J_c degradation of less than 4% and a filling factor of 87%. Several Nb_3Sn cables have just been fabricated and the quality is to be tested. Wind and React method will be adopted for winding our Nb_3Sn coils.[12,13] Copper cables will be used to make several dummy coils firstly, to explore winding process, heat reaction and epoxy impregnation technologies. After mastering all the technologies, we will use real Nb_3Sn cables to try all these procedures.

3. A Subscale Magnet Fabricated with Only Nb₃Sn Superconductor

After the first subscale magnet has been fabricated and the characteristics of Nb_3Sn coils have been well known to us, a 12-T subscale magnet will be fabricated with only Nb_3Sn superconductors to test the stress management method for Nb_3Sn coils. The iron yoke manufactured for the first subscale magnet will be reused both in this magnet and the magnet in step 3. The second magnet is a subscale magnet including two flat racetrack Nb_3Sn coils which have been built for the first magnet, and two other flat racetrack Nb_3Sn coils. For each layer of coil, there are still 20 turns of cables. The coil width is 51.6 mm, and the coil height is 36 mm in the first quadrant. To increase efficiency of Nb_3Sn strands utilization, graded coil configuration has been designed. Broader cable IHEPW3 will be adopted for the inner coil and narrower cable IHEPW1 will be used for the outer two coils. The main parameters of these two kinds of cables are shown in Table 5. Table 6 shows the required amount of superconductors per meter of such magnet: the required length of the IHEPWCJC strand (Nb_3Sn) is 9.12 km in total. We can get a coil field of 12.4 T with an operating margin of 20% at 4.2 K corresponding to an operating current of 6250 A, as shown in Fig. 4. When the current reaches 8050 A, we can get the maximum coil field of 15.4 T with 100% load line ratio at 4.2 K.

Large Lorentz force will be induced in coils when the field reaches 12 T. After excitation, the racetrack coils in both window frames tend to move outward. Figure 5 shows the detailed Lorentz force distribution in coils for the first quadrant. The total force is 3.35 MN/m in horizontal direction and 0.26 MN/m in vertical direction. The peak stress in coil is around 103 MPa. As Nb_3Sn superconductor is a stress-sensitive material, high level stress (higher than 150 MPa) may cause J_c degradation.[14] For such level, stress is sustainable for Nb_3Sn superconductors.[15] Shell-based mechanical support structure will be adopt in this dipole magnet: with an aluminum shell outside of the iron yoke to provide pre-stress for coils.[16] And bladders will be used to supply a pre-stress about 60 MPa.[17] Manufacturing the bladders is a challenge as it requires high quality welding techniques. Typically, we need bladders to sustain more than 100 MPa water pressure without any leakage.

Table 5.   The main parameters of the cables.

Cable	Height	Width-I	Width-o	Ns	Strand	Filament	Insulation
IHEPW1	5.8	1.5	1.5	14	IHEPWCJC	Nb_3Sn	0.15
IHEPW3	1.2	1.5	1.5	29	IHEPWCJC	Nb_3Sn	0.15

Table 6.   The required amount of superconducting per meter of such magnet.

Cable	Cable turns (1/4)	Cable turns (all)	Strand length (km)
IHEPW1	80	320	4.48
IHEPW3	40	160	4.64

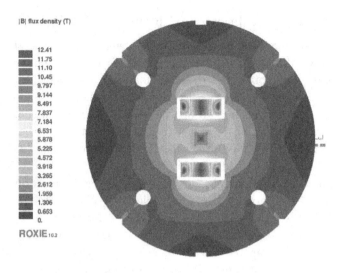

Fig. 4. The field distribution in the cross-section of coil and yoke.

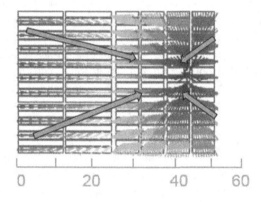

Fig. 5. Lorentz force distribution in coil (in first quadrant).

4. A Common Coil Magnet Fabricated with HTS (YBCO) and Nb$_3$Sn Superconductors

Finally, a 12-T common-coil dipole magnet with two apertures will be fabricated with HTS and Nb$_3$Sn superconductors, to test the field optimization, fabrication and quench protection methods for HTS coils. The cross-section of this magnet is shown in Fig. 6. The magnet will reuse the four Nb$_3$Sn flat racetrack coils in second subscale magnet but have two apertures (top and bottom). The clear bore diameter is temporarily set to be 15 mm and the inter-aperture spacing is 156 mm. There are eight layers of coils with HTS (YBCO) for the inner two layers and Nb$_3$Sn for the outer six layers. The YBCO insert coils are fabricated with 4-mm width and 0.2-mm thickness YBCO tape. The main parameters are listed in Table 7. Table 8 shows the required amount of superconductors per meter of such magnet. The main

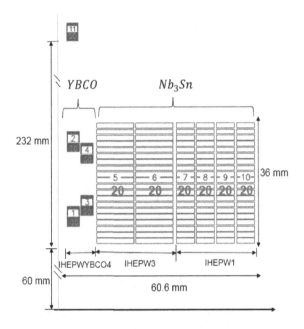

Fig. 6. First quadrant cross-section of a 12-T common-coil dipole magnet.

Table 7. The main parameters of the YBCO tape.

Strand	Size	cu/sc	RRR	Tref	Bref	Ic@BrTr	dIc/dB
YBCO4	4×0.2 mm^2	—	—	4.2	12	400	15

Table 8. The required amount of superconducting per meter of such magnet.

Cable	Cable turns (1/4)	Cable turns (all)	Strand length per 1 m coil (km)
IHEPW1	80	320	4.48
IHEPW3	40	160	4.64
IHEPWYBCO4	150	600	0.6

field with 100% load line ratio at 4.2 K is 14.6 T, corresponding to an operating current of 358 A in YBCO tapes and 9930 A in Nb$_3$Sn cables. The peak field is 14.7 T in YBCO coil and 14.5 T in Nb$_3$Sn coils. Or we can get a main field of 12 T with an operating margin of 19% at 4.2 K, corresponding to an operating current of 280 A in YBCO tapes and 7900 A in Nb$_3$Sn cables. The peak field is 12.2 T in YBCO coil and 12 T in Nb$_3$Sn coils. Figure 7 shows the field distribution in coils with an operating margin of 19% at 4.2 K.

Mastering quench protection method for HTS coils is one of the main objectives of this step. Coupling-loss-induced quench (CLIQ) technology seems to be effective, but more verification tests are required.[18]

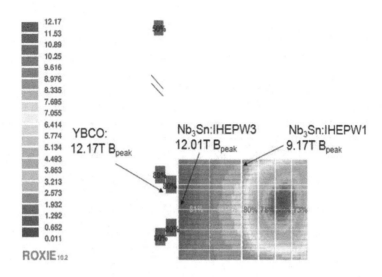

Fig. 7.   Field distribution in the coil (in first quadrant).

The field quality has been optimized at the main field of 12 T for this 2D electromagnetic design. All the high order harmonics are less than 1 unit after optimization. As YBCO superconductor is tape-shaped material, persistent current problem[19] is to be solved as it would distort the field quality. Many research has been carried out to learn the distribution of persistent current in YBCO tapes but there is still no effective solution to deal with it.[20,21] This is a challenge but also a chance for us to make it clear.

5. Conclusion

The 3-step R&D steps to realize a 12-T common-coil dipole magnet are presented in this paper. We first fabricate a 12-T subscale magnet with Nb3Sn and NbTi superconductors; after we have fully understood the characteristics of the Nb3Sn superconductor and fabrication procedures of the coils, we will fabricate a 12-T subscale magnet with only Nb3Sn superconductor. Finally, a 12-T common-coil dipole magnet with HTS and Nb3Sn superconductors will be fabricated to test the characteristics of HTS coils and field optimization method of such type of magnets. Lots of work have been carried out to investigate key technologies required for completing this project.

Acknowledgment

The authors would like to thank Dr. GianLuca Sabbi of LBNL (Lawrence Berkeley National Lab, Berkeley, USA), Dr. Ramesh Gupta of BNL (Brookhaven National Lab, Upton, NY, USA) and Dr. Tiina Salmi of TUT (Tampere University of Technology, Tampere, Finland) for their helpful discussions and suggestions.

References

1. M. Carena and H. E. Haber, *Prog. Part. Nucl. Phys.* **50**, 63 (2003), arXiv:0208209 [hep-ph].
2. CEPC-SPPC Preliminary Conceptual Design Report (2015), available at http://cepc.ihep.ac.cn/preCDR/volume.html.
3. J. D. Wells, Lectures on Higgs Boson physics in the standard model and beyond (2009), arXiv:0909.4541 [hep-ph].
4. LHC Design Report: The LHC Main Ring, Chapter 7: Main magnets in the arcs (2004), available at http://ab-div.web.cern.ch/ab-div/Publications/LHC-DesignReport.html.
5. L. Rossi, *IEEE Trans. Appl. Supercond.* **14**, 153 (2004).
6. R. Gupta *et al.*, *IEEE Trans. Appl. Supercond.* **11**, 2168 (2011).
7. G. Sabbi, S. E. Bartlett and S. Caspi, *IEEE Trans. Appl. Supercond.* **15**, 1128 (2005).
8. L. Brouwer, D. Arbelaez and S. Caspi, *IEEE Trans. Appl. Supercond.* **24**, 1 (2014).
9. Q. Xu *et al.*, *IEEE Trans. Appl. Supercond.* **26**, 4000404 (2016).
10. K. Zhang *et al.*, *IEEE Trans. Appl. Supercond.* **26**, 4003705 (2016).
11. N. Andreev *et al.*, *IEEE Trans. Appl. Supercond.* **17**, 1027 (2007).
12. P. He *et al.*, *J. Supercond. Nov. Magn.* **25**, 1805 (2012).
13. K.-H. Mess, P. Schmüser and S. Wolff, *Superconducting Accelerator Magnets* (World Scientific, Singapore, 1996).
14. D. R. Dietderich *et al.*, *Cryogenics* **48**,331 (2008).
15. E. Barzi, T. Wokas and A. V. Zlobin, *IEEE Trans. Appl. Supercond.* **15**, 1541 (2005).
16. P. Ferracin, *Cheminform* **41** (2009).
17. M. Juchno, *et al.*, *IEEE Trans. Appl. Supercond.* **25**, 1 (2015).
18. E. Ravaioli *et al.*, *IEEE Trans. Appl. Supercond.* **26**, 1 (2016).
19. C. C. Rong *et al.*, *IEEE Trans. Appl. Supercond.* **25**, 1 (2015).
20. N. Amemiya and K. Akachi, *Supercond. Sci. Technol.* **21**, 40 (2008).
21. Y. Iwasa *et al.*, *Appl. Phys. Lett.* **103**, 052607 (2013).

Multi-objective Dynamic Aperture Optimization for Storage Rings

Yongjun Li and Lingyun Yang

Energy Sciences Directorate, NSLS-II Department,
Brookhaven National Laboratory, Upton, NY-11973, USA

We report an efficient dynamic aperture (DA) optimization approach using multi-objective genetic algorithm (MOGA), which is driven by nonlinear driving terms computation. It was found that having small low order driving terms is a necessary but insufficient condition of having a decent DA. Then direct DA tracking simulation is implemented among the last generation candidates to select the best solutions. The approach was demonstrated successfully in optimizing NSLS-II storage ring DA.

Keywords: Multi-objective optimization; dynamic aperture; storage ring.

1. Introduction

In designing storage ring-based X-ray light sources, accelerator physicists are pushed very hard to achieve low emittance pursuing bright photon beam. Low emittance means strong focusing quadrupoles, which requires strong sextupoles (installed at in dispersive region) to compensate chromatic aberrations. These strong sextupoles result in small dynamic aperture (DA), making it difficult to realize efficient off-axis injection, and maintain sufficient Touschek lifetime. Based on modern accelerator theory, improvement of DA can be realized by minimizing various nonlinear driving terms (NDT), which can be computed to a certain order.[1] But there are already several tens terms even we only compute them up to the second-order. In the meantime, the designers have to face the difficulty of specifying weights for numerous NDTs based on their experience, if the conventional optimization method (single merit function) is used.

Recently, people used a new method of simultaneously optimizing dynamic and momentum aperture,[2] or beam lifetime[3,4] with multi-objective genetic algorithm (MOGA). This method is very successful. The optimization heavily relies on a powerful and reliable tracking simulator to directly probe the dynamic and momentum apertures (or beam lifetime). Usually the optimization driven by the element-by-element symplectic-tracking simulator[5] is very time-consuming, especially when ring scale is big. This time-consuming objective function computation significantly degrades optimization efficiency although powerful parallel computers are available.

Based on previous genetic optimization experiences and results, we observed that: (1) the driving terms reduce automatically once DA is set as one of the objectives for optimization;[3] (2) the optimal candidates always have small driving terms, while small driving terms do not always imply a good DA.[2] In another word, having small driving terms is a necessary but insufficient condition for ensuring a good DA. This observation hints an efficient optimization method. First, we can apply MOGA to optimize various driving terms simultaneously to avoid blindly weights on them. Then we implement tracking simulation as a final filter to select the best solutions among the last generation candidates. The reason of why this new strategy becomes efficient is that the optimization driven by analytical formulae computation is much faster than direct tracking simulation. We can further extract the correlation between DA and driving terms from the raw data of simulation to identify which resonances dominate DA. The correlations can help us to better understand the nonlinearity from the view of beam dynamics.

2. Multi-Objective Genetic Algorithm

MOGA[6] has been widely adopted in storage ring lattice design recently. Ring designer used it to find low emittance lattice for a given magnet layout,[7,8] or to optimize DA by varying sextupole configuration. The method we are using is an elitist multi-objective evolutionary algorithm proposed by Deb.[6] It is a population-based evolution algorithm to find the Pareto optimum iteratively. First, a fixed number of candidates is initialized as the first generation, and they are randomly chosen and uniformly distributed in their allowed ranges. Then one pair of them are randomly chosen as parents to cross over to generate two new children according to a certain probability density function. The process of crossover is repeated until the population is doubled. Next, all children mutate randomly also with a certain probability density function. The objective functions and constraints are evaluated for each of these new children. The whole population, including the parents, is then sorted according to their dominance relations. Since parents are included into sorting, elitists are kept once they are found, which can speed up the performance of optimization significantly by preventing the loss of good solutions. Candidates not dominated by anyone are in the first rank. Only half of the better candidates are kept by dropping out the candidates with larger rank. Within same rank, candidates in a high population density region have lower priority to be selected. Up to this point, the population is kept the same but the overall qualities in terms of objective functions and constraints are evolved not worse than the previous population before crossover and mutation. The population is evolved generation by generation until it converges or the maximum number of iterations is reached.

3. Nonlinear Driving Terms

For storage rings, NDTs can be obtained by concatenating individual maps into one-turn-map via the similarity transformation and BCH formula in Lie algebra

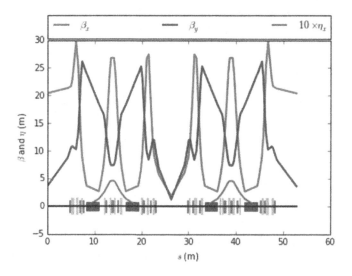

Fig. 1.   NSLS-II optics in one super-cell.

language.[1] Wang[9] explicitly derived all second-order driving terms due to sextupoles and chromatic effects of quadrupoles. We combined them with first-order driving terms as our optimization objectives. Of course, h_{11001} and h_{00111} are excluded to control linear chromaticity. In the meantime, three terms, h_{11110}, h_{22000} and h_{00220} are imposed with some extra constraints to reduce tune dependence on amplitude.

4. Applications

4.1. *NSLS-II lattice*

NSLS-II[10] lattice has 30 double bend achromatic (DBA) cells. Two DBA cells with mirror symmetry have low- and high-beta functions at short and long straight sections. The linear optics for one cell is illustrated in Fig. 1. The whole bare lattice has 15-fold symmetry. Each DBA cell has nine sextupoles. Three of them are sitting in the dispersive region for chromaticity correction. The other six sextupoles are geometric sextupoles.

4.2. *Chromaticity +7/+7 lattice*

To effectively suppress beam transverse instabilities at high beam current, high positive linear chromaticity is preferable. High positive linear chromaticity means stronger sextupoles, and then a smaller DA. The DA optimization is therefore demanding. In this case, the linear chromaticity is corrected to +7/+7 with three chromatic sextupoles (Fig. 2). Then the free parameters left for DA optimization are the strengths of 6 families of geometric sextupoles in the nondispersive straights. Our objective functions are 30 low order NDTs totally. The searching space of

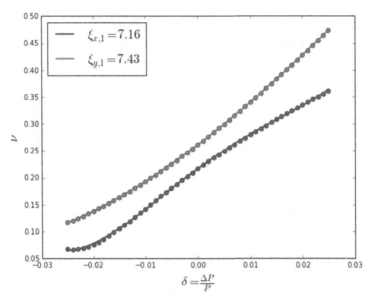

Fig. 2.   Tune versus energy deviation for linear chromaticity +7/+7.

free variables (i.e. sextupole strengths) is limited within the ranges limited by the engineering specification and power supply polarities. Three tune-dependency-on-amplitude terms were imposed with some extra constraints to boost the convergence. We use a population of 4000, and run for 100 generations. With 100 Xeon 2.33 GHz CPUs in a Sun Grid Engine cluster, it takes 6–12 hours to finish the optimization. Finally, we use a symplectic tracking code to check the DAs for all these 4000 candidates in the last generation. Based on tracking results, we can choose some of the best solutions among them.

The frequency map analysis was carried out for one of solutions as shown in Figs. 3 and 4. This sextupole configuration has been successfully tested at the NSLS-II ring. DA is proved to be sufficient for a decent injection efficient (80–90%) and up to 18 h lifetime was observed with 10 mA stored beam in 100 bunches.

4.3. Preliminary study for low-alpha lattice

Low-alpha lattice with short bunch longitudinal length can provide coherent synchrotron radiation in the THz range. Such kind of lattice has been tested on most of modern synchrotron radiation facilities. This preliminary study focuses on the feasibility of low-alpha mode on the NSLS-II storage ring. The DA optimization is more difficult than the previous case, because we need to control linear chromaticity, DA and longitudinal stability simultaneously with sextupole configuration. There is no dispersion-free section, thus we cannot separate chromaticity correction and DA optimization. Once the zeroth-order momentum compactor is small enough, the higher order terms will play important roles in the longitudinal motion stability.

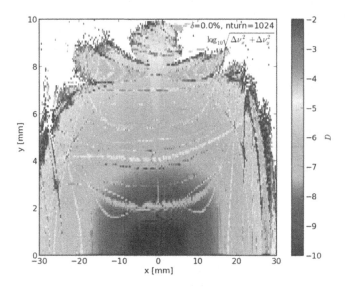

Fig. 3.   Dynamic aperture in x–y plane.

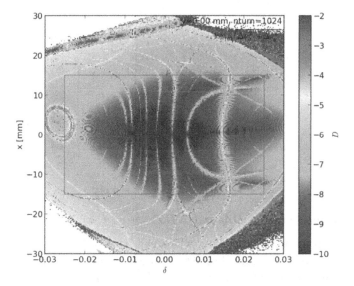

Fig. 4.   Energy acceptance in x–δ plane.

And finally, we still need sufficient DA and energy acceptance to achieve decent injection efficiency and lifetime. All these optimization objectives need to be satisfied simultaneously with a given sextupole configuration. The preliminary results on linear optics, DA, energy acceptance and stable RF bucket are shown in Figs. 5–8, which are positive and promising.

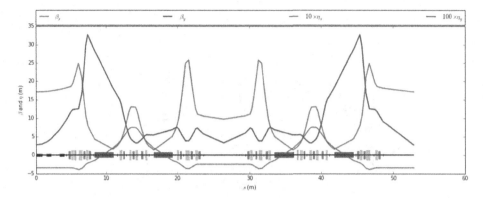

Fig. 5.   Linear optics for low-alpha configuration.

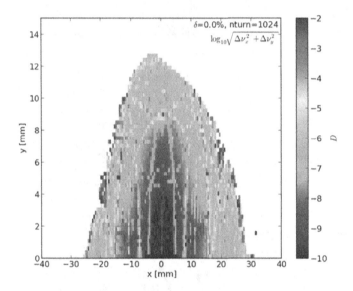

Fig. 6.   Dynamic aperture in x–y plane.

4.4. Correlation between DA and NDTs

MOGA produces numerous raw data, which can be used to study the correlations between DA and NDTs. Usually, inside the last generation, none of candidate in the first rank is dominated by any others. From the view of multi-objective optimization, they are equally good if only considering NDTs. But from the aspect of DA, they are quite different. We use DA tracking simulator to compute their DAs for all 4000 candidates, and illustrate their correlation with the summation of NDTs in Fig. 9. By observing the correlation, we can conclude that having small low order NDTs is necessary but insufficient condition for having a good DA (as marked with red A). The necessity is demonstrated by observing that all candidates with large DAs must have small NDTs without exception.

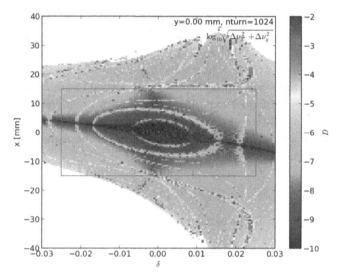

Fig. 7.   Energy acceptance in x–δ plane.

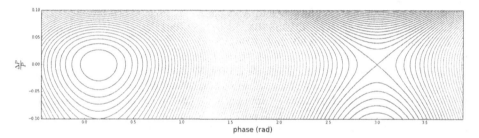

Fig. 8.   Stable bucket for low alpha configuration.

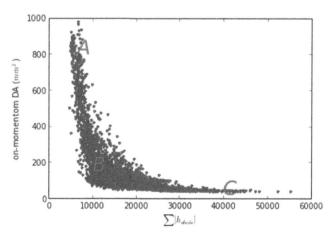

Fig. 9.   (Color online) Correlation between DA and NDTs.

5. Summary

An efficient approach of using MOGA driven by NDTs computation to optimize storage ring DA is discussed in this paper. The optimization produces numerous raw data, which can satisfy the necessary condition to have a good DA at the first step. Among them we can select some good solutions with the DA tracking code. Applying this approach on the NSLS-II bare lattice to achieve a high positive chromaticity was demonstrated experimentally. And a preliminary study on low-alpha lattice is quite positive also. So far, we applied this method to a ring-based light source. But it will be interesting to explore how to implement this algorithm on high-energy ring collider design.

Acknowledgment

This work was supported by Department of Energy with Contract No. DE-AC02-98CH10886.

References

1. A. Dragt, *AIP Conf. Proc.* **87**, 147 (1982).
2. L. Yang *et al.*, *Phys. Rev. ST Accel. Beams* **14**, 054001 (2011).
3. M. Borland, L. Emery, V. Sajaev and A. Xiao, *Conf. Proc.* **C110328**, 2354 (2011).
4. L. Wang *et al.*, *Conf. Proc.* **C1205201**, 1380–1382 (2012).
5. H. Yoshida, *Phys. Lett. A* **150**, 262 (1990).
6. K. Deb, *Multi-Objective Optimization Using Evolutionary Algorithms* (Wiley, 2001).
7. L. Yang *et al.*, *Nucl. Instrum. Methods A* **609**, 50 (2009).
8. C. Sun *et al.*, *Phys. Rev. ST Accel. Beams* **15**, 054001 (2012).
9. C. X. Wang, ANL/APS/LS-330 (unpublished) (2012), https://www1.aps.anl.gov/ icms_files/lsnotes/files/APS_1429490.pdf.
10. BNL, NSLS-II Conceptual Design Report (CDR) (2013), https://www.bnl.gov/nsls2/ project/CDR/.

Experiment

Detectors and Experiments

John Hauptman

Department of Physics and Astronomy,
Iowa State University, Ames, IA 50011, USA
hauptman@iastate.edu

The talks in the Program and the Conference parallel sessions make clear that high quality pixel vertex chambers are presently well developed and with continuing improvements (M. Caccia,[1] X. Sun,[2] M. Stanitzki,[3] J. Qian[4]); that there are at least two major tracking chambers that are well studied, a TPC and silicon-strip chambers (H. Qi,[5,6] C. Young,[7,8] A. de Roeck[9,10]); that the energy measurement of photons and electrons is generally very good (H. Yang,[11] S. Franchino[12]); and, that the last remaining detector that has not yet achieved the high precision required for good e^+e^- physics is the hadronic calorimeter for the measurement of jets, most importantly, jets from the decays of W and Z to quarks (S. Lee,[13,14] M. Cascella,[15] A. de Roeck[16]). The relationship of the detectors to physics and the overall design of detectors was addressed and questioned (Y. Gao,[17] M. Ruan,[18] G. Tonelli,[19] H. Zhu,[20] M. Mangano,[21] C. Quigg[22]) in addition to precision time measurements in detectors (C. Tully[23]).

Keywords: Pixel vertex chambers; tracking chambers; calorimeters; detectors.

1. Introduction

The detectors and experiments of high energy physics hold the field together by providing the bridge between actual particle interactions in the colliding machines and the theoretical understanding of those interactions, once expressed by a student:

> *"Now I see. The experimentalist connects the nut and bolt to the Feynman diagram."*
> — Sung Keun Park (student)

An important question about the capabilities and physics reach of detectors was asked by Michelangelo Mangano

> *"How far can we push the detector technology to maximize the Higgs measurement capabilities?"*
> — Michelangelo Mangano[21]

which is a complex and vital question considering the importance of good measurements and the timescales and costs of big experiments that force early decisions on detector technologies: when do you stop improving your technology and start building it? The question of good-enough measurements has been answered within the ILC community in three parts: (i) a vertex chamber with an impact parameter resolution of $\sigma_b \approx 10\ \mu$m, (ii) a tracking chamber momentum resolution of $\sigma_p/p \approx 3 \times 10^{-5}\,p(\text{GeV}/c)$ and (iii) a hadronic calorimeter energy resolution of $\sigma_E/E \approx 30\%/\sqrt{E(\text{GeV})}$.

A further question that is seldom addressed by experimentalists was asked in different ways by John Ellis and Chris Quigg. In the grand strategies of physics about what collider to build next and what detectors to approve, one can ask a deep question:

> *"Should one be considering exotic signatures that might have been missed, or just push on hopefully to higher masses?"*
>
> — John Ellis[24]

A philosopher who looks at present-day particle physics might be concerned that we seem to predict new particles (charm, bottom, top, W, Z, Higgs) within the context of the standard model, we proceed to look for them with conventional detectors that have changed little in 30 years (excepting resolutions), and we find them, thus confirming the standard model. Is there something we have missed?

> *"How are we prisoners of conventional thinking?"*[a]
>
> — Chris Quigg[22]

These are profound questions: the detectors and colliders we build, and the way we search for all "the usual suspects" has not changed much from those searches at PEP, LEP, Tevatron and now the LHC. One way to understand this is the pleasant fact that the standard model is so successful (the "unreasonable effectiveness" of the standard model[22]) that an aspiring physicist is not encouraged to propose a detector far outside the conceptual bounds of the standard model with prospects of seeing nothing.

The inventions of detectors, from the alcohol mist in Wilson cloud chambers to the 85 million silicon channels in the LHC tracking chambers, and from simple shower counters to measurements of jet four-vectors, portend future detectors that will more successfully exploit all the known particles of the standard model and thereby measure all possible processes whose final decay products are simple hadrons ($\pi^\pm$, $K^\pm$, K^0, p, n), electromagnetic particles ($e^\pm$, γ) and the strong and weak force carriers (W, Z, gluon).

We believe that achieving a four-vector resolution of 1–2% on hadronic particles, that is, on jets and gluons, will "close the gap" with all the other detectors of a big experiment such that each and every particle in Fig. 1 will be measured with high precision and that whole event reconstructions will be so precise that backgrounds in general will be greatly reduced.

---

[a]Chris Quigg does not ask "Are we prisoners ...," but instead "How are we prisoners ..."

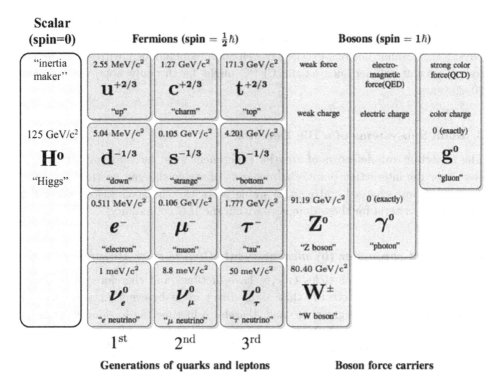

Fig. 1.   Standard model fermions shadowed by their boson force carriers. The Higgs presumably couples to everything.

An examination of papers at the Elba and Vienna instrumentation conferences shows that extraordinary instrumentation capabilities are in our future, but a collider detector requires a little conservatism to guarantee an actual working detector. In every experiment, the design of the detector defines the search limits and may preclude easy observation of "exotic signatures" not anticipated in the standard model.

2. The Importance of an e^+e^- Collider

Everyone is here at this wonderful meeting to discuss future colliders, specifically electron–positron colliders, at which the experiments are generally capable of achieving an event efficiency of nearly 100% due to the low interaction rate and the uncluttered events with only one interaction per event readout. This is compared to events at the LHC with tens of events (mostly low-p_T) per readout and an overall event efficiency of 0.1%. One merely needs to compare an LHC event to an LEP event to see this difference. One might be tempted to say, "so, we do not need a good detector, a mediocre one with do." This is true, but the CEPC can open a window to exceptionally excellent detectors with 2% or better energy–momentum resolution on all particles of the standard model, excellent particle identification such that mis-assignments are less than 10%, and timing resolutions at the sub-nanosecond level on all the particles of the event. It is difficult to imagine physics that cannot be done excellently well and unambiguously with such a detector. A circular collider such as the CEPC might be the only hope for the next 20–30 years.

3. Main Subsystems of a Big Detector

The important sub-detectors of a major experiment must measure impact parameters near the interaction point, the momenta of stable charged particles, and the energies of both charged and neutral stable particles. The goals below follow the well-studied criteria for these detectors within the ILC community.

3.1. Impact parameter (b) measurement: Goal is $\sigma_b \approx 10~\mu m$

Three standard model particles (τ, c, b) have lifetimes and therefore mean decay lengths $(\gamma\beta c\tau)$ in the detector that allow direct event-by-event measurement of their laboratory decay flight paths and, thereby, a lifetime tag for identification of the τ lepton and the D and B mesons.

The mean decay lengths for $\tau^\pm$, D^0 and B^0 of momentum 10 GeV/c inside the detector are all around one-half millimeter, Fig. 2, an easily measured flight distance with a vertex chamber impact resolution of 10 μm. In fact, due to the Lorentz boost of these particles, the impact parameter for each is approximately $c\tau$ itself.

The two main parameters of a vertex chamber are its inner radius and the spatial resolution at the inner radius. A pixel size that is 1/30 of that in CMS and ATLAS is designed to achieve an impact parameter resolution of

$$\sigma_{ip} \approx 5~\mu\text{m} \oplus 10~\mu\text{m}/p \cdot \sin^{3/2}\theta$$

for a track of momentum p at polar angle θ with respect to the beam. The main number (5 μm) is determined very geometrically: it is the spatial resolution of the first layer, times the lever arm from large radius down to the vertex.

Dramatic advances[1] in silicon detector technologies now allow two-dimensional measurements with small (25 μm × 25 μm) pixels.

Particle, mass (GeV/c^2)	$c\tau$ (mm)	$\gamma\beta c\tau$ (mm) at 10 GeV/c
$\tau^\pm$, 1.78	0.087	0.49
$c \to D^0$, 1.87	0.123	0.66
$b \to B^0$, 5.28	0.457	0.87

Fig. 2.   The mean decay lengths of 10 GeV/c $\tau^\pm$, D^0 and B^0.

Collider	a (μm)	b (μm GeV/c)
LEP	25	70
SLC	8	33
Belle II	15	15
LHC	12	70
RHIC-II	13	19
ILC	< 5	< 10

Fig. 3.   Impact parameter resolution constants a, b for several recent detectors.

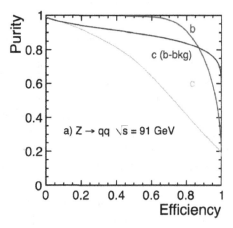

Fig. 4. c and b tagging in the ILD vertex chamber design, ILD DBD.

The primary measurement[b] of a vertex chamber is the impact parameter of charged tracks, and the resolution can be written as

$$\sigma_{ip} = a \oplus b/p \cdot \sin^{3/2}\theta \,,$$

and several achieved values of a and b are shown in Fig. 3. A simulation of c and b quark tagging efficiency and purity in the ILD design is shown in Fig. 4 along with the b contamination of the c sample.

There seems to be an embarrassment of riches in silicon pixel vertex detectors[1] of several technologies with differing features: (i) MAPS: Monolithic Active Pixel Sensors, (ii) DEPFET: DEPleted Field Effect Transistor, (iii) Fully depleted SOI: Silicon On Insulator and (iv) CCD: Charge Coupled Devices. This variety allows detector designers to match the time response, time recovery, efficiency, etc., of a pixel vertex detector to a given detector and a given machine subject to costs and risks.

---

[b]One secondary purpose is to assist the main tracker.

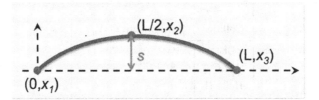

Fig. 5. Momentum measurement: sagitta of a circular track in a magnetic field.

The tagging of D and B mesons containing c and b quarks is critical to Higgs decay measurements and t ensemble purities, in addition to the discovery potential of these heavy quarks as a window to new physics.

3.2. Momentum measurement: Goal is $\sigma_p/p \approx 3 \times 10^{-5} p(\text{GeV}/c)$

The only known means to measure the momentum of a charged particle is by bending the particle in a magnetic field and measuring the radius of curvature, as illustrated in Fig. 5, by measuring the sagitta s.[c] The best discussion of all aspects of momentum measurement is by Gluckstern.[26] There are two popular paths to this resolution: a large radius time projection chamber (TPC) (plus a silicon strip blanket) as in the ILD design for the ILC, and a 5-layer silicon strip system (plus vertex chamber) as in the SiD design for the ILC, with both immersed in a 3–5 Tesla magnetic field.

The essentials of track momentum resolution are seen in this simple example, Fig. 5. Three equally spaced points making a track length L (m) and with space point resolution σ_x (m) in a uniform magnetic field B (T) with track momentum p (GeV/c) perpendicular to $\mathbf{B}$, the sagitta is $s = x_2 - (x_1 + x_3)/2$ and is related to the radius of curvature r by $s = L^2/8r$. The sagitta error is $\sigma_s \approx \sqrt{3/2}\sigma_x$. Since the radius of curvature is proportional to the momentum, $p = 0.3Br$, the momentum resolution is

$$\frac{\sigma_p}{p} = \frac{\sigma_s}{s} ,$$

or

$$\frac{\sigma_p}{p} = 8\sqrt{3/2}\left(\frac{\sigma_x p}{0.3BL^2}\right) .$$

This degradation with increasing momentum is in contrast with calorimeters that improve with increasing energy as $1/\sqrt{E}$ (in the absence of constant terms).

One concern with a tracking system is the multiple scattering material which degrades the momentum resolution at low momenta and presents a converter-radiator

---

[c]One could measure the intensity of transition x-ray radiation as a measure of $\gamma = E/m$, but the forward-going x-rays are too weak and obscured by the passing charged track. This technique was given up shortly after it was attempted. A recent interesting talk on using meta-materials and Čerenkov light might provide for a momentum measurement.[25]

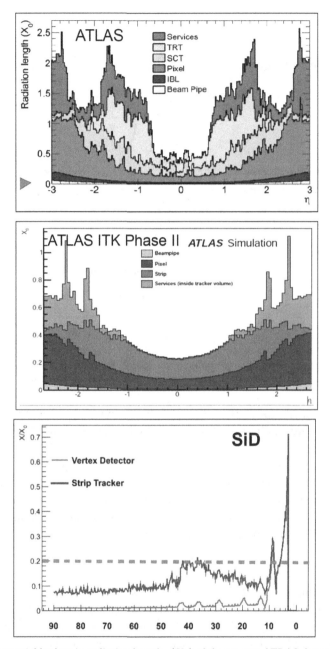

Fig. 6.   The material budget in radiation lengths (X_0) of the present ATLAS detector, its planned upgrade and the designed SiD detector for the ILC.

for both high and low energy γ's. In Fig. 6, the material budget in the current ATLAS detector (top), the planned material in a future ATLAS upgrade (middle) and the material in the SiD design for the future ILC (bottom) are compared.

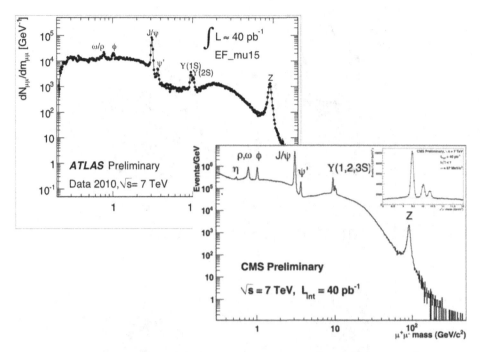

Fig. 7.   The invariant masses of $\mu^+\mu^-$ pairs in the ATLAS ($k = 3 \times 10^{-4}$ $(\text{GeV}/c)^{-1}$) and CMS ($k = 1 \times 10^{-4}$ $(\text{GeV}/c)^{-1}$) silicon-strip tracking detectors. Plot from Jianming Qian.

The main contributor is the mass of cables leaving the detector around $\eta \approx$ 1.5 ($\theta \approx 40°$) where the barrel meets the end cap. Someone should find a solution to this problem.

For N equally spaced points, the constant in front is $\sqrt{720/(N+4)}$ in place of $8\sqrt{3/2}$. The important factors are that the momentum resolution scales proportional to σ_x, and inversely proportional to BL^2. Hence, the "easy way" to achieve good momentum resolution is to build a large radius tracking system. Overall, the momentum resolution degrades linearly in the momentum, p.

$$\frac{\sigma_p}{p} = k \cdot p. \tag{1}$$

The most important Higgs final state is shown in Fig. 8 (left) as the "Higgsstrahlung" diagram in which a Z^0 radiates a Higgs. A detector event is shown in Fig. 8 (middle), and the ensemble of recoil masses from this process is shown in Fig. 8 (right). This event depends entirely on the quality of the tracking system for the precision measurement of the $\mu^+\mu^-$ pair and their (presumably) constrained fit to the central Z^0 mass.[d]

---

[d]If the $\mu^+\mu^-$ mass is better measured than the natural width of the Z^0, then you do not do this fit.

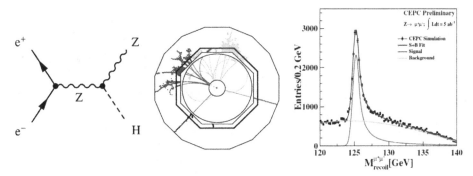

Fig. 8. Primary benchmark process for any detector at a Higgs factory through the decay $Z^0 \rightarrow \mu^+\mu^-$.

Momentum resolution is obvious in the $\mu^+\mu^-$ mass plot measured in the CMS tracking system and the same measured in the ATLAS tracking system, both shown in Fig. 7. For CMS, the resolution constant $k = 1 \times 10^{-4}$ $(\text{GeV}/c)^{-1}$, whereas for ATLAS $k = 3 \times 10^{-4}$ $(\text{GeV}/c)^{-1}$, that is, k is a factor of three smaller for CMS than for ATLAS, and this difference is evident in the sharpness and clarity of the particles revealed in these plots.

This is a missing mass experiment in which the known initial state and the measured Z^0 state are used to calculate the mass of the missing recoil particle,

$$M^2_{\text{recoil}} = (\sqrt{s} - E_Z)^2 - |\mathbf{p}|^2 \, .$$

This procedure scans the entire mass spectrum coupling to the $Z^0 Z^0$ vertex and produces all the final states of the Higgs. Of course, any decay mode of the Z^0 can and should be used in this study, including $Z^0 \rightarrow e^+e^-$ and $Z^0 \rightarrow q\bar{q}$ which depend on the tracking and EM calorimeter, and on the hadronic calorimeter, respectively.

3.3. *Choice of silicon-strip chamber or time projection chamber*

The choice between a silicon-strip tracker and a TPC tracker[?] is not simple, shown in Fig. 9. The cylindrical shells of silicon strips on wafers with a 10–20 μm single point spatial resolution and effective 4π coverage looks impressive, but the excellent point resolution is only effective if the calibration has comparable or better spatial fidelity. The large volume TPC can be easily calibrated with its ten times worse point resolution compensated by 100 times more spatial point measurements. An interesting but misleading comparison of the silicon and TPC measurements is shown in Fig. 10 of the same simulated event. The TPC produces essentially continuous tracks and excellent pattern recognition, whereas not a single track is evident in the silicon system. Yet, overall, both trackers achieve (in simulations) a momentum resolution constant of $k \approx 3 \times 10^{-5}$ $(\text{GeV}/c)^{-1}$ and nearly 100% event reconstruction efficiencies, yielding a Higgs missing-mass measurement like that in Fig. 8. Therefore, both tracking options are acceptable subject to rates and backgrounds.

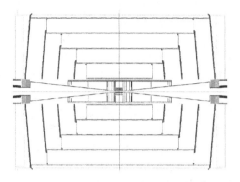 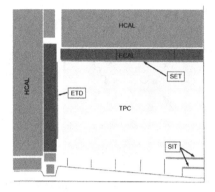

Fig. 9.   (Color online) A 5-layer silicon-strip tracker from the SiD concept at the ILC (left), and the (featureless, yellow colored) gas volume of a Time Projection Chamber (TPC) (right).

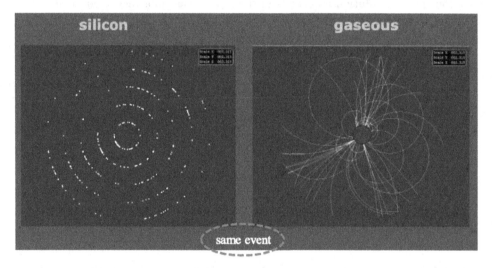

Fig. 10.   Images of the same simulated event as it would be recorded by a silicon-strip system (left) and a TPC (right).

In summary, the strengths of a TPC are (i) 100–200 spatial points leading to nearly continuous tracks and excellent pattern recognition at nearly 100% efficiency, (ii) easy to calibrate with most available tracks in events and by UV laser lines, (iii) particle identification below 10 GeV by specific ionization, (iv) relatively inexpensive per unit volume, and (v) very high efficiency even for large track densities in jets. The weaknesses of a TPC are (a) slow electron drift leading to long drift times and overlapping events at high luminosity (e.g. at the Z^0), (b) positive ion production both by primary track ionization and electron multiplication on the end planes, after which the ions are drawn back into the tracking volume producing a radial component to the axial drift electric field, which will distort the drift trajectories and (c) the end planes with wires, micromegas, or other structures, introduces a lot of material before the calorimeter, and the inner radius E-field defining cage introduces multiple scattering material for the charged tracks. To the extent that this multiple scattering is large, the coupling between the vertex chamber and the TPC is lost and the effective L^2 gain in momentum resolution is degraded.

The strengths of a silicon-strip tracker are (i) excellent spatial point resolution, (ii) uniformity of silicon technology over all layers, (iii) fast response and readout of tracks hits on a time scale shorter than the beam crossing, and recovery for further hits on a comparable scale. The weaknesses of silicon strips are (a) only five point measurements will lose some $K^0 \to \pi^+\pi^-$ decays if the K^0 decay point is beyond the third layer; however, its proponents have shown that the tracking efficiency is nearly 100% with algorithms that form every combination of points as tracks, (b) the current drawn in the silicon yields Joule heating which can be cooled by a liquid resulting in too much multiple scattering, or by "power pulsing" the currents, i.e. turning the currents off during the 0.2 s period between ILC bunch trains, (c) calibration might be very difficult with too few tracks available to fix the five (or more) positioning parameters per wafer, (d) occupancy in the inner layer and (e) conversion of both high and low energy γ's in the high-Z silicon. All of these issues can and are being addressed in beam tests within the ILC groups.

3.4. *EM energy measurement: Goal is $\sigma/E < 10\%/\sqrt{E}$*

An electromagnetic calorimeter is both easy to build and its resolution is easy to predict, as shown in Fig. 11 (right frame) for many diverse technologies, sampling fractions f_{sample}, and sampling frequencies d. A simple goal is $(\sigma/E)\sqrt{E} \approx 0.10$, although better or worse is often acceptable. Crystals are nonsampling media and have intrinsically better energy resolution.[29] The simple reason is that in

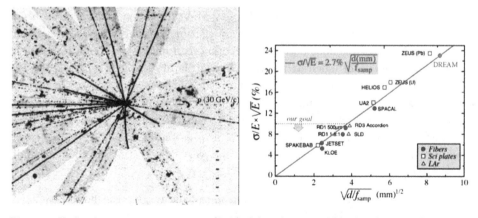

Fig. 11. Hadronic energy measurement: (Left) *debris* from a nuclear break-up, as happens in a calorimeter due to hadronic particle interactions; (right) electromagnetic energy resolution is easily described by a simple formula.

an electromagnetic shower, the electrons and positrons mainly emit photons by bremsstrahlung, and the photons mainly pair-produce electron–positron pairs, and both of these processes occur in roughly a radiation length of material. Therefore, the development of electromagnetic showers is very uniform.

3.5. *Hadron energy measurement: Goal is $\sigma/E = 30\%/\sqrt{E}$*

This goal is about a factor of 2–4 times better than the resolutions in ATLAS and CMS at the LHC, and is achievable with dual-readout calorimeters[29] and possibly with particle-flow calorimeters.[16] Figure 11 displays the break-up of a nucleus like those in a hadronic calorimeter, where it is necessary to measure the energy of the incoming particle (a 30 GeV proton in this case) by summing up the energies of all the particles that leave the collision, including the ones that are invisible in this emulsion photograph.

Hadronic particle energy measurements are much more difficult[27] and the reason why is illustrated in Fig. 11. Each hadronic interaction is different and widely varying in the numbers and types of particles produced, for example:

- energetic charged pions, $\pi^{\pm}$: the products of ordinary hadronic scattering, relatively few in number and produce relatively small calorimeter signals;
- energetic target nucleons, p, n: usually only one or two, producing only small signals;
- energetic neutral pions, π^0: decay to two photons, $\pi^0 \to \gamma\gamma$, essentially immediately on the spatial scale of a detector, yielding two electromagnetic showers which produce large calorimeter signals;
- spallation protons, p: low energy protons up to 100 MeV from the broken nucleus carrying away the electrostatic stored energy of the nuclear protons; generally stop (range out) within the absorber material or give large signals in the active elements of the sampling calorimeter;
- Fermi-energy neutrons, n: neutrons of a few MeV kinetic energy liberated from the broken-up nuclei, with ranges of tens of centimeters, and which give a small signal in a nonhydrogenous active material and can give a very large signal in a hydrogenous medium (such as an organic scintillator) through repeated elastic $np \to np$ scatters and the subsequent ionization of the recoil protons;
- nuclear excitation and gamma de-excitation, γ's: an excited nucleus can decay in its final stages through the emission of γ's in the MeV range;
- muons from pion and kaon decay, $\pi^{\pm} \to \mu^{\pm}, \ldots$ and $K^{\pm} \to \mu^{\pm}$: muons from hadron decay are generally low energy and very penetrating, leaving a small calorimeter signal.

In summary, the $e^{\pm}$ from the γ showers from π^0 decay give the largest signals, followed by np scatters in an organic scintillator.

The main difficulty for good energy resolution is that fluctuations in the numbers and the composition of these diverse contributions to the total calorimeter signal

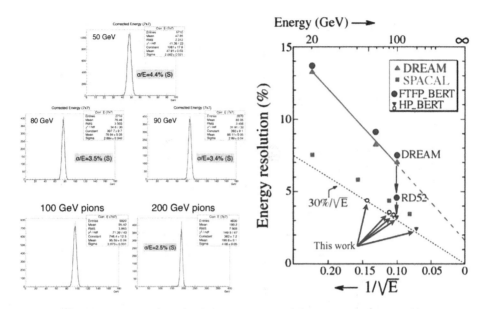

Fig. 12. The energy resolutions of fiber dual-readout calorimeters as simulated with the high-precision version of GEANT4 (FTFP_BERT_HP) at π^- beam energies of 50, 80, 90, 100 and 200 GeV. The DREAM and RD52 calorimeters are copper-fiber and SPACAL is lead-fiber. The dotted line on this plot of resolution versus $1/\sqrt{E}$ is a calorimeter with an energy resolution of $30\%/\sqrt{E}$.

are large and, to make matters worse, non-Poisson. Successful solutions to this problem have been:

- compensation, $e/h = 1$: arrange the absorbing material and the active material of a calorimeter to be, on average, such that the electromagnetic response to $e^\pm$, called "e", be equal to the nonelectromagnetic response to $p, n, \pi^\pm, \ldots$, etc. called "h", or that $e/h = 1$. This works very well for Pb-absorber of thickness t combined with plastic scintillator of thickness $t/4$;
- preferential weighting: weight the electromagnetic-like signals relative to the non-electromagnetic signals (by local energy loss density, for example) to achieve an approximate compensation;
- energy flow or particle flow: use the momentum and direction of measured charged tracks to assist in the assignment of energies observed in the calorimeter. The ultimate result of this is essentially the particle-flow analyses of the CALICE calorimeters;
- dual-readout: measure more-or-less directly the electromagnetic ($\pi^0 \to \gamma\gamma$) content of a hadronic shower relative to all the rest and, in addition, measure the $np \to np$ scatters in the scintillating fibers as a means of measuring the fluctuations of the invisible binding energy losses. The simulated energy resolution of a copper-fiber dual-readout calorimeter (RD52) is shown on the left panel of

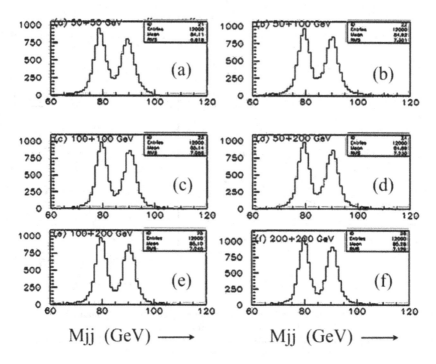

Fig. 13.   The di-jet invariant mass distribution reconstructed from leakage suppressed DREAM data events. There were three beam energies, 50, 100 and 200 GeV, and therefore six W, Z energies of 100, 150, 200, 250, 300 and 400 GeV.

Fig. 12 at π^- beam energies of 50, 80, 90, 100, and 200 GeV. These five points are plotted on the right-side panel of Fig. 12 and fall close to the dotted line for $\sigma/E = 30\%/\sqrt{E}$.

An essential measurement at any future collider, e^+e^- or pp, is the reconstruction of the hadronic decays of the weak bosons, $W \to q\bar{q}$ and $Z \to q\bar{q}$. Using DREAM data with leakage suppressed[28] events from beam data files at 50, 100 and 200 GeV, these decays were simulated by combining two beam data events per each $W \to q\bar{q}$ and $Z \to q\bar{q}$, using fully the fluctuations in the S and C signals and including the spatial reconstruction fluctuations by taking the (x, y) centroids of the showers in the DREAM module. The results are shown in Fig. 13 from which it is seen that the mass resolution is good enough to distinguish W from Z at the 95% level, and that this resolution is nearly independent of the W, Z energies.

The energy resolution of a digital CALICE module, shown in Fig. 14 (top), for different thresholds on the readout, and event images from this 0.5 M channel module, are shown in Fig. 14 (bottom). The energy resolution of about 7% at 100 GeV is not as good as that obtained in dual-readout, but it does not have to be as good considering the spatial imaging capabilities which can enhance certain measurements.

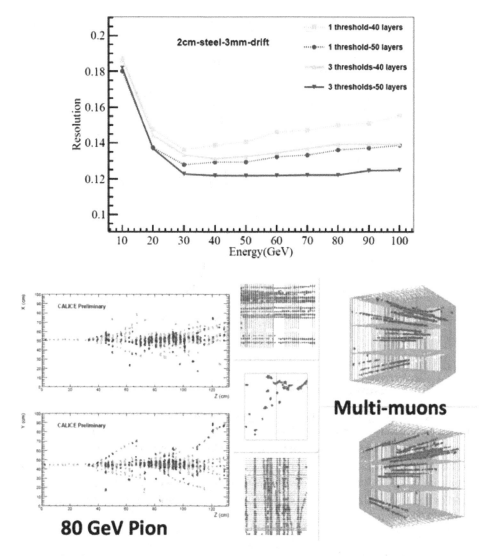

Fig. 14. (Top) The energy resolution of a CALICE hadronic calorimeter for different digital thresholds; (bottom) event images for this 0.5 M channel module.

The characterization of a calorimeter is not as simple as the characterization of a tracking chamber which might require, for example, only three points as in Fig. 5. In a calorimeter test, the full ensemble of beam particles must be used and the full volume of the calorimeter must be used. Restricting either one of these requirements leads to an ensemble bias and, therefore, a nontest of the calorimeter. Only a full and complete pulse height distribution, from which the mean μ and Gaussian width σ are extracted, can be considered an actual test of a calorimeter.

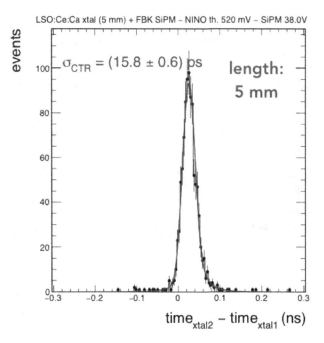

Fig. 15.   Timing resolution from two small crystals.

3.6. *Time measurement: Goal is $\sigma_t \approx 1$ ps*

The time domain is now achieving resolutions near 10 psec (see Fig. 15), or 3 mm spatial resolution, we think largely as a result of H. Frisch's pursuit of the large area picosecond photo-converters.[30] In the RD52 calorimeters,[29] the time domain is used to slice the z-position (shower depth coordinate) within the calorimeter volume into approximately 4 cm spatial slices, subject to the bandwidth of the photo-converter and the digitizer. This has been shown to be very valuable in $e^- - \pi^-$ discrimination and identification.[29]

4. The Rest of the Detector

A big detector is more than vertex, tracking and calorimetry. The tagging of muons outside the calorimeter is usually accomplished with massive (10 kton) iron plates (which also serve to partially return the magnetic flux). There must be a data acquisition system (DAQ) and a trigger. However, the heart of any big detector are the three detector systems considered here.

5. Matters We Did Not Discuss, But Should Have

A detector at a frontier collider should be prepared for the unknown, e.g. particles not on the list of SM particles in Fig. 1 that might have highly unusual lifetimes or interaction characteristics. Let us repeat:

There has always been a fundamental philosophical problem in physics. We often look for what is expected, for example, for SUSY particles with prescribed decays. In detectors, we build triggers that will only find these particles. Obviously, we will not see the unexpected unless we are lucky. Recall C. Quigg: *"How are we prisoners of conventional thinking?"* The design of the detector with its capabilities and limitations is the critical decision.

What might be some examples? (a) super-strongly interacting particles that do not survive the residual gas in the beam pipe or the beam pipe material itself, (b) non-Dirac magnetic monopoles, (c) long-lived slightly weakly interacting neutral particles or (d) a decay chain of neutral weakly interacting particles which result in charged particles appearing in the detector a millisecond after the primary interaction. These exotics would require re-designed detectors, at least over a portion of their solid angle, and DAQ readout to include long-time monitoring and, at least in the innermost detectors, ultra-low-mass materials.

Acknowledgments

We all thank the hospitality at HKUST brought to us by Henry Tye, Tao Liu and Prudence Wong for a most fruitful and interesting meeting.

Appendix A. Speakers in the Experimental Sessions

Speaker	Institution	Speaker	Institution
Albert De Roeck	CERN	Cia-ming Kuo	Nat'l Central U.
Xuai Zhuan	IHEP, Beijing	Yuanning Gao	Tsinghua U.
Marcel Stanitzki	DESY	Sehwook Lee	Kyungpook Nat'l U.
Huirong Qi	IHEP, Beijing	Massimo Caccia	INFN, U. dell'Insubria
Hongbo Zhu	IHEP, Beijing	Xiangmin Sun	Central China Normal U.
Richard Talman	Cornell U.	John Hauptman	Iowa State U.
Charles Young	SLAC	Silvia Franchino	U. Heidelberg
Chris Tully	Princeton U.	Haijun Yang	Shanghai, Jiaotong U.
Guido Tonelli	U. Pisa	Michele Cascella	University College London
Manqi Ruan	IHEP, Beijing	Jianming Qian	U. Michigan
Xin Chen	Tsinghua U.	Shin-shan Yu	National Central U.
Aurelio Juste	IFAE		

References

1. M. Caccia, Pixel detectors for an experiment at the ILC, http://iasprogram.ust.hk/hep/2016/conf.html.
2. X. Sun, CEPC Vertex/Si tracker, http://iasprogram.ust.hk/hep/2016/conf.html.
3. M. Stanitzki, Future trends in silicon trackers, http://iasprogram.ust.hk/hep/2016/conf.html.

4. J. Qian, Detector requirements for precision Higgs boson physics, http://iasprogram. ust.hk/hep/2016/conf.html.
5. H. Qi, Status of TPC hybrid detector module for the circular collider, http:// iasprogram.ust.hk/hep/2016/conf.html.
6. H. Qi, TPC vs. silicon tracking, http://iasprogram.ust.hk/hep/2016/program.html.
7. C. Young, 750 GeV di-photon resonance: Experimental overview, http://iasprogram. ust.hk/hep/2016/conf.html.
8. C. Young, The importance of tracking: TPC vs. silicon, http://iasprogram.ust.hk/ hep/2016/program.html.
9. A. de Roeck, Status of the studies for a FCC-hh detector, http://iasprogram.ust.hk/ hep/2016/conf.html.
10. A. de Roeck, Detector challenges at future colliders, http://iasprogram.ust.hk/hep/ 2016/program.html.
11. H. Yang, Status report about the CEPC calorimeters, http://iasprogram.ust.hk/hep/ 2016/conf.html.
12. S. Franchino, Crystal and fiber dual-readout calorimeters: Building and understanding them, http://iasprogram.ust.hk/hep/2016/program.html.
13. S. Lee, Everything you always wanted to know about the original DREAM module, http://iasprogram.ust.hk/hep/2016/program.html.
14. S. Lee, Energy resolution and particle identification of the dual-readout calorimeter, http://iasprogram.ust.hk/hep/2016/conf.html.
15. M. Cascella, Time structure in dual readout calorimeter, http://iasprogram.ust.hk/ hep/2016/conf.html.
16. A. de Roeck, Calorimetry, http://iasprogram.ust.hk/hep/2016/program.html.
17. Y. Gao, CEPC detector, http://iasprogram.ust.hk/hep/2016/conf.html.
18. M. Ruan, Higgs measurements at CEPC, http://iasprogram.ust.hk/hep/2016/conf. html.
19. G. Tonelli, Higgs physics at LHC and beyond, http://iasprogram.ust.hk/hep/2016/ conf.html.
20. H. Zhu, Machine detector interface at electron colliders, http://iasprogram.ust.hk/ hep/2016/conf.html.
21. M. Mangano, Precision Higgs physics at 100 TeV, http://iasprogram.ust.hk/hep/ 2016/conf.html.
22. C. Quigg, Scientific overview, http://iasprogram.ust.hk/hep/2016/conf.html.
23. C. Tully, A new experimental perspective on long-lived particles and beam backgrounds, http://iasprogram.ust.hk/hep/2016/conf.html.
24. J. Ellis, talk at *International Symposium on Lepton Photon Interactions at High Energies*, 17–22 August 2015, University of Ljubljana, Slovenia.
25. I. Kaminer, Photonic crystals, graphene, and new effects in Čerenkov radiation, CERN EP Seminar, 15 April 2016.
26. R. L. Gluckstern, *Nucl. Instrum. Methods* **24**, 381 (1963).
27. R. Wigmans, *Calorimetry: Energy Measurement in Particle Physics* (Oxford University Press, 2001).
28. J. Hauptman, The importance of high-precision hadronic calorimetry to physics, http://iasprogram.ust.hk/hep/2016/program.html.
29. M. Cascella, S. Franchino and S. Lee, Fiber and crystal dual-readout calorimeters, http://iasprogram.ust.hk/hep/2016/program.html.
30. B. W. Adams, *et al.*, A brief technical history of the Large-Area Picosecond Photodetector (LAPPD) collaboration, arXiv:1603.01843.

Conceptual Design Studies for a CEPC Detector

S. V. Chekanov[*] and M. Demarteau[†]

HEP Division, Argonne National Laboratory,
9700 S. Cass Avenue, Argonne, IL 60439, USA
[*] chekanov@anl.gov
[†] demarteau@anl.gov

The physics potential of the Circular Electron Positron Collider (CEPC) can be significantly strengthened by two detectors with complementary designs. A promising detector approach based on the Silicon Detector (SiD) designed for the International Linear Collider (ILC) is presented. Several simplifications of this detector for the lower energies expected at the CEPC are proposed. A number of cost optimizations of this detector are illustrated using full detector simulations. We show that the proposed changes will enable one to reach the physics goals at the CEPC.

Keywords: e^+e^-; jets; Monte Carlo; CEPC.

1. Introduction

The Circular Electron Positron Collider (CEPC) project is currently planned in China as a Higgs factory. Operating at the center-of-mass (CM) energy of 250 GeV (or above), the CEPC experiment will take advantage of the clean environment of e^+e^- collisions needed for high-precision measurements of the Higgs boson. CEPC experiments can significantly strengthen our understanding of the fundamental processes at the electroweak sector of the Standard Model (SM), and can lead to discoveries of new physics through the precision measurements of the SM.

In order to achieve the physics goals at the CEPC, a detector should be well optimized for measurements of e^+e^- annihilation. In particular, the studies of physics processes in the Higgs sector are considered to be the primary goal of the new experiment. A promising approach for a detector at the CEPC can be based on the Silicon Detector (SiD) concept[1] developed for the International Linear Collider (ILC).[2,3] The design of this detector has a long history, and the experience gained during the R&D phase of this detector can be extremely valuable during the preparation to the CEPC concept.

The abbreviation "SiD" stands for "silicon detector" — a compact general-multipurpose detector designed for high-precision measurements of e^+e^- annihilation at a CM energy of 500 GeV, which can be extended to 1 TeV. Radiation hardness of silicon sensors used for vertexing, tracking and Electromagnetic Calorimeter (ECAL) ensures the required robustness to beam background or beam loss, while a high-granular calorimeter is well suited for the reconstruction of separate particles.[1] Some key characteristics of the SiD detector are:

(1) 4π solid angle coverage for reconstructed particles.
(2) Full five-layer silicon tracking system with 50 μm readout pitch size.
(3) Silicon pixel detector with 20 μm readout pitch size. The detector has excellent flavor-tagging capabilities through a measurement of displaced vertices.
(4) Superconducting solenoid with a 5 Tesla (T) field.
(5) Highly segmented silicon-tungsten ECAL with the transverse cell size of 0.35 cm.
(6) Highly segmented Hadronic Calorimeter (HCAL) with a transverse cell size of 1×1 cm. The depth of the HCAL in the barrel region is about 4.5 interaction length[a] (λ_I). The calorimeter has 40 longitudinal layers in the barrel and 45 layers in the endcap region; Resistive-Plate Chambers (RPCs) are the baseline for the active media, although other options, such as Gas-Electron Multipliers (GEM), micromegas, scintillator/SiPM are being considered.
(7) High-resolution muon detector based on scintillator strips with SiPM readout. Bakelite RPC is an alternative option for the muon-detector readout.

Both ECAL and HCAL calorimeters are finely segmented longitudinally and transversely. This is required for "imaging" capabilities of the calorimeter system: Together with the efficient tracking, the fine segmentation of the calorimeter system optimizes the SiD detector for particle-flow algorithms (PFAs) which allow identification and reconstruction of separate particles. The PFA objects can be reconstructed using the Pandora PFA.[4,5]

The response of the SiD detector to physics processes is simulated using the "Simulator for the Linear Collider (SLIC) software package"[6] developed for the ILC project. The main strength of this software lies in the fact that it can easily be configured using XML option files, and it has a platform-independent reconstruction which can be easily deployed on computers with different operating systems.

The M&S cost of the baseline design of the SiD detector is estimated to be around \$320M,[1] with 32% being allocated for the calorimeter, and 37% for the magnet (estimated in 2009).

---

[a]Nuclear interaction length, (λ_I), is the average distance traveled by a hadronic particle before undergoing an inelastic nuclear interaction.

2. SiD for the CEPC

For the CEPC physics goals, the SiD detector is over-designed. For example, the cost can be substantially reduced by simplifying the calorimeters and by reducing the magnetic field of the solenoid. Due to the lower CM energy of 250 GeV at the CEPC, a number of optimizations of the SiD detector are proposed:

(1) 5 T solenoid field can be reduced to 4 T;
(2) 40 layers of HCAL can be reduced to 35 by removing five outer HCAL layers in the SiD design. The remaining 35 layers of the steal absorber correspond to about 4.1 nuclear interaction length;
(3) 45 layers of the HCAL endcap can be reduced to 35 layers. This makes the CEPC detector fully uniform from the point of view of the HCAL depth.

The reason for the reduction of the solenoid field lies in the fact that the typical track momentum measured at CEPC is a factor of two (four) smaller compared to the 500 (1000) GeV e^+e^- collisions at the ILC. The magnetic field could be further reduced, but this will require a more detailed study than shown in this paper. Similarly, the reduction of the calorimeter depth is motivated by the fact that the maximum jet transverse momentum at the CEPC is 125 GeV, which is a factor two (four) smaller than for the 500 (1000) GeV e^+e^- machine. In terms of the HCAL interaction length, the proposed $4\lambda_I$ calorimeter is similar to that of the OPAL experiment.[7] The total absorber (steal and tungsten) of the ECAL and HCAL calorimeter systems corresponds to about $5.1\lambda_I$.

In order to explore the possibility of optimization of the SiD detector to a lower CM energy, we use the HepSim Monte Carlo repository[8] with several benchmark processes for e^+e^- collisions. The e^+e^- events at the 250 GeV CM energy were generated using the PYTHIA6 model.[9] The following processes were generated:

- Fully inclusive QCD dijet process;
- Z-boson production with the decays $Z \to e^+e^-$, $Z \to \mu^+\mu^-$, $Z \to \tau^+\tau^-$, $Z \to b\bar{b}$;
- Higgs production (Z^0H) with the decays $H \to 4l$, $H \to \gamma\gamma$, $H \to \tau^+\tau^-$, $H \to b\bar{b}$. The Higgs mass was set to 125 GeV.

The events were simulated using the SiD detector geometry, and reconstructed using the SLIC package with Pandora PFA. The simulation and reconstruction steps were performed using the Open-Science Grid.[10] Events before and after the simulation of the detector response were registered in the HepSim data catalogue.

In the following, the original SiD detector geometry will be called SiDloi3. The number of reconstructed events after the SiDloi3 detector simulation and reconstruction was about 10,000. Most representative observables which are expected to be sensitive to the tracking and calorimeter performance of the SiD detector were analyzed. The obtained results (not shown) were found to be within the specification of the SiD detector.[3]

The same data samples were simulated and reconstructed using the CEPC-optimized geometry discussed in the beginning of this section, i.e. with the solenoid field changed from 5 T to 4 T, and the HCAL calorimeter depth reduced from 40 (45) to 35 layers. In the following, the SiD geometry after such modifications called SiDcc1. Full details of this detector geometry are available from the HepSim repository. To reduce computation time, the number of simulated and reconstructed events for the SiDcc1 detector were a factor two smaller than for the SiDloi3 simulation.

The distributions of several observables which are particularly sensitive to the change in the strength of the solenoid field and the HCAL absorber depth is shown in Figs. 1 and 2. The distributions were reconstructed from the PFA objects

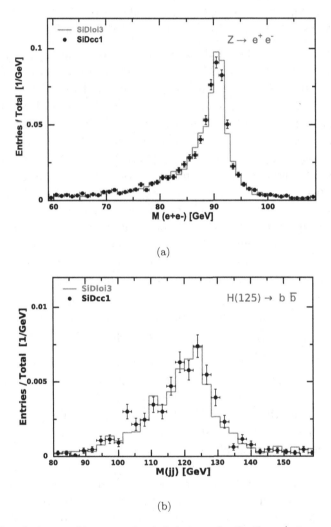

(a)

(b)

Fig. 1. The invariant mass of two reconstructed electrons for the $Z \to e^+e^-$ process (a) and the invariant mass of two jets for the process $H(125) \to b\bar{b}$ (b). The distributions were reconstructed from the PFA objects. The figure shows the original SiD setup (SiDloi3) and a CEPC optimized version of the SiD detector (SiDcc1). The distributions of the latter setup are shown as solid dots with statistical uncertainties.

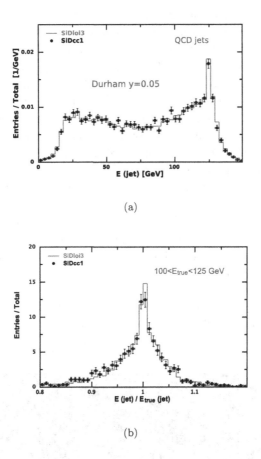

Fig. 2. The distribution of QCD jets in e^+e^- collisions using the Durham algorithm with $y_{cut} = 0.05$ (a), and the jet energy response for jets with energy close to the kinematic peak of 125 GeV. The jets were reconstructed from the PFA objects. The figure shows the original SiD setup (SiDloi3) and a CEPC optimized version of the SiD detector (SiDcc1). The distributions of the latter detector setup are shown as solid dots with statistical uncertainties.

which combine the information from four-momenta of tracks and calorimeter energy deposits. For example, the Z boson mass reconstructed from the invariant mass of two electrons (Fig. 1(a)) is sensitive to the performance of tracking system to high-momentum tracks (e^+/e^-). The energy distribution of hadronic jets reconstructed from the PFA objects is sensitive to both the performance of the tracking system, and to the HCAL longitudinal segmentation. Figure 1(b) shows the invariant mass of two jets for the process $H(125) \to b\bar{b}$. The jets were reconstructed with the Jade algorithm[11] by forcing two jets per event, and requiring the transverse momentum of jets to be above 20 GeV.

To take a closer look at the hadronic jets, Fig. 2 shows the distribution of the jet transverse momentum for inclusive QCD processes in e^+e^- at 250 GeV.

The jets were reconstructed using the Durham algorithm[12] with the parameter $y_{\text{cut}} = 0.05$. Unlike the Jade algorithm which uses the invariant mass-squared of the pair of particles for the clustering procedure, the Durham algorithm uses the relative transverse momentum-squared. Such a choice for two-particle distance measure better reflects the nature of jet evolution. As before, the input to this algorithm are the PFA objects. Jets were selected with a minimum transverse momentum of 20 GeV. Figure 2(b) shows the jet energy response by taking the ratio of the reconstructed jet energy to the energy of jets reconstructed from stable particles, which are defined if their lifetime τ are smaller than $3 \cdot 10^{-10}$ s. Neutrinos were excluded from consideration. As expected, the distributions for this ratio peaks at one, indicating that no energy leakage is observed for both the SiDloi3 and SiDcc1 detectors.

Figure 3 illustrates a typical $Z \rightarrow e^+ e^-$ event in the Jas3/Wired4 event display. A prominent feature of this event is the energy deposits in the ECAL corresponding to the electrons from the Z decay. The space between the outer layer of the HCAL and the solenoid is due to the removal of 5 HCAL layers from the original design of the SiD detector.

In summary, this paper suggests that the SiD detector (or its sub-detectors) can be repurposed for the CEPC. We have illustrated a few directions to optimize the SiD detector for lower CM energies. The results obtained with the SiDloi3

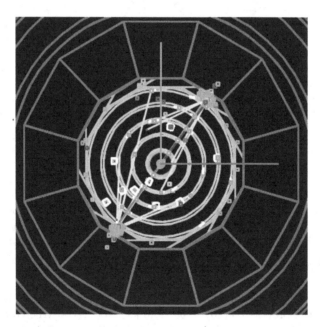

Fig. 3.   (Color online) An event display of an $Z \rightarrow e^+ e^-$ event using the optimized SiDcc1 setup. Clusters of the green points on the surface of the ECAL correspond to the reconstructed e^+/e^- from the Z decay. The space between the outer layer of HCAL (shown with the trapezoid segments) and the inner radius of the cylindrical solenoid is due to removed 5 outer HCAL layers of the original SiD detector.

and SiDcc1 detector concepts show good agreements (within statistical errors), thus the optimized SiDcc1 detector will enable to reach the physics goals at the CEPC. It should be noted that the changes to the SiD concept listed above are just a few possible options to reduce the cost of a detector for the CEPC energy, without compromising the physics goals at the CEPC. It is very likely that a more substantial optimization can be made after dedicated performance studies.

Acknowledgments

This research was done using resources provided by the Open Science Grid, which is supported by the National Science Foundation and the U.S. Department of Energy's Office of Science.

The submitted manuscript has been created by UChicago Argonne, LLC, Operator of Argonne National Laboratory (Argonne). Argonne, a U.S. Department of Energy Office of Science laboratory, is operated under Contract No. DE-AC02-06CH11357. Argonne National Laboratory's work was supported by the U.S. Department of Energy under contract DE-AC02-06CH11357. The U.S. Government retains for itself, and others acting on its behalf, a paid-up nonexclusive, irrevocable worldwide license in said article to reproduce, prepare derivative works, distribute copies to the public, and perform publicly and display publicly, by or on behalf of the Government.

References

1. H. Aihara *et al.*, SiD Letter of Intent, presented to ILC IDAG, arXiv:0911.0006 [physics.ins-det].
2. C. Adolphsen *et al.*, The International Linear Collider Technical Design Report — Vol. 3, II: Accelerator Baseline Design, arXiv:1306.6328 [physics.acc-ph].
3. H. Abramowicz *et al.*, The International Linear Collider Technical Design Report — Vol. 4: Detectors, arXiv:1306.6329 [physics.ins-det].
4. M. J. Charles, PFA performance for SiD, in *Proc. Int. Linear Collider Workshop, LCWS08 and Int. Linear Collider Meeting, ILC08*, Chicago, USA, 16–20 November 2008 (2009), arXiv:0901.4670 [physics.data-an].
5. J. S. Marshall and M. A. Thomson, Pandora particle flow algorithm, in *Proc. Int. Conf. on Calorimetry for the High Energy Frontier (CHEF 2013)* (2013), pp. 305–315, arXiv:1308.4537 [physics.ins-det].
6. N. Graf and J. McCormick, *AIP Conf. Proc.* **867**, 503 (2006).
7. OPAL Collab. (K. Ahmet *et al.*), *Nucl. Instrum. Methods A* **305**, 275 (1991).
8. S. Chekanov, *Adv. High Energy Phys.* **2015**, 136093 (2015), available at http://atlaswww.hep.anl.gov/hepsim/.
9. T. Sjöstrand, S. Mrenna and P. Z. Skands, *J. High Energy Phys.* **0605**, 26 (2006).
10. R. Pordes *et al.*, *J. Phys. Conf. Ser.* **78**, 012057 (2007).
11. JADE Collab. (W. Bartel *et al.*), *Phys. Lett. B* **123**, 460 (1993).
12. Y. L. Dokshitzer, *J. Phys. G* **17**, 1537 (1991), Contribution in Report of the Hard QCD Working Group, in *Proc. Workshop on Jet Studies at LEP and HERA*, Durham, December (1990).

Fast Timing for Collider Detectors

Christopher G. Tully

Jadwin Hall Physics Department, Princeton University,
Princeton, New Jersey 08544, USA
cgtully@princeton.edu

Advancements in fast timing particle detectors have opened up new possibilities to design 4π collider detectors that fully reconstruct and separate event vertices and individual particles in the time domain. The applications of these techniques are considered for the physics at CEPC.

Keywords: Fast; timing; precision; collider; detectors.

1. Introduction

The developments and innovations of 4π collider detectors have been pivotal to decades of discovery in elementary particle physics, most recently with the discovery of the Higgs boson.[1] As the beam energies are pushed to higher and higher energy, the corresponding wavelengths that set the scale of the cross-sections for high-energy interactions drop. At the same time, increasing beam energies for a fixed-mass process produces a steep increase in the parton luminosities or equivalent Weizsäcker–Williams approximations to luminosities in scattering processes. This situation results in the search for new physics in a rising sea of low-energy backgrounds.

In the LEP era of high-energy e^+e^- collisions, the average number of beam collisions per beam crossing was much less than unity. This was true not only for the hard center-of-mass interaction, but also for forward Bhabha scattering, two-photon processes and beam-related backgrounds from beam gas or photon radiation into the detector. In these conditions, an almost total reliance on the beam collision timing was developed for the readout of the 4π colliders where typical integration times for a calorimeter were 5 μs with a beam separation in 4-bunch operation of 22 μs. In 1995, the time structure of the beams was altered to include bunch trains.[2] The time spacing of the bunchlets within the trains was approximately 250 ns, with four bunchlets over a total time interval of 750 ns. This time spread was significant enough to warrant bunchlet tagging techniques to correct calorimeter integration times and time-zero markers for drift detectors.

In the preparations for the HL-LHC era, the number of proton–proton interactions per 25 ns bunch crossing is anticipated to reach a mean value of 200. The time spread of the colliding beams, depending on the machine optics, is expected to be between 50–160 ps. While one is apt to conclude that these collisions are essentially simultaneous, it is no longer a luxury to record purely spatial information in such a dense environment. Of particular importance are the discovery modes for the Higgs boson and the further pursuit of those decay modes to search for new physics processes that involve the Higgs boson. The most challenging of these is the Higgs boson decay to diphotons. Indeed, this decay mode was a driving factor in the original design of the LHC collider detectors and pushed the electromagnetic calorimeter resolutions down to constant terms of 0.5% or below.

The calorimetric measurement of the photon energy is a defining constraint in the design of the LHC detectors. However, the need to identify the correct production vertex to within 1 cm along the colliding beam axis is required to preserve the invariant mass resolution from the reconstruction of the angular separation of the photons. Multivariate algorithms used to select the diphoton primary vertex are based on the sum of the p_T^2 of the tracks associated to the spatially reconstructed vertices and other kinematical variables. These variables encounter steady degradation from the finite probability of pileup vertices to appear harder than the signal vertex and more rapid degradation from merged pileup vertices and, in particular, pairs of vertices whose spatial separation drops below approximately 0.3 mm along the beam direction. At these separations, the track reconstruction methods lose vertex separation power when confined to purely spatial separation. The degradation of the z-coordinate intercept resolution for forward tracks is much more rapid.

Initial studies of 4D vertex reconstruction at the HL-LHC show that the problems of spatial merging of vertices are significantly reduced.[3] Initial simulation results show that for vertex separation performance, the 4D algorithms for 200 pileup events achieve performances comparable to the Run 1 LHC pileup levels of 20–25. An example event with low pileup is shown in Fig. 1 for a simulation of a Higgs diphoton decay in a fast-timing enhanced simulation of the CMS detector to illustrate how photon timing is used in conjunction with the 4D vertex algorithm to resolve the signal vertex.[3] These simulation studies were done to demonstrate that at the HL-LHC the number of compatible primary vertex candidates using 4D vertices and photon timing constraints can be brought down to Run 1 LHC pileup levels before being further subjected to multivariate kinematical separation techniques. To achieve these performance benchmarks, the concept of hermetic fast timing was implemented into the simulation of the CMS 4π collider detector.

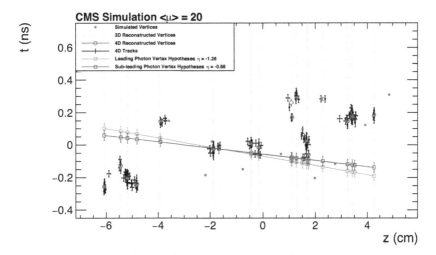

Fig. 1. (Color online) This figure is part of the collection of figures included in the CMS detector performance summary (CMS-DP-2016-008) on fast timing simulation studies.[3] The space–time diagram shows the 4D vertex algorithm for a low-pileup ($\langle\mu\rangle = 20$) event, including the overlay of the spatial coordinates (vertical dotted lines) from the corresponding 3D algorithm. Overlaid on top of the 4D vertices are two sets of photon timing information (in green) to demonstrate how the photon arrival times at the calorimeter are translated through time-of-flight corrections to the list of spatially reconstructed vertices to assess the timing compatibility of the diphotons and the charged track vertex. The triple coincidence, seen at (-2 cm, -0.02 ns), of the two photons and a track vertex in space–time indicates uniquely the signal vertex.

2. Physics at CEPC

The L3 detector at LEP implemented a low-threshold single photon trigger and operated at thresholds down to 1 GeV. The low threshold of the single photon trigger made it possible to directly count the number of Z boson decays into neutrinos through the direct detection of the initial-state radiation (ISR) photon when operating near the Z peak.[4] At the higher center-of-mass energies of LEP2, the single photon spectrum was a uniquely sensitive probe to models with missing energy.[5] The recoil mass distribution from single and multiple photons recorded by the L3 detector at LEP is shown in Fig. 2.[5] These measurements and searches were performed in the absence of primary vertex information.

At CEPC, the collision luminosities will surpass LEP2 by over two orders of magnitude.[6] With this leap in luminosity, further constraints will be needed to separate the spectrum of single photons from low-energy interactions and beam backgrounds from those coming from high-energy interactions with missing energy. Similarly, the possibility for low-energy processes to overlap a hard scatter interaction, including forward Bhabha scattering, becomes nonnegligible. In this environment, one can consider the implementation of hermetic fast timing techniques into the calorimeters and a MIP-sensitive fast timing layer surrounding, or incorporated into, the central tracking system. A purely conceptual design of a fast timing layer with MIP-sensitivity is shown in Fig. 3.

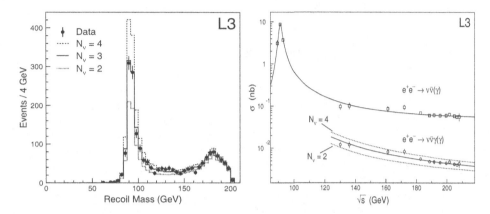

Fig. 2. This figure is part of the collection of figures from the L3 paper on single and multiple photons with missing energy.[5] The distribution of the recoil mass (on the left) computed with single and multiple photons using the beam constraint information shows a clear Z peak at 91 GeV from the decay into neutrinos. Comparisons with standard model couplings provide a clear counting of the $N_\nu = 3$ light neutrino families in the recoil mass spectrum and in the cross-section normalization (on the right). Anomalous structure in the recoil mass spectrum is uniquely sensitive to new physics interactions with a dark sector and will be a powerful probe at CEPC with a 4π collider detector that can efficiently trigger and cleanly separate the single photon signal from out-of-time backgrounds.

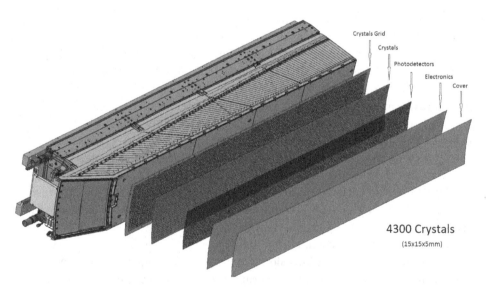

Fig. 3. A conceptual drawing of a MIP-sensitive fasting layer based on the short crystal plus photodetector technologies. The short crystals are placed in the transition region between a total absorption crystal calorimeter, as shown here, and an inner tracking system. Test beam prototypes of this system achieve 10–15 ps timing resolution for individual MIPs at nearly 100% detection efficiency.[7] (Credit: A. Galdames Perez, A. Conde, E. Auffray, EP-CMX CERN).

At the trigger level, the beam crossing will have nominal constraints from the machine that will translate to a timing window of roughly 400–500 ps. Individual particle timing at a level that is an order of magnitude smaller than the timing window of the beam spot provides discrimination power to separate vertices within the same bunch crossing. Fast timing information at this level of precision provides two layers of discrimination. Independent of explicitly reconstructed charged particle 4D vertices, a single photon whose time is recorded by the calorimeter can be tested for compatibility with the timing window of the beam spot, taking into account the time-of-flight path for a photon originating from the corresponding spatial distribution of the beam spot. This level of compatibility provides robust rejection that can be implemented at the trigger level to operate efficiently at low thresholds. At a second level, with a specific list of 4D vertices reconstructed from the tracking information and fast timing layer, the compatibility of a single photon with other photons or charged particles can be tested within the 4D domain of an individual bunch crossing to eliminate overlapping low-energy or forward processes. The hermiticity of the fast timing layer in conjunction with the calorimeter and tracking systems will provide high efficiency for 4D vertex reconstruction and remove blind regions that have no discrimination against out-of-time particles.

3. Conclusions

The past decades of collider detectors have served to push the field to new discoveries with unprecedented sensitivity. The study of Higgs physics at CEPC opens a portal to new physics processes which, if exist, may be the only path to interact nongravitationally with the dark matter sector. We must expect that any breakthrough in this area will produce signatures that are not fully visible in the detector, the most extreme being the single photon triggered events where the entire final state interaction goes undetected. The fast timing constraints on single photons and particle-flow event reconstruction, in general, provide a powerful level of discrimination that would otherwise not be available and provide a vital tool to limit the impact from the growing rate of background processes produced at high luminosity colliders. The fast timing layer and concept of hermetic timing detectors are a clear path forward to continue the search for new physics at high sensitivity at CEPC.

Acknowledgments

Many important contributions to these ideas were developed in the context of the CMS Fast Timing Working Group. The author would like to thank the current co-leaders of this group, Tommaso Tabarelli de Fatis and Lindsey Gray, for their collaboration in the working group and for their feedback on this paper.

References

1. E. M. Henley and S. D. Ellis, *100 Years of Subatomic Physics* (World Scientific, Singapore, 2013).

2. W. Herr, B. Goddard, E. Keil, M. Lamont, M. Meddahi and E. Peschardt, Experience with bunch trains in LEP, in *Proc. 1997 Particle Accelerator Conf.* (1997), pp. 309–311.
3. CMS Collaboration, Initial Report of the Fast Timing Working Group (March 2016), CMS-DP-2016-008, http://cds.cern.ch/record/2143491.
4. B. Adeva *et al.*, *Phys. Lett. B* **275**, 209 (1992).
5. P. Achard *et al.*, *Phys. Lett. B* **587**, 16 (2004).
6. C.-S. S. Group *et al.*, CEPC-SPPC Preliminary Conceptual Design Report, Volume II-Accelerator, Tech. Rep., IHEP-AC-2015-01, March (2015).
7. A. Benaglia, S. Gundacker, P. Lecoq, M. T. Lucchini, A. Para, K. Pauwels and E. Auffray, *Nucl. Instrum. Methods A* **830**, 30 (2016), doi:10.1016/j.nima.2016.05.030.

The Importance of High-Precision Hadronic Calorimetry to Physics

John Hauptman

Department of Physics and Astronomy,
Iowa State University, Ames, IA 50011, USA
hauptman@iastate.edu

The reconstruction and high-precision measurement of the four-vectors of W and Z decays to quarks, which constitute nearly 70% of their decay branching fractions, are critical to a high efficiency and high quality experiment. Furthermore, it is crucial that the energy resolution, and consequently the resolution on the invariant mass of the two fragmenting quarks, is comparable to the energy–momentum resolution on the other particles of the standard model, in particular, electrons, photons, and muons, at energies around 100 GeV. I show that this "unification of resolutions" on all particles of the standard model is now in sight, and will lead to excellent physics at an electron–positron collider.

Keywords: Hadronic calorimeters; detector precision; W, Z hadronic decays

1. Isn't This Just Some Technical Stuff?

Yes, it is highly technical and you need to understand nuclear physics and the details of interactions of low energy particles in materials, subjects which are no longer taught in physics schools. Personally, I have been working on calorimeters, off and on, since the TPC/PEP4 experiment at SLAC on simulations with EGS and GEANT of electromagnetic calorimeters at PEP and LEP, on hadronic calorimeter studies for SDC/SSC, and on electron test beams. When I first started to work on dual-readout calorimetry (DREAM) at CERN, I thought I knew everything. It was a rude shock to realize that the half of calorimetry that I "knew" was wrong, and the other half I did not know existed.

Hadronic calorimetry is difficult and expensive, requiring a minimum of 1 ton of detector to even begin to claim that you are making calorimeter measurements, and requiring a detailed physics understanding of interactions and nuclei, which most of us lack. One small design error or oversight can yield a mediocre calorimeter performance, and there is no "reset" button. A prominent European physicist once remarked to me that

> "Calorimetry was the low point of my life."

This reflects the frustration and stress inherent to hadronic calorimetry in funding necessities, tonnage of detector mass, difficulties of testing, and the harsh fact that even a small mistake can leave you with a poor result from which there is no recovery.

The building of new instruments to watch nature has progressed remarkably over 100 years. Tracking chamber precisions have progressed from cm-precision with Geiger counters, mm-precision with cloud chambers, to μm-precision with silicon pixel detectors. Hadronic calorimeters have progressed from near-nothing to big detectors with energy resolutions of 5–7% at 100 GeV. Considering also the time domain in both tracking and calorimetry, the capabilities of these detectors have improved by many orders of magnitude. The goal of 2% energy–momentum resolution on all particles of the standard model at 100 GeV is a worthy goal for new detectors at a machine like the CEPC.

2. What We Know and How We Know It: pp *versus* e^+e^- Machines

The experimental history of the standard model, as opposed to the theoretical history, is more arduous, more interesting, and a lot more expensive. Some particles were discovered by accident (the $\mu^\pm$, $K^0 \to \pi^+\pi^-$ and $\Lambda \to p\pi^-$) in cosmic rays, many quark states ($\Delta, \Sigma, \Xi, \Omega, \rho, \omega, \ldots$) were uncovered in bubble chambers, and some were found by deliberate searches (the J at BNL, the ψ at SPEAR, the b, t quarks at the Tevatron, the W, Z bosons at the Sp$\bar{p}$S collider, and the Higgs at the LHC). None of these discoveries depended critically on calorimetry. The point of this paper is to argue that future discoveries and precision measurements will depend critically on the quality of the hadronic calorimeters in experiments, specifically, in the reconstruction of $W \to q\bar{q}$ and $Z \to q\bar{q}$ decays with precisions comparable to those for $e^\pm$, $\mu^\pm$ and γ.

When the Stanford Positron Electron Asymmetric Rings (SPEAR) was proposed by B. Richter at SLAC, the proposal was denied several times on the argument that proton machines will scan the entire mass range for vector particles in one experiment, whereas at an e^+e^- machine the luminosity and event rates are small and the beam spread is so narrow that scanning in mass is too time consuming. These were true statements, and still true today. In fact, the J was found at a proton machine much earlier than the ψ was found at SPEAR, the $W^\pm$ and Z^0 were discovered at a proton machine, and the Higgs was found at a proton machine. Only afterwards are these particles studied with high precision at electron machines.

The ψ at SPEAR was missed months earlier than November 1974 due to the narrow beam spread. Repeated scans in the 3.0–3.2 GeV region gave conflicting event rates due to sitting at different beam energies below the ψ and on the radiative tail above the ψ. It took a careful time-consuming precision energy scan to hit the ψ.

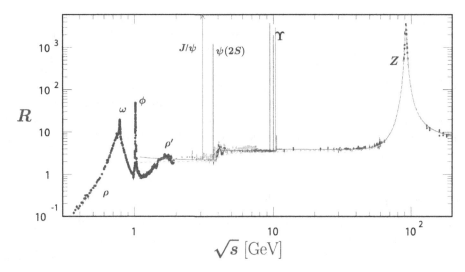

Fig. 1. Measurements of the $e^+e^- \to$ hadrons cross-section relative to $e^+e^- \to \mu^+\mu^-$ up to 200 GeV. The high precision in the definition of particles in this plot is due to the machine precision, not detector precision. From PDG.[1]

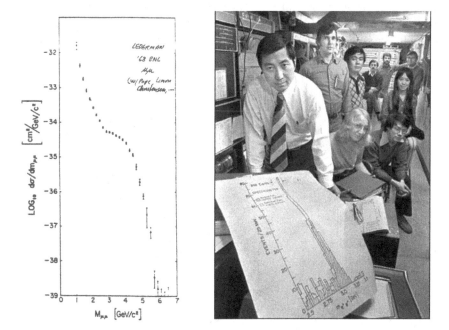

Fig. 2. On the left, the $\mu^+\mu^-$ mass in pN collisions; on the right, the e^+e^- mass in similar pN collisions at BNL.

It is important to realize that the precision in the masses of the particles evident in Fig. 1 is mostly determined by the machine and not the detectors. The converse is true at a proton–proton collider where you depend entirely on the detector precision.

The importance of raw, direct resolution is illustrated in Fig. 2 showing, on the left, a failed attempt in pN collisions at BNL to see the $J/\psi \to \mu^+\mu^-$ decay and, on the right, a successful attempt to see the $J \to e^+e^-$ in pN collisions at 28.5 GeV also at BNL. The difference between the two is momentum resolution on the leptons. In the di-muon $(\mu^+\mu^-)$ experiment, an iron absorber was used to filter out the pions from the target resulting in substantial multiple scattering of the muons. In the di-electron (e^+e^-) experiment, no material filter was present and a bag of helium gas was placed downstream of the target before the tracking chambers to reduce multiple scattering.

3. Measuring Hadronic Particles and Jets, e.g. Quarks and Gluons

Nearly 70% of the decay branching fractions of Z^0 and $W^\pm$ bosons is to quark-antiquark, that is, two jets of hadrons and photons striking the detector. The four-vectors of $W \to q\bar{q}$ and $Z \to q\bar{q}$ are measured well only when all the particles from both jets are measured, and this demands a calorimeter, combined with a nearly-4π tracking system for the low energy particles from the jets.

As far as I know, the current best hadronic reconstruction of $W \to q\bar{q}$ is by the ATLAS collaboration shown in Fig. 3 in which it is evident that it will be difficult to use these data in physics analyses with discrete W decays tagged at high efficiency since the background under the W signal is about 100 times larger than the signal.

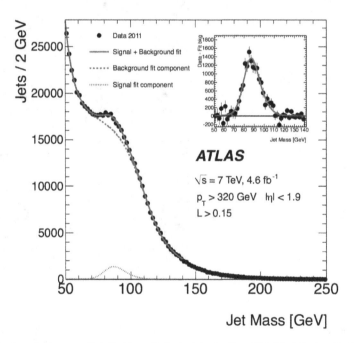

Fig. 3.   The invariant mass distribution of single jets in the ATLAS detector that have been selectively enriched in jets that are consistent with $W \to jj$ events.[2]

The reconstructed mass of a two-jet system such as $W \to q\bar{q}$ is $M^2 = 2E_1E_2(1 - \cos\theta_{12})$ for jet energies of E_1 and E_2 with angle θ_{12} between them. Apart from the angle θ_{12} which depends on the measurements of the centroids of the two jets, the mass resolution depends directly on the jet energy resolution.

It seems to me[3] that the best strategy for $W \to q\bar{q}$ reconstruction is to use all the characteristics of $W \to q\bar{q}$ events, viz., two 40 GeV quark-jets, each with lower particle multiplicity than gluon jets, and the di-jet decay angular distribution in the W center-of-mass is flat, whereas the gluon–gluon background di-jet events will be strongly peaked at $\cos\theta \approx \pm 1$. For highly energetic $W \to q\bar{q}$ events, the two jets will co-mingle in the calorimeter, and therefore a sensible degree of transverse segmentation is necessary.

3.1. *Jet energy measurement*: *Why is it so difficult?*

Since a jet consists of a collection of pions, kaons, electrons, photons, protons and neutrons, a calorimeter must be prepared to measure all of these. Absorbing these particles into a calorimeter is the easy part. Measuring the energies with the same response for each is difficult[4] and requires, as we shall see, new techniques in hadronic calorimetry, either particle-flow or dual-readout (or a combination of both).

A nuclear break-up recorded in emulsion is shown in Fig. 4 due to a 30 GeV proton entering from the right. Most of the particles leaving the nuclear break-up are low energy particles: spallation protons, Fermi-energy neutrons and other nuclear *debris* such as α particles and fission fragments. The energetic particles such as $\pi^\pm$ leave small signals, but proceed onwards to break up more nuclei, yielding more low energy particles. Thus, the main signal from hadronic calorimeters comes from a large number of low energy particles. In addition, there are binding energy losses of about 8 MeV for each liberated nucleon, usually called *invisible energy*.

After the absorption of all particles and the effective equalization of the responses of all particle types, there are second-order effects which compromise jet energy measurements at the 1–2% level. Wigmans[4,7] discusses these in detail: (i) the response of a calorimeter to γ differs from the response to $e^\pm$ in a beam test, (ii) the response to a jet of particles differs from the response to a single $\pi^\pm$ in a beam test, (iii) the response to a proton differs from the response to a π^+ (by about 10% in both mean and width) and (iv) hadrons below 5 GeV act like *mips* and give a larger signal that expected, etc.

Whatever signal sensors are embedded in a calorimeter (scintillator, liquid Argon, silicon, gaseous chambers, etc.) it is generally always true that electrons and positrons inside a hadronic shower generate larger signals than the hadronic particles, essentially because every $e^\pm$ develops a signal, whereas not all protons, neutrons, or nuclear fragments do. For a specific calorimeter, the average values of these responses are called the "electromagnetic component, e" and the "non-electromagnetic component, h" and their ratio is here defined as

$$\eta = h/e\,.$$

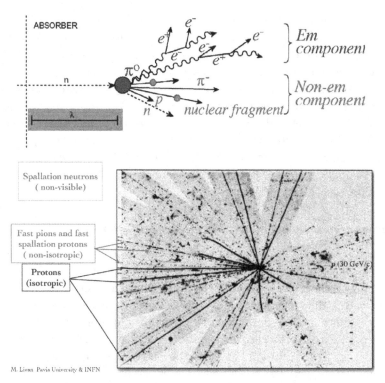

Fig. 4.   An illustration of "EM component" (mostly $\pi^0 \to \gamma\gamma$) and the "non-EM component" from the interaction of a hadron with a nucleus. The mean spatial scale is one nuclear interaction length (λ_{Int}). An emulsion image of a real such interaction from a 30 GeV proton entering from the right. The MeV-energy neutrons are not visible; the α particles and other nuclear debris have very short range; the heavily ionizing spallation protons and lightly ionizing charged pions are visible. From M. Livan, INFN Pavia.

In a nuclear interaction such as Fig. 4, charged pions $(\pi^\pm)$ and neutral pions (π^0) are produced approximately in the ratio two-to-one and most importantly to realize, there are large fluctuations both in the numbers of pions produced and the fractions of neutral and charged pions. A neutral pion immediately delivers all its energy to electromagnetic particles as $\pi^0 \to \gamma\gamma$ and this energy is dumped locally in the calorimeter. The charged pions continue to break up more nuclei as in Fig. 4. Since the mean electromagnetic response is almost always larger than the hadronic response, $e > h$ and fluctuations in the numbers of $\pi^\pm$ and π^0 result in fluctuations in the total calorimeter signal.

The solution to this problem[4] was "compensation," that is, build a calorimeter that gives the same response to electromagnetic and hadronic energies, $h/e \approx 1$. This is possible in a number of ways, but possibly the easiest and best way is a lead-scintillator calorimeter[8] in which the neutrons from the nuclear break-ups scatter elastically from the protons in the scintillator and these recoil protons give a large scintillation signal. This gives compensation because the kinetic energy of

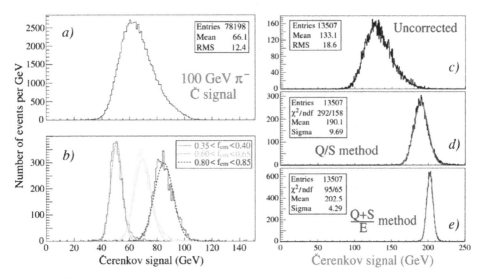

Fig. 5.   (a) Čerenkov signal for 100 GeV π^-; (b) Čerenkov signal distributions for three selections on the EM fraction, f_{EM}; (c) Čerenkov signal distribution for 200 GeV π^-; (d) distribution of the solution E from Eqs. (1) and (2); (e) distribution of E from Eqs. (1) and (2) for $f_{EM} \approx C/E_{beam}$.

the neutrons is strongly correlated with the number of broken-up nuclei and their binding energy losses. The success of the Bernardi *et al.*, experiment[8] led Wigmans[9] to build SPACAL, a 20 ton lead-scintillating fiber module that still holds the world record for energy resolution,

$$\sigma/E \approx 30\%/\sqrt{E} \oplus 1\%, \quad \text{(SPACAL energy resolution)}$$

shown plotted in Fig. 6 as the green square symbols.[a] The dotted line in this plot of σ/E versus $1/\sqrt{E}$ is a calorimeter with $\sigma/E = 30\%/\sqrt{E}$ and a zero constant term. The intercept at $1/\sqrt{E} = 0$ is the constant term.

The problem with this lead-scintillator compensating calorimeter is that the scintillator fraction had to be small, about 2%, which limited the electromagnetic energy resolution. Instead of compensating (in hardware) to get $h/e \approx 1$, if the electromagnetic part could be measured directly and separately, then $h/e \approx 1$ could be achieved in software by appropriately adding the e and h parts. This was proposed by Wigmans[10] and was the beginning of the DREAM and RD52 projects. Dual-readout can be called "flexible compensation" since the actual design of the calorimeter does not matter much, e.g. the scintillator fraction can be anything you like and the absorber can be any high density medium.

Data from the DREAM module are shown in Fig. 5. The distribution of the measured Čerenkov signal for 100 GeV π^- is shown in Fig. 5(a) for all "electromagnetic fractions," f_{EM}, and in Fig. 5(b) for three selections on f_{EM}: 0.35–0.40,

---

[a]This is one representation of the SPACAL data. There are others, as discussed at length by Wigmans.[4]

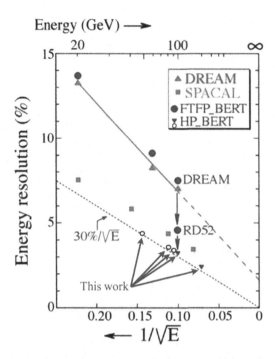

Fig. 6.   (Color online) Energy resolution data from SPACAL (compensating calorimeter) and DREAM and RD52 (dual-readout) calorimeters. The dotted line depicts a calorimeter with $\sigma/E\,30\%/\sqrt{E}$ with no constant term, or 3% resolution at 100 GeV.

0.60–0.65 and 0.80–0.85. The mean of these three sub-ensembles is very nearly linear in f_{EM}, leading to Eqs. (1) and (2) below for the separate S and C expected mean signals. Knowing f_{EM} each event gives you the narrow Gaussian distributions in Fig. 5(b) instead of the wide non-Gaussian distribution of 5(a). This is the essence of dual-readout calorimetry.

In a dual-readout calorimeter, the mean response of the scintillating fibers is S and the mean response of the Čerenkov fibers is C,

$$S = E\left[f_{\mathrm{EM}} + \eta_S \cdot (1 - f_{\mathrm{EM}})\right] \quad \eta_S = (h/e)_S\,, \tag{1}$$

$$C = E\left[f_{\mathrm{EM}} + \eta_C \cdot (1 - f_{\mathrm{EM}})\right] \quad \eta_C = (h/e)_C\,, \tag{2}$$

where η_S and η_C are the mean relative hadronic-to-electromagnetic responses, which are constant. The Čerenkov response for a 200 GeV π^- is shown in Fig. 5(c) and the distribution of E from Eqs. (1)–(2) is shown in Fig. 5(d) which is narrower, Gaussian and at the right energy.

This simple representation of the S and C signals as linear functions of the electromagnetic fraction, f_{EM}, is justified by the linear dependence of the mean response versus f_{EM} in Fig. 5(b). Someday, when we start to probe the 1–2% resolution level in a big calorimeter, we may find that the responses in S and C are not perfectly linear, but so far we have not seen this.

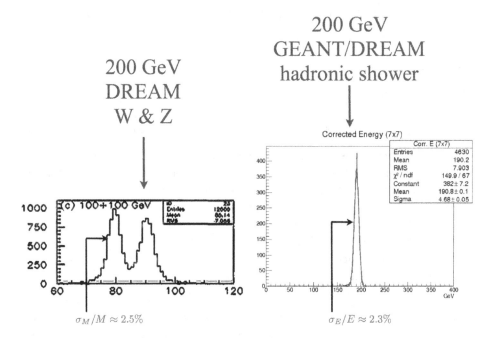

$\sigma_M/M \approx 2.5\%$ $\qquad\qquad\qquad$ $\sigma_E/E \approx 2.3\%$

Fig. 7. Di-jet invariant mass of $W \to q\bar{q}$ and $Z \to q\bar{q}$ from leakage-suppressed DREAM jet data (Subsec. 3.2) and GEANT4 simulation of RD52 calorimeter with physics list FTFP_BERT_HP. The mass and energy resolutions are comparable at 2–3%. See contribution to this conference, "Energy Resolution and Particle Identification of the Dual-readout Calorimeter," Sehwook Lee.[5]

3.2. *Suppressing leakage fluctuations in DREAM data*

The response functions shown in Fig. 5(e), Fig. 7 (left figure) and Fig. 9 are DREAM data events for which the leakage fluctuations of 4% have been suppressed by a re-estimation of the electromagnetic fraction, $f_{\rm EM}$. From Eq. (2), the EM fraction $f_{\rm EM}$ is

$$f_{\rm EM} = (C/E - \eta_C)/(1 - \eta_C) \quad \sim C/E\,.$$

The $(h/e)_C$ mean hadronic response relative to the mean EM response is small, $\eta_C \approx 0.2$ in the DREAM module and E is the unknown shower energy. Substituting $f_{\rm EM}$ into Eq. (1) gives the shower energy E which, overall, fluctuates by about 4% due solely to leakage fluctuations (and not intrinsic calorimeter resolution). By estimating $f_{\rm EM}$ as

$$f_{\rm EM} = (C/E_{\rm beam} - \eta_C)/(1 - \eta_C) \quad \sim C/E_{\rm beam}\,,$$

these leakage fluctuations are suppressed, while leaving in place the fluctuations in the two direct measurements, S and C, to determine the intrinsic calorimeter resolution. This procedure has been criticized, but I believe that it is not only legitimate, but also consistent with all other measurements (DREAM/RD52 modules) and simulations (GEANT4/FTFP_BERT_HP and FLUKA) of these dual-readout calorimeters.

4. How To Get There

It has been a long 40-year experimental journey from the early electromagnetic "shower counters" to precision hadronic calorimeters and we are not there yet. There are two current paradigms for achieving excellent hadronic energy resolution:

Particle-flow: A jet consists of charged particles (measured very well by the tracking system) and photons (measured very well by the electromagnetic calorimeter). This is most of the jet's particles. If a hadronic calorimeter behind the electromagnetic calorimeter catches the neutral hadrons (n, K_L^0) each jet would be completely measured. This has been shown to improve the energy resolution on CDF and CMS, but excellent jet energy resolutions near 2% at high energies have not yet been achieved experimentally, although simulations suggest that 4% resolution at high energies is possible.[11,12] I will not discuss particle-flow further since I am neither an expert nor a proponent.

Dual-readout: A jet consists of widely fluctuating numbers of π^0 and $\pi^\pm$ with differing responses and therefore, a dual-readout calorimeter measures the electromagnetic fraction, f_{EM}, each event (which is approximately the π^0 fraction) and properly weights the different EM and hadronic responses. The next largest fluctuation is the nuclear break-up energies which we measure through the neutrons that elastically scatter from protons in the scintillating fibers. Reducing both of these to the 1% level (at high energies) allows an overall jet energy measurement of 2%.

The results from RD52 dual-readout studies, both data and GEANT4 (FTFP_BERT_HP) simulations, imply that $\sigma/E \approx 2\%$ is achievable in a copper-fiber dual-readout calorimeter at high energies, say, above 300 GeV. Both GEANT4 simulations and leakage-suppress DREAM data (Subsec. 3.2) are shown in Fig. 7.

Simulations at 50, 80, 90, 100 and 200 GeV in GEANT4 (high-precision code) are shown in Fig. 6 as the blue-colored points labeled with triangles and open circles. These simulations are very close to the dotted line representing $\sigma/E \approx 30\%/\sqrt{E}$.

The leakage-suppressed DREAM data events, which are also consistent with an energy resolution of $\sigma/E \approx 30\%/\sqrt{E}$, were used to form $W \rightarrow q\bar{q}$ and $Z \rightarrow q\bar{q}$ decays. The procedure was (i) to sample the W Breit–Wigner for M_W, (ii) choose two data events randomly from two beam data files of energies E_1 and E_2, (iii) set the space angle α between the two beam particles for the two events such that $M_W^2 = 2E_1 E_2 (1 - \cos\alpha)$. Then, the measured energies E_1^* and E_2^* are taken from the data and the shower centroids in each module are calculated from the two data events to get α^* and the di-jet mass is calculated as $M_W^2 = 2E_1^* E_2^* (1 - \cos\alpha^*)$ and plotted. The same procedure is followed for the Z and both are superposed in Fig. 9 for all six combinations of beam data files at 50, 100, and 200 GeV. It is

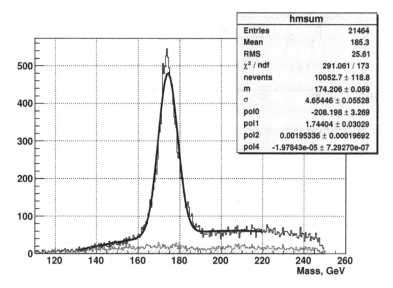

hmsum	
Entries	21464
Mean	185.3
RMS	25.61
χ^2 / ndf	291.061 / 173
nevents	10052.7 ± 118.8
m	174.206 ± 0.059
σ	4.65446 ± 0.05528
pol0	-208.198 ± 3.269
pol1	1.74404 ± 0.03029
pol2	0.00195336 ± 0.00019692
pol4	-1.97843e-05 ± 7.29270e-07

Fig. 8. The top mass fitted in a 20 K event sample of $e^+e^- \to t\bar{t} \to$ 6-jets with full ILC backgrounds.

interesting that the W–Z mass discrimination is almost the same from 100 GeV W/Z's up to 400 GeV W/Z's and also that the mass discrimination is so good that the mis-identification rate between W and Z is only about 5%.

4.1. $t\bar{t} \to 6 - jet$ top mass resolution

The process $e^+e^- \to$ 6-jets is good for accessing the jet resolution. The fourth concept dual readout simulations used the FLUKA code and demonstrated excellent resolution on the top quark mass shown in Fig. 8. The plot includes all physical backgrounds. This work was done by F. Ignatov (BINP, Novosibirsk).[6]

4.2. *Mistakes in calorimeter data analysis*

When the decay particles from a $W \to q\bar{q}$ or $Z \to q\bar{q}$ leave the interaction point, pass through the tracking system and enter the calorimeter, all of these particles must be measured and their energies and directions summed into an overall four-vector of the W or Z. Anything less than this constitutes a nonmeasurement, or a measurement that has nothing to do with calorimetry. There is an absolute scientific requirement in the assessment of calorimeters from beam test data:

You must use the *whole data set* and the *whole calorimeter*.

When calorimeter data analysts remove particles (e.g. beam pions) from the ensemble for whatever reason, the resulting data sample is biassed. When particles are selected to only hit a certain part of the calorimeter, or selected to interact in only a small region of the calorimeter, the data sample is biassed. A biassed ensemble

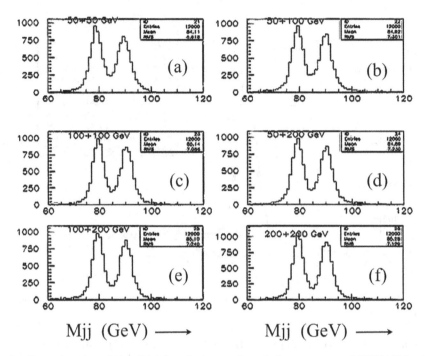

Fig. 9. Reconstruction of the di-jet invariant mass from leakage-suppressed DREAM best test data (Subsec. 3.2) with two beam events per W or Z. Beam taken at 50, 100 and 200 GeV are paired into $W \to jj$ and $Z \to jj$ events at (a) 100, (b) 150, (c) 200, (d) 250, (e) 300 and (f) 400 GeV. The overlap between W and Z is about 5% and the competition between jet energy resolution and inter-jet angular resolution approximately cancels over this wide energy range.

has a wrong mean, a wrong width and a wrong shape, all of which are misleading when attempting to assess the physics quality of a calorimeter.

The most common bias is driven by the need to handle leakage from a calorimeter. All calorimeters leak energy.[b] Given the cost of even a small 1 ton test module, often times a beam test module leaks substantial energy, thus destroying the measurable energy resolution. What to do? Throw out events with a lot of energy near the end of the module? Only keep events that interact in the front and no energy at the end of the module? Both of these selections severely bias the resulting ensemble in events that are predominantly electromagnetic. For example in a π^- beam, events in which the first interaction (Fig. 4) produces high energy π^0's followed by $\pi^0 \to \gamma\gamma$ deposit their full energy locally without leakage. The resulting response distribution reflects the calorimeter resolution for electromagnetic particles, e.g. as if an e^- beam were used.

---

[b]There are five types of calorimeter leakage: (1) EM albedo from the front face of a calorimeter, (2) leakage of ν's from $\pi^\pm$ and $\mu^\pm$ decay, (3) $\mu^\pm$ leakage from $\pi^\pm \to \mu^\pm$ decay, (4) lateral leakage out the sides of a narrow calorimeter (usually these are neutrons) and (5) longitudinal leakage out the back of a shallow calorimeter (usually high energy pions, protons or neutrons).

The second most common bias is to use only one part of the calorimeter, for example, just the hadronic section. In most calorimeters, the depth is divided into an "EM" section at the front for measuring $e^\pm$ and γ, and a "hadronic" section afterwards for measuring the more penetrating hadronic particles $\pi^\pm$, p, n, K_L^0, $K^\pm$, etc. Due to their differing purposes, these two sections have different sampling fractions, absorbers, etc. Inter-calibrating these two sections presents problems that are too complex to discuss here, but suffice it to say that in many tests the incident beam particles are selected to interact only in the front part of the hadronic section. This is convenient for assessing the resolution and other responses of the hadronic section, but $W \to q\bar{q}$ and $Z \to q\bar{q}$ decays send particles into both sections.

Both of these exercises may be interesting or useful for some purposes, such as validating the corresponding simulation codes, but they have nothing to do with calorimetry to measure the four-vectors of $W \to q\bar{q}$ and $Z \to q\bar{q}$ decays.

5. "You Pays Your Money and You Takes Your Choice"

Given the high costs and difficulties in building and testing calorimeters, the temptation to degrade performance or quality in order to lower the overall cost of an experiment is strong.

5.1. *Degrading the calorimeter*

The decisions that experimental collaborations make for their calorimeters is very complex and it is not our purpose to criticize or second-guess these decisions. However, since the calorimeter is often the most costly of all the subsystems of a big detector (comparable to the large superconducting solenoid), it is an easy target for balancing the budget of a detector late in its construction stage. Designing a good calorimeter is hard enough, but re-designing a calorimeter for the purpose of lowering its cost is harder and can be costly for physics.

5.2. *"Unification of experimental resolutions" near the 2% level*

In the physics reconstruction of an event, for example, an $e^+e^- \to Z^0 Z^0$ event,

$$e^+e^- \to Z^0 Z^0 \to \mu^+\mu^- + q\bar{q} \to \mu^+\mu^- + \text{jet} - \text{jet},$$

the overall energy–momentum balance of the event depends on the tracking system for the $\mu^\pm$ and the hadronic calorimeter for the jets and the quality is limited by the weaker link, usually the calorimeter. Given the expectations for dual-readout achieving $\sigma/E \approx 30\%/\sqrt{E}$, this event class at a $\sqrt{s} = 500$ GeV collider would have comparable resolution on the muons (1.2%) and the jets (2.7%). At higher energies, the jets become better measured than the muons.

Excellent hadronic energy resolution leads to the prospect for detectors with comparable energy and momentum resolutions on all the particles of the standard model, Fig. 10, a kind of "unification of experimental resolutions" in which

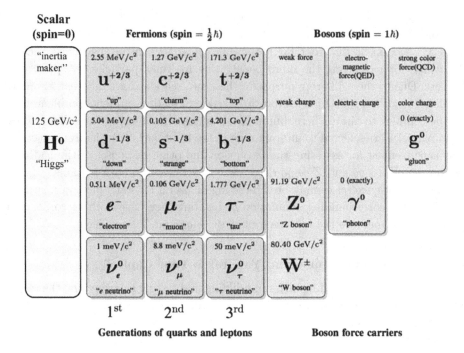

Fig. 10. "Unification of experimental resolutions" on all particles of the standard model.

Table 1. Particles of the standard model and achievable energy–momentum resolutions at 100 GeV.

SM parton at 100 GeV	Detector system	Resolution (%) σ_p/p, σ_E/E
u, d, s, g	Hadronic calorimeter	~3.0
e^-, γ	EM calorimeter	1–2
μ^-	Tracking system	~1.0
$W \to q\bar{q}$, $Z \to q\bar{q}$	Hadronic calorimeter	~3.0
$Z \to ee$	EM calorimeter	1–2
$Z \to \mu\mu$	Tracking system	~1.0
$t \to$6-jet	Hadronic calorimeter	~3.0
c, b	Vertex chamber/track/calor	1–4
ν's	Track/calor (missing $\mathbf{p}$)	1–5
$H^0 \to e$, μ, c, b, W, etc.	All systems	1–5

theoretical final states are treated and tested equally. This is summarized in the table for partons at energies of 100 GeV, Table 1.

Other than extreme exotics, all conceivable theoretical proposals will always result in events decaying in the detector as standard model particles in Fig. 10 and be reconstructed with comparable four-vector resolutions. It is also likely to be the case that an excellent detector will be well prepared to detect and reconstruct exotic particles not conceived or anticipated.

Acknowledgments

I thank the organizers and leaders for this very delightful conference and program at Hong Kong University of Science and Technology. Henry Tye, Tao Liu, and Prudence Wong that fostered communication and the free flow of ideas.

References

1. Particle Data Group (K. A. Olive *et al.*), *Chin. Phys. C* **38**, 090001 (2014).
2. Atlas Collab., *New J. Phys.* **16**, 113013 (2014), arXiv:1407.0800 [hep-ex].
3. H.-U. Bengtsson, J. Hauptman, S. Linn and A. Savoy-Navarro, *Phys. Rev. D* **40**, 1465 (1989).
4. R. Wigmans, *Calorimetry*: *Energy Measurement in Particle Physics*, 2nd edn. (Oxford University Press, 2000).
5. RD52 Collaboration (S. Lee), Energy resolution and particle identification of the dual-readout calorimeters, IAS Program on High Energy Physics (2016).
6. Letter of Intent from the Fourth ("4th") Detector Collaboration, March 2009, available at www.4thconcept.org/4LoI.pdf.
7. R. Wigmans, On wrong and correct ways, to analyze and interpret calorimeter test beam data, April 03 (2014); R. Wigmans, Hadron calorimetry options for future experiments in particle physics, April 02 (2014); Detector physics — trends and challenges, 31 March–11 April (2014), Bethe Forum, Bonn, Germany.
8. E. Bernardi *et al.*, *Nucl. Instrum. Methods A* **262**, 229 (1987).
9. R. Wigmans, *Nucl. Instrum. Methods A* **259**, 389 (1987); Acosta *et al.*, *Nucl. Instrum. Methods A* **316**, 184 (1992).
10. R. Wigmans, Prospects for hadron calorimetry at the one percent level, in *Proc. 7th Int. Conf. on Calorimetry in High Energy Physics*, Tucson, Arizona (World Scientific, Singapore, 1998), p. 182.
11. M. A. Thomson, *Nucl. Instrum. Methods A* **611**, 25 (2009).
12. M. Ruan and H. Videau, arXiv:1403.4784 [physics.ins-det].

Fiber and Crystals Dual Readout Calorimeters

Michele Cascella*

University College London, Gower Street, London WC1E 6BT, UK
m.cascella@cern.ch

Silvia Franchino*

Kirchhoff-Institut for Physics, Heidelberg University, Germany
silvia.franchino@cern.ch

Sehwook Lee*

Department of Physics, Kyungpook National University,
80 Daehak-ro, Buk-gu, Daegu 41566, Republic of Korea
seh.wook.lee@cern.ch

The RD52 (DREAM) collaboration is performing R&D on dual readout calorimetry techniques with the aim of improving hadronic energy resolution for future high energy physics experiments. The simultaneous detection of Cherenkov and scintillation light enables us to measure the electromagnetic fraction of hadron shower event-by-event. As a result, we could eliminate the main fluctuation which prevented from achieving precision energy measurement for hadrons. We have tested the performance of the lead and copper fiber prototypes calorimeters with various energies of electromagnetic particles and hadrons. During the beam test, we investigated the energy resolutions for electrons and pions as well as the identification of those particles in a longitudinally unsegmented calorimeter. Measurements were also performed on pure and doped $PbWO_4$ crystals, as well as BGO and BSO, with the aim of realizing a crystal based dual readout detector. We will describe our results, focusing on the more promising properties of homogeneous media for the technique. Guidelines for additional developments on crystals will be also given. Finally we discuss the construction techniques that we have used to assemble our prototypes and give an overview of the ones that could be industrialized for the construction of a full hermetic calorimeter.

Keywords: Dual-readout; energy resolution; calorimeter.

1. Introduction

The role of calorimeter in modern high energy physics experiments has been identifying particles as well as measuring their energy and momentum. Such experiments,

---

*On behalf of the RD52 Collaboration.

the calorimeters have provided the key information on the measured energy and momentum of particles identified as the electromagnetic particles and hadrons by itself, and have become the heart of particle physics experiments. From this fact, we can infer that the measurement of particle energy with excellent calorimeter energy resolution ends up with high quality particle physics experiments and good physics results. After the Higgs discovery, it is believed that a future lepton collider will focus on understanding the Higgs mechanism, and study of the Higgs boson properties. To accomplish this, we will need a dedicated Higgs factory with excellent detectors which can measure the energy of jets from hadronic decays of W's and Z's in the same precision as the energy measurement of electrons and gammas.

However, while the detection of electrons, photons and other particles that develop electromagnetic showers can be performed with high precision, the same is not true for hadrons.

This is primarily due to the fact that

(1) most calorimeters generate a larger signal per unit deposited energy for the electromagnetic shower component (e) than for the hadronic one (h); that is $e/h > 1$,

(2) the fluctuations in the energy sharing between these two components are large and non-Poissonian.

As a result, in typical instruments the hadronic response function is non-Gaussian, the hadronic signals are nonlinear.

Several approaches have been proposed to deal with this problem:

Compensating calorimeters are designed to deliver equal response to the em and non-em shower components: $e/h = 1$. This can be achieved by boosting the response to the hadronic component, for instance in calorimeters with a hydrogenous active medium that is very sensitive to the soft neutrons abundantly produced in hadronic shower development. Such calorimeters require a precisely tuned sampling fraction. Since this sampling fraction is typically small (2.3% in lead/plastic-scintillator calorimeters), the em energy resolution of such devices is in practice limited to $15\%/\sqrt{E}$.[1]

Off-line compensation is a technique applied in devices with $e/h \neq 1$, in which signals from different sections of the calorimeter are re-weighted according to some scheme to reconstruct the shower true energy. This is one of the most common approach, however, it requires attention to avoid introducing nonlinearities in the response.

Energy Flow combines the calorimeter information with measurements from a tracking system, to improve the performance for jets. This method, usually deployed in combination with the previous one, has been successfully applied in several modern collider experiments.

Particle Flow extends the previous technique by trying to reconstruct each individual particle in a jet using highly segmented calorimeters and sophisticated reconstruction algorithms.

Dual Readout Method that is the topic of this paper and will be fully described in the next section.

1.1. *The dual readout method*

Since the resolution is determined by fluctuations in the electromagnetic fraction of the shower (f_{em}), measurement of the f_{em} value, event by event, is the key to improving the hadronic energy resolution of an intrinsically noncompensating calorimeter. In our method, f_{em} is measured by comparing the shower signals produced in the form of Scintillation light (S) and Cherenkov light (C) in the same detector.

In dual-readout calorimetry each hadronic shower is measured in two nearly independent ways. While the scintillation channel is sensitive to both components of the shower, the signals from the hadronic fraction are strongly dominated by spallation hadrons produced in nuclear reactions, these are usually not sufficiently relativistic to produce Cherenkov light. The Cherenkov channel is, for all practical purposes, only sensitive to the em shower component ($e/h \gg 1$).

This can be realized in different ways:

- In a fiber calorimeters, scintillating fibers and clear fibers can used to measure the S and C channels separately.
- In a crystal calorimeter, both scintillation and Cherenkov light are generated in the same optical volume, and the necessary separation of the two kinds of light is accomplished using time structure, direction, wavelength spectrum, and polarization.

The e part is calibrated to unit response, and the resulting average response of the h part is denoted by $\eta = h/e$, which is less than unity for most calorimeters. If E is the hadronic energy (either single hadron or jet), and the responses expected in the two channels are

$$S = E\left[f_{\text{em}} + (1 - f_{\text{em}})\eta_S\right],$$

$$C = E\left[f_{\text{em}} + (1 - f_{\text{em}})\eta_C\right].$$

These equations can be inverted to measure both E and f_{em} event-by-event

$$E_{\text{meas}} = \frac{S - \chi C}{1 - C} \text{ with } \chi = \frac{1 - \eta_S}{1 - \eta_C}, \tag{1}$$

where E_{meas} is the corrected energy which is a function of the scintillation and Cherenkov signals, and the factor χ, which, in turn, is determined by the physical characteristic of the calorimeter.

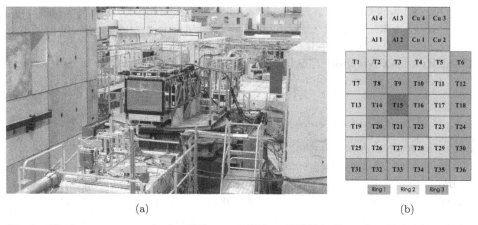

Al 4	Al 3	Cu 4	Cu 3
Al 1	Al 2	Cu 1	Cu 2

T1	T2	T3	T4	T5	T6
T7	T8	T9	T10	T11	T12
T13	T14	T15	T16	T17	T18
T19	T20	T21	T22	T23	T24
T25	T26	T27	T28	T29	T30
T31	T32	T33	T34	T35	T36

Ring 1   Ring 2   Ring 3

(a)   (b)

Fig. 1. The beam test setup in the H8C area of SPS at CERN in November 2012 (a), and the tower map of the Pb- and Cu-fiber calorimeters (b). The front end of the Cherenkov fibers in the left copper module have been aluminized on the front end to increase the light output.

2. The RD52 Fiber Calorimeter

The RD52 collaboration has undertaken an extensive experimental campaign to validate the dual readout method with fiber calorimeters. We have a semi-permanent setup at H8 beam line in the CERN SPS North Area. The experimental setup for the November 2012 beam test is shown in Fig. 1 and all experimental results described in this paper have been obtained using a substantially unmodified set of detectors.

A number of auxiliary detectors are installed on the beam line:

Wire chambers (DC) the two wire chambers are used to constrain the particle position and divergence to obtain a well 10×10 mm^2 beam spot.
Preshower detector (PSD) a scintillator placed right after a 5 mm thick Pb plate. Electrons start developing showers in this device, while muons and hadrons typically produced a signal characteristic for a minimum ionizing particle (mip) in the scintillator plate.
Tail catcher (TC) a 20×20 cm^2 scintillation counter placed right after the calorimeter ($10\lambda_{\mathrm{int}}$), used to identify escaping pions and muons.
Muon counter (MuC) a 50×50 cm^2 scintillator placed right after the tail catcher behind an additional $8\lambda_{\mathrm{int}}$ of iron.

Elimination of the hadron (electron) and muon contamination in the electron (hadron) beams is performed using these ancillary detectors.

The RD52 fiber calorimeters consisting of nine Pb-fiber and two Cu-fiber modules were housed by the rectangular box, which is shown in the middle of Fig. 1(a).

Each module is 2.5 m long, corresponding to $\sim 10\lambda_{\mathrm{int}}$ and its cross-section is 92×92 mm^2. Each tower is a square of 46×46 mm^2 and has one Cherenkov and one scintillation channels; one module consists of four towers. Figure 1(b) shows how

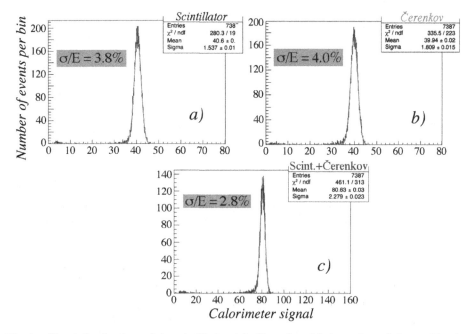

Fig. 2.   Signal distributions of the scintillation (a), Cherenkov (b) channels, and the combination of two types of signals (c) for 40 GeV electrons.

the nine Pb-fiber modules are arranged in a 3×3 matrix and the relative position of the two Cu-fiber modules that are placed on top of the Pb matrix. Tower 15 (T15) is defined as the central tower of the matrix, surrounded by three rings. The results described here were obtained with beams aimed at T15 (for the Pb matrix) or Al2 (for the Cu-fiber calorimeter).

The cross-section of 3×3 matrix is 27.6×27.6 cm^2. It is small compared to the average lateral size of an hadronic shower, so shower leakage affects the performance of the RD52 calorimeters. To try and count for that, 20 leakage counters, made of $50 \times 50 \times 10$ cm^3 of plastic, are installed around the RD52 fiber calorimeters.

2.1. *Electromagnetic performance*

Each tower of the RD52 calorimeter contains an equal number of scintillating and clear fibers (where only Cherenkov light is produced). Both channels of every individual tower are calibrated using the response to 20 GeV electrons.

Figure 2 shows the response of the aluminized Cu-fiber calorimeter to 40 GeV electrons. In both scintillation and Cherenkov channels, the electron energy is measured with a resolution of σ_E/E of 3.8% and 4.0%, respectively. The signal distributions of the scintillation and Cherenkov channel is plotted in Figs. 2(a) and 2(b).

Since the same shower is sampled by the Cherenkov and scintillation fibers at the same time the two signals can be simply combined giving an improved energy resolution of $\sigma_E/E = 2.8\%$.

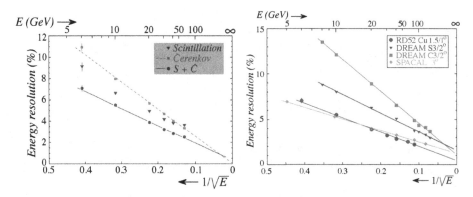

Fig. 3. (Color online) The energy resolution of the Cu-fiber calorimeter for electrons as a function of $1/\sqrt{E}$ (right). Comparisons of the em energy resolutions of the prototype DREAM, RD52 calorimeters and SPACAL (left).

To investigate the electromagnetic performance, the Cu-fiber module was exposed to various electron energies. Figure 3 shows the energy resolution as a function of $1/\sqrt{E}$. The blue inverted triangles and the red squares are the energy resolutions for the scintillation and Cherenkov channels, respectively. The Cherenkov channel is well described by the straight line, and this suggests a very small constant term. The scintillation channel, on the other hand, deviates from the linear behavior by 2% to 3%. This deviation comes from the response difference between particle hitting the absorber first or the scintillation fibers first.

Combining the two signals, the energy resolution is clearly improved as the green circles in Fig. 3 show. The green straight line fit results in a resolution

$$\frac{\sigma_E}{E} = \frac{13.9\%}{\sqrt{E}}$$

with constant term of less than 1%.

It is useful to compare the electromagnetic performances of the RD52 calorimeter with that of other fiber calorimeters such as SPACAL[1] and the old DREAM[2] (the first prototype dual-readout calorimeter). Figure 3 shows that the RD52 calorimeter has better energy resolution than DREAM (thanks to the improved sampling fraction). Moreover the energy resolution of our copper calorimeter comparable to SPACAL and better than that for energies larger than 20 GeV.

2.2. *Hadronic performance*

Figures 4(a)–4(c) show the Pb-fiber calorimeter responses to 20 GeV, 60 GeV and 100 GeV pions, respectively. The top three plots show the scintillation (blue) and Cherenkov (red) signal distributions for those three energies. These are typical noncompensating calorimeter response to hadrons. Their response functions are non-Gaussian and the average energies measured are lower than the beam energies. The energy resolutions in the scintillation channel are 20%, 15% and 13%, respectively.

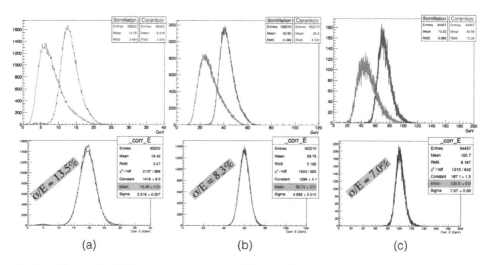

(a)   (b)   (c)

Fig. 4.   (Color online) The response functions of the scintillation and Cherenkov channels to pions before (upper) and after (lower) the dual-readout method correction for 20, 60 and 100 GeV.

However, we can use the dual-readout method described in Eq. (1) to reconstructs the corrected beam energies with Gaussian response functions, and significantly improve the energy resolution. After the dual-readout correction the resolutions become 13.5%, 8.3% and 7.0%, respectively.

Figure 5(a) shows the calorimeter responses as a function of pion energy for the scintillation, Cherenkov and combined channels. The responses of the scintillation and Cherenkov channels are, on average, 70% and 45%, of that of electrons of the same energy. But the dual-readout method correction restores the electron energy scale and the linearity of the response.

As we have already mentioned the RD52 calorimeter lateral size is smaller than the average lateral size of the typical hadronic shower and its resolution is negatively affected by shower leakage. To see how a bigger module can improve the hadronic performance, we simulated the performance of a 65×65 cm^2 RD52 fiber matrix with GEANT4[3] for 50, 80, 90, 100 and 200 GeV pions. We used two physics lists such as FTFP_BERT and FTFP_BERT_HP,[4] the second one being much slower but with a high precision neutron model used for neutrons below 20 MeV.

In Fig. 5(b), we compare the results obtained with the RD52 calorimeter and SPACAL, with the simulation. SPACAL with a stochastic term of $30\%/\sqrt{E}$ still holds the world record for the hadron energy resolution. It was a relatively larger detector than the RD52 calorimeter, so that the leakage fluctuation has a negligible contribution to the hadronic energy resolution.

First we simulated a dual readout Pb-fiber calorimeter the same size of the real RD52 Pb matrix (27.6×27.6 cm^2) using the FTFP_BERT physics list. The performance of this detector with 100 GeV pions is shown with a blue circle in Fig. 5(b) and is consistent with the experimental results.

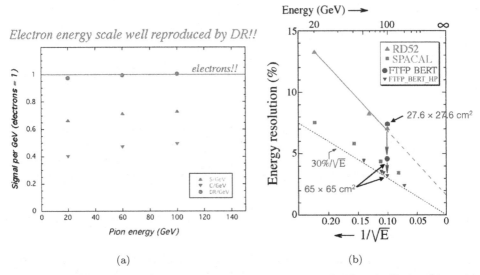

(a)   (b)

Fig. 5. (Color online) The RD52 calorimeter responses to pions for the scintillation (blue triangles) and Cherenkov (red inverted triangles) channels, and after the dual-readout correction (green circles) (a). Comparison of the hadron energy resolutions of the RD52 experimental data (red triangles), the RD52 simulation (blue circles for FTFP_BERT and blue inverted triangles for FTFP_BERT_HP), and SPACAL (green squares) (b).

Following this result, we investigated the hadron performance of a larger version of the RD52 calorimeter (65×65 cm^2, equivalent to a 4×4 matrix) using the FTFP_BERT_HP for 50, 80, 90, 100 and 200 GeV pions (see Fig. 5(b)). These simulations indicate that a dual readout calorimeter large enough (and in a collider experiment it certainly would be) to have negligible lateral shower leakage would have an hadronic resolution of about $30\%/\sqrt{E}$ and a very small constant term.

Our simulation predicts that the dual-readout method calorimeter can achieve the high-quality hadron and jet energy measurement with a much better electromagnetic resolution than that of a compensating calorimeter.

2.3. *Particle identification*

Identification of isolated electrons, pions and muons is of particular importance in particle colliders, for the study of the decay of Higgs bosons into pairs of τ leptons. In the following we discuss the possibility of doing particle separation with the RD52 Pb-fiber calorimeter. Such an ability would be a great asset for an experiment at a future Higgs factory.

The particle beams in the H8 beam line typically contain a mixture of electrons, pions and muons. Using the auxiliary detectors described above we can select the following samples:

Electrons are identified as particles that produced a signal larger than that of two minimum ionizing particles (mips) in the PSD. We also require a signal

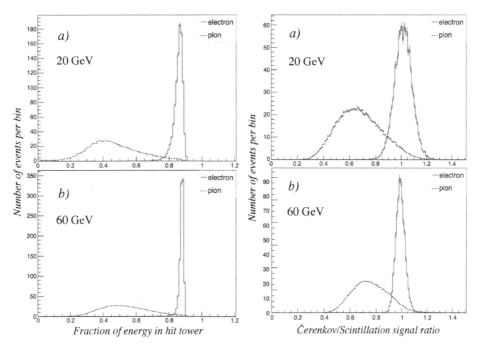

Fig. 6. Distribution of the energy fraction deposited in the hit tower by electrons and pions of 20 GeV (top left) and 60 GeV (bottom left). Distribution of the C/S signal ratio in the hit tower for 20 GeV (top right) and 60 GeV (bottom right) electrons and pions.

compatible with electronic noise in the TC and MuC. The total scintillation signal in the calorimeter should be larger than 15 GeV for the 20 GeV beam and larger than 50 GeV for the 60 GeV beam.

Pions are particles that produced a signal compatible with a mip traversing in the PSD, and a signal compatible with noise in the MuC. The total scintillation signal in the calorimeter should be larger than 7 GeV.

We have developed several techniques to identify the nature of a particle using our Pb-fiber calorimeter.

The first separation method uses the lateral shower size to distinguish between em and hadronic showers. The calorimeter towers have a lateral size of 1.6 Moliere radii, or 0.2 nuclear interaction lengths. Our measurements show that electrons hitting the center of a tower deposit typically 85% of their energy in that tower, while hadrons typically deposit only 40–50%; this is shown in Fig. 6.

A unique aspect of a dual readout calorimeter is the fact that two types of signals are produced: scintillation (S) signals and Cherenkov (C). The ratio of the two types of signals, C/S, is typically around 1 for electron showers, while it is smaller than 1 for hadron showers. Figure 6 shows the distribution of the C/S signal ratio for electrons and pions, at energies of 20 and 60 GeV. The width of the electron distribution shrinks because of the reduced event-to-event fluctuations,

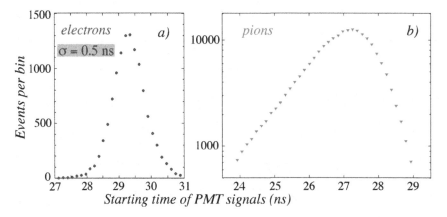

Fig. 7.   The measured distribution of the starting time of the calorimeter's scintillation signals with respect to the trigger signal produced by 60 GeV electrons (a) and 60 GeV pions (b).

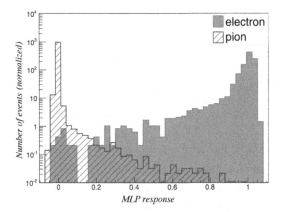

Fig. 8.   Results from the multivariate analysis of the electron/pion separability at 60 GeV, using the lateral shower profile, the C/S signal ratio and the starting time of the PMT signals together to separate electrons from pions.

while the average value of the pion distribution increases because of the increased em shower fraction.

Measurements of the average depth at which the energy is deposited inside the calorimeter provide a powerful tool for particle identification. In longitudinally unsegmented calorimeters, this depth can be measured using the fact that the light in the optical fibers travels at a lower speed (~ 17 cm/ns for polystyrene based fibers) than the particles that generate this light (close to $c \sim 30$ cm/ns). As a consequence the deeper inside the calorimeter the light is produced, the earlier the PMT signal. The depth of the light production can be determined for individual events with a precision of ~ 20 cm. Figure 7 shows the measured TDC distribution for 60 GeV pions and electrons; pions peak 1.5 ns earlier than electrons. The effect of the PSD on this measurements is estimated to be smaller than 0.02 ns.

A high degree of e/π separation can be achieved by combining different methods: using a multi-layer perception can identify 99.8% of all electrons with 0.2% pion contamination (see Fig. 8).

More information on these studies, including the treatment of another method based on the PMT pulse width, can be found in Ref. 5.

3. Construction of the RD52 Fiber Prototypes

The tested calorimeter prototypes were built in 2012 inside the INFN mechanical workshops of Pavia for the lead modules[15] and of Pisa for the copper ones.

The geometry chosen for the lead and copper profiles represented the best compromise between the limitations given by the production technologies and the need to maximize the sampling fraction and the sampling frequency of the calorimeter. One of the most challenging aspects of building this type of calorimeter is the problem of how to get very large numbers of optical fibers embedded in a uniform way in the metal absorber structure.

In case of the lead absorber, a regular structure with equidistant fibers was chosen (Fig. 9(b)), while for the copper absorber another solution was adopted: grooves were created only on one side of each absorber plate, due to the difficulty to machine both sides of the copper plates. In this case the Cherenkov and scintillating fibers were alternated in a single layer (Fig. 9(a)). The sampling fraction was calculated to be 4.5% and 5% in the case of copper and lead geometries, respectively.

3.1. *Choice of fibers*

Before the beginning of the prototypes construction, we have chosen the type of fibers more suitable for our application, in term of numerical aperture and absorption length. A test bench was built at the INFN Pisa in order to measure the fiber absorption length. A led and the light produced by the Ru^{106} β decay (above the Cherenkov threshold) were used. For the scintillation signal fibers Kuraray doped

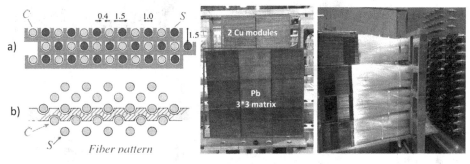

Fig. 9. (Color online) Left: Fiber pattern used in our prototypes for the copper absorber (a) and lead absorber (b). The two different colors represent the two different kind of fibers. Right: Transversal and lateral view of the Pb-fiber and Cu-fiber RD52 calorimeter prototypes as installed at the CERN test beam area.

SCSF-78 were chosen. Before the PMT each S channel was also equipped with a yellow filter to eliminate effect of self-absorption, higher at shorter wavelengths. For the Cherenkov light, Mitsubishi Polymethyl Methacrylate Resin (PMMA)[7] based fibers SK40 were chosen. Their attenuation length was measured to be ~ 6 m.

In order to further increase the Cherenkov light yield, clear fibers used in one of our copper prototypes have been aluminized on one side with the spattering technique at Fermilab.

3.2. *The Pb-fiber matrix*

The nine lead modules were arranged for the test beam in such a way to form a 3×3 matrix as shown in Fig. 9 (right), on top of them the two copper prototypes were laid. For the construction of the nine lead modules, lead plates were produced by an Italian company by the cold extrusion process; after few months of development, the mechanical tolerances were measured to be suitable for our needs: nominal thickness of 1 mm with a tolerance of 50 μm, a fiber-to-fiber distance of 2 mm and a total width of 9.3 cm. Extruded lead was shipped by the producer rolled up on a reel; during the assembly phase each plate was cut at the nominal length (2.5 m), stretched with a roller (Figs. 10(a) and 10(b)) and placed on a worktable (Fig. 10(e)), where the stacking process occurred. Each layer of fibers was created, by laying carefully 46 fibers of a given type in each of the lead grooves. The stacking continued, alternating layers of Cherenkov and scintillating fibers, until the height of the module reaches 9.3 cm.

For each module, eight fiber bundles (4 scintillating and 4 Cherenkov) depart from the rear end and are glued to a support and are milled before being connected to a PMT (Figs. 10(c) and 10(d)).

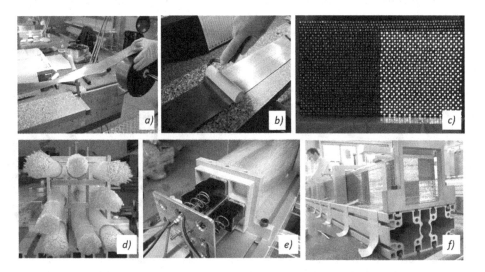

Fig. 10. Pictures taken during different phases of the assembly of lead/fibers RD52 modules.

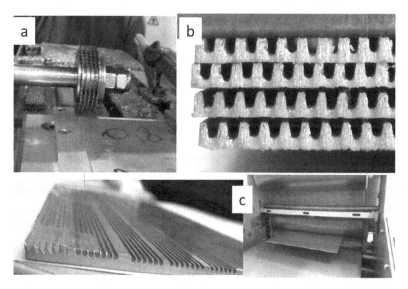

Fig. 11. Different methods for the fabrication of the RD52 copper absorbers: (a) The saw scraper method used at the INFN Pisa for building the first two copper modules prototypes. (b) First results of grooves created in the copper planes with a water jet (Iowa University). (c) Grooves in the copper plane created with chemical milling (CERN).

3.3. *The Cu-fiber modules*

For the copper prototypes construction has turned out that copper itself is a particular difficult material to work with such high density aspect ratio. We have tried many different ways (e.g. rolling, extrusion), but so far only machining grooves in thin copper plates has provided the desired quality. This is why the design was slightly modified in order to etch grooves only on one side of the copper plates (see Fig. 9(a)). For the first two prototypes, copper planes were grooved with a rotating saw (water cooled) as shown in Fig. 11(a). The rotating saw was fabricated specifically for this application, with four parallel lames on a common rotating tool. Once the copper planes were prepared (50 cm long), the assembly procedure was similar to the one used for the lead module, with some differences, like the fact that copper plates were laid one after the other in order to reach the 2.5 m of total longitudinal length, and that between one layer and the other some glue was used and pressed during the night.

3.4. *Industrialization of the construction process*

Copper is a better material than lead for the dual readout calorimetry because of the much better electromagnetic resolution. However, in order to be able to produce in the next future a real scale, full containment copper calorimeter with such high sampling frequency as the one we tested so far, we need to find a less time consuming and an industrial-compatible technique for making grooves in copper.

The method used with rotating saws is time consuming if is needed to be applied in large scale. We have then tried other ways and from these studies the more promising ones seem to be chemical milling and water grooving. After having started the rough grooving process with one of the two methods one could finish the grooving process with rolling to better define the precision of the grooves. First trials of chemical milling were done at CERN using the classic photolithography technique (Fig. 11(c)) and tuning the width of the mask and the exposure time to the chemical bath. First trials of water grooving were done at the Iowa State University (Fig. 11(b)) tuning the water pressure and the jet speed.

Finally, toward a fully projective fiber calorimeter structure, the possible geometry has been studied years ago by the RD1 collaboration[6] and this geometry could be still applied for the dual readout fiber calorimeter.

3.5. *Future directions*

In order to study the possibility to use a dual readout, not longitudinally segmented calorimeter, for a future collider experiment, the RD52 collaboration has recently equipped, a small copper prototype with Silicon photomultipliers. This readout way offers the possibility to eliminate the forests of optical fibers that stick out at the rear end and also to choose a transversal segmentation at a fiber level, if it is needed. With PMT readout, fiber bunches (Fig. 10) occupy precious space and act as antennes for particles that comes from other sources than the showers developing in the calorimeters. The analysis of the data collected using this new readout is currently ongoing but preliminary results are promising.

4. Dual Readout Calorimetry with Crystals

Using high-Z crystals as dual-readout calorimeters offers potential benefits, since one could then in principle eliminate or greatly reduce two remaining sources of fluctuations that dominated the hadronic resolution of the fiber calorimeter: sampling fluctuations and fluctuations in the Cherenkov light yield. In case of using homogeneous media, the light signals generated in the crystals into scintillation (S) and Cherenkov (C) components need to be decomposed. In recent years, we have developed four different methods to separate these two components, exploiting their different properties:

- differences in the angular distribution of the emitted light,
- differences in the spectral characteristics,
- differences in the time structure of the signals,
- light polarization in case of Cherenkov.

These methods were experimentally investigated and optimized for three different types of crystals: bismuth germanate ($Bi_4Ge_3O_{12}$, BGO),[8] bismuth silicate ($Bi_4Si_3O_{12}$, BSO),[9,10] and lead tungstate ($PbWO_4$).[8] The latter crystal was also doped with small amounts of impurities (Molybdenum and Praseodymium) to further improve its dual-readout characteristics.[11] However, of these four methods only the differences in the spectrum and in time structure are easily applicable in hermetic detectors needed for 4π experiment at particle colliders.

4.1. *Undoped PWO crystals*

The first tests that proved the feasibility of separation of Cherenkov from scintillation light in homogeneous media were done on a single $PbWO_4$ crystal[8,12] and exploited the directionality of Cherenkov light. Results are shown in Fig. 12(a), where one can see the average time structure from the crystal oriented at $+30°$ and $-30°$ with respect to the incoming electron beam. The difference of the two signals is an indication of the Cherenkov component, that makes the leading edge of the forward signal much steeper than the backward one.

After these promising first studies, an intense R&D program of Dual Readout calorimetry with crystals started. In order to exploit the difference in spectral emission of scintillation (that has a characteristic emission spectrum, depending from the type of crystal) from Cherenkov light (with a $1/\lambda^2$ behavior), four optical transmission filters were used in our studies, all were 3 mm thick and made of glass; each of them had 90% of transmission for different wavelengths: UG11: $\lambda < 400$ nm, UG330: $\lambda < 410$ nm, UG5: $\lambda < 460$ nm, GG495: $\lambda > 495$ nm. For the isolation of the Cherenkov component, a cut toward shorter wavelengths ended up with the effect to have less contamination of scintillation light but, on the other end, Cherenkov signals obtained with this filter were rather small and strongly depended on the distance the light had to travel to the PMT. Different combination of filters were used in order to find the right compromise.

4.2. *BGO crystals*

The importance of optical transmission filters for our purpose was first demonstrated with BGO crystals[8] and then successfully applied to Mo-doped $PbWO_4$ ones.[11] Figures 12(b) and 12(c) show the typical signal shapes for events in which beam particles traversed a single crystal placed perpendicular to the beam line. One side of the crystal was equipped with a "yellow" (GG495) transmission filter, the other side with a UV (UG11) one. The UV filter absorbed more than 99% of the scintillation light, while a large fraction of the Cherenkov light was transmitted. As a result, the Cherenkov component of the light produced by the crystal became clearly visible, in the form of a prompt peak superimposed on the remnants of the scintillation component, which has a longer decay time (25 ns for Mo-$PbWO_4$ and 300 ns for BGO).

4.3. *Doped PWO crystals*

In Ref. 11, one can read about the studies we did in order to optimize $PbWO_4$ crystals for dual readout applications, introducing small amounts of molybdenum

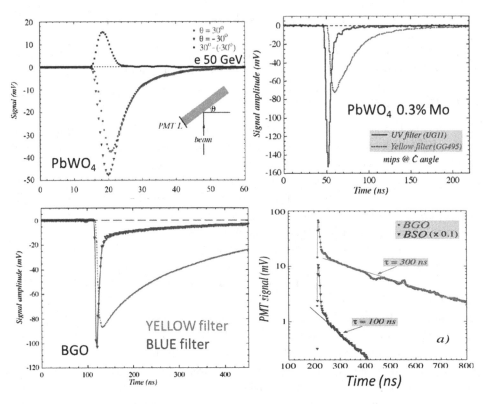

Fig. 12.   (Color online) Results from the single crystals measurements. (a) Time structure from a PMT for two orientations of the PbWO$_4$ crystal with respect to the incoming beam; the difference between the two distributions is an indication of the presence of Cherenkov light in the forward direction (green squares). (b) Mo-doped PbWO$_4$ pulse shapes for the crystal side equipped with yellow filter and the UV one. (c) BGO pulse shapes for yellow and blue filter. (d) Comparison of scintillation amount and time scale for BGO and BSO crystals.

and praseodymium doping. The best solution found was 0.3% Mo-doping, which had two important effects for the separation of Cherenkov from scintillation light: the shift of the emission spectrum toward longer wavelengths, making possible to use of optical filters, and the increase of the decay time of the scintillation process from ~ 10 ns to ~ 25 ns, making easier the separation of the prompt Cherenkov component from the delayed scintillation one. In Fig. 12(b), one can see that for the Mo-doped PbWO$_4$ crystals, we could achieve an almost complete separation of the two optical components.

4.4. Time structure of the light signal

In order to exploit the different time structures between prompt Cherenkov light and scintillation, the pulse shape of the signals produced in these crystals irradiated with electron beams were recorded by means of an high sampling frequency oscilloscope

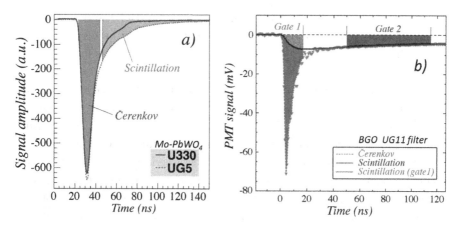

Fig. 13. The time structure of a typical shower signal measured in the Mo-PbWO$_4$ equipped with two optical filters (a) and of a BGO crystal equipped with an UG11 filter (b). In order to measure the relative contributions of scintillation and Cherenkov light, the time spectrum is integrated in two different gates.

(4 ch) or by 32 channels digitizer based on the DRS-IV chip, with a time resolution up to 200 ps.

The time structure information of the signals was used to determine their scintillation and Cherenkov components, integrating, event by event, the pulse shape in two different gates, as it can be seen in Fig. 13. The Cherenkov signal is defined to be the result of the integral of the pulse in the first gate, while the scintillation signal is defined by the integral on the second and more delayed gate. These integrated charges were then converted into deposited energy using appropriate calibration constants (see later for more details).

Using the time structure of the signal has the advantage that the C and S components are extracted by the same readout channel, while for the optical filter case two different channels, equipped with different filters are needed. However, as it can be seen from Fig. 13, not always the use of time structure from only one readout channel is easy. The result will depend on how much scintillation light is passing from the optical filter bandwidth and on how far in time is the scintillation component from the prompt Cherenkov one. In the case of BGO crystal (Fig. 13(b)), because of a longer scintillation decay time, one can see that the definition of two well separated time gates is more easy than in the other case. A much higher scintillation light yield is also helpful in the BGO case because the signal filtered by the UV filter is much more "contaminated" by scintillation photoelectrons, that give the characteristic long tail to the time distribution.

4.5. Comparison of BGO and BSO crystals

Another type of crystal that has also been studied for dual readout applications[9] was the bismuth silicate (BSO), which has the same crystal structure as BGO (Bi$_4$Si$_3$O$_{12}$), with silicon atoms replacing the germanium ones.

We have performed a systematic comparison of the BGO and BSO crystals relevant properties and we found that the BSO crystal could be very promising if one would like to invest more studies in the optimization of crystals for dual readout calorimetry.

First the purity of the Cherenkov signals that can be obtained with UV filters was studied. We found the contamination of scintillation light in the Cherenkov signals to be smaller by about a factor of two in the BSO crystal. Then we studied the number of Cherenkov photoelectrons detected per unit of deposited energy; we measured this yield to be about a factor two to three larger for the BSO crystal. The light attenuation length of Cherenkov light has been found to be approximately the same as the BGO. The cheaper price of BSO with respect to BGO could also be an advantage. In Fig. 12(d) a comparison between the emission spectrum of BGO (~ 300 ns) and the BSO (~ 100 ns) is shown. The slightly faster signal of BSO could be an advantage for high energy applications, the time scale of hundred of nanoseconds is still ideal for using informations coming only from one readout channel and, as in the case of BGO, using the integral of pulse shape in two well-defined time windows.

4.6. *Electron showers in dual-readout crystal calorimeters*

The application of the dual readout technique to an hybrid system made by a crystal matrix (BGO or Mo-doped $PbWO_4$) as electromagnetic section and the original DREAM fiber calorimeter as the hadronic one was evaluated.[13]

Thanks to the experience gained with previous single crystal studies on the way to separate the Cherenkov from the scintillation light components, we have also applied to the crystal matrices the same methods: the optical filters and time structure to select the desired type of light.

The BGO matrix was made by 100 BGO crystals, 24 cm long, from a projective segment of the L3 experiment. The segment was placed perpendicular to the beamline (25 X_0 deep), as shown in Fig. 16(b), in front of the DREAM calorimeter. The readout was made in different ways, the most promising results were obtained with 16 square PMTs (Photonis XP3392B) arranged in a way such that each PMT collected light produced by clusters of at least nine adjacent crystals. Each PMT was equipped with an UG11 optical filter.

The $PbWO_4$ matrix (Fig. 16(a)) consisted of seven custom made 0.3% Mo-doped crystals, 20 cm long, arranged in a matrix and placed in the beam line with the beam entering in the central crystal (22.5 X_0 longitudinal dimensions). Each crystal was equipped with one PMT on each side (Hamamatsu R8900-100); both faces of the matrix were covered with a large optical transmission filter. Several filter combinations were used during our tests. A calibration procedure was done for both matrices using electron beams.

4.6.1. *Calibration*

For the PbWO$_4$ one a narrow beam was steered into the center of each of the seven crystals constituting the matrix. According to GEANT4-based Monte Carlo simulations 93% of the energy was deposited in the entire matrix. After the signals from the crystal were disentangled into Cherenkov and scintillation components, the integrated charge in each of these components was determined, equalization constants were found and applied before summing the contribution of each crystal to get the total inter-gate charge. The average value of that total charge for beam particles traversing the center of the matrix was equated to 93% of the beam energy, and this yielded the conversion factor between the normalized integrated charge and the deposited energy.

In the case of the BGO matrix, calibration constants had to be assigned to each of the four PMTs that read out the four longitudinal segments of the matrix into which the showers developed. This calibration procedure was carried out in two steps: first, the gains of all 16 PMTs were equalized, by means of an LED signal with an amplitude comparable to that of a typical electron shower signal. In the second step, 100 GeV electrons were sent into each of the four columns and the HV values of the four PMTs in the hit column were varied, in an iterative procedure, until the energy resolution for the summed signals reached a minimum value. Because of the size of the BGO matrix, size leakage was considered negligible, and we assumed that the integrated charge collected by the 16 PMTs was a good measure of the deposited energy. On that basis, the integrated charge measured in each individual PMT contributing to the signal could be converted into GeV as well.

4.6.2. *Linearity of the response*

We have investigated the linearity of the observed signals in both crystal matrices and results are shown in Fig. 14. In the case of BGO the calorimeter is linear within a 3% for both the C and S components (Fig. 14(c)). In the case of Cherenkov signal in PbWO$_4$, the linearity is restored only if the signal extracted from the PMTs placed on both sides of the crystals are summed (Fig. 14(b)). This effect is the result of the strong attenuation of the UV light in these kind of crystals, hence as the energy increases, the shower maximum is closer to the downstream PMT and its response is higher, vice versa for the downstream one, as can be seen in Fig. 14(a). In order to obtain a linear Cherenkov response we have equipped both sides of the matrix with a low pass wavelength filter (U330). The main disadvantage of this readout geometry is that the U330 filters transmit almost no scintillation light, as can be seen from Fig. 13(a). In an alternative setup, we therefore replaced the downstream U330 filter by a UG5 one, which also transmits light with wavelengths in the region around 500 nm, where scintillation dominates. This led to a usable scintillation signal.

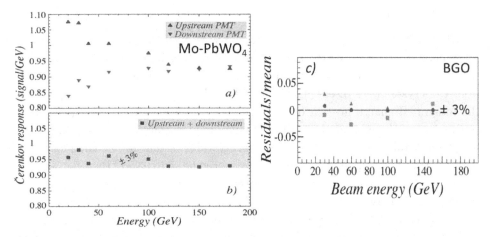

Fig. 14. Signal linearity of electron detected with the crystals matrices. Shown is the response as a function of the electron energy for the C signal in a Mo-PbWO$_4$ matrix equipped with U330 filters on both sides ((a) and (b)); S, C and Σ signals in case of BGO matrix equipped with UG11 filters (c). The signals measured at both ends are both shown separately (a) and added together (b).[14]

4.6.3. Energy resolution

The energy resolution obtained for the Cherenkov channel is shown as a function of the electron energy in Fig. 15(a) for these two sets of experimental data. The stochastic fluctuations that dominate the measured energy resolutions were found to contribute $20\%/\sqrt{E}$ and $28\%/\sqrt{E}$ for these two filter configurations. No evidence was found for an energy independent contribution to the energy resolution. The resolutions measured at the low-energy end of each data set deviate from the

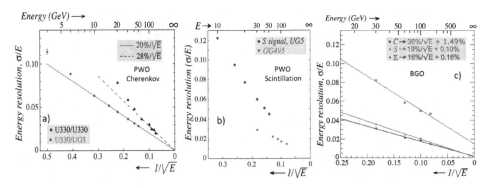

Fig. 15. (Color online) Energy resolution for electrons in the crystal matrix: MO-doped PbWO$_4$ in (a) and (b) figures and BGO in (c). (a) Resolution measured Cherenkov signals, derived from UV-filtered light detected at both ends of the crystal matrix. (b) Scintillation signal measured with two different filters. (c) Results for the total charge collected by the PMTs (Σ) and for Cherenkov (C) and scintillation (S) components of the signal. Results from the fit of experimental points are also shown.[14]

straight lines. These deviations are consistent with the contribution of the signal baseline fluctuations to the measured energy resolution. The energy resolution is strongly dominated by fluctuations in the Cherenkov light yield, this can be seen by the fact that the Cherenkov energy resolution improves if UG5 filter, that transmit a larger fraction of Cherenkov light, is used.

Even though the UG5 filter led to usable scintillation signals, the energy resolution for electron showers measured in the scintillation channel (as an integral of the tail of the signal) was somewhat worse than in the Cherenkov channel, and about a factor of two worse compared to resolutions measured with the yellow (GG495) filter (Fig. 15(b)). This is of course due to the very small fraction of the scintillation light that was detected in this setup. From the sloped of the lines drawn in Fig. 15(a), the Cherenkov light yield was estimated to be 25 photoelectrons per GeV in case of UG5 filter and 13 for the U330.[a]

In case of BGO matrix the energy resolution measured with electrons in the range 10–150 GeV is shown in Fig. 15(c):

$$\frac{\sigma E_C}{E} = \frac{36\%}{\sqrt{E}} + 1.5\%, \quad \frac{\sigma E_S}{E} = \frac{19\%}{\sqrt{E}} + 0.1, \quad \frac{\sigma E_\Sigma}{E} = \frac{16\%}{\sqrt{E}} + 0.16\%. \quad (2)$$

The resolution obtained for the scintillation component, as well as that for the total collected charge (Σ) i.e. the integral over the entire waveform, are well described by $E^{=\frac{1}{2}}$ scaling, while the energy resolution measured for the Cherenkov component exhibits a deviation, with a constant term $b \sim 1.5\%$. The resolution for this component is affected by significant non-Poissonian fluctuations. This could be due to the fact that in order to use crystals for dual readout calorimetry, we had to apply cuts in the wavelength and in the amount of emitted light; in particular for the C component we selected a very small fraction of the total light produced in the crystals, in a wavelength range in which that light is strongly attenuated.

In the case of the BGO matrix we have also studied the performance of the whole calorimeter with pion beams.[13] The information collected with the two sections are combined to obtain an overall evaluation of the electromagnetic fraction of the shower. In order to perform such a measurement we selected only those events in which pions start showering in the crystal section.

In Fig. 16(c), the distribution of the total Cherenkov signal of the hybrid system BGO + DREAM fiber calorimeter is shown. This signal is broad, asymmetric and centered around a value of only 110 GeV, whereas the total jet energy was 200 GeV. In Fig. 16(d), three different subsets of events, selected on the basis of the measured C/S signal ratio are shown. These three distributions are narrower, well described by Gaussian fits and centered at a value that increases roughly proportionally with

---

[a]From this small amount of Cherenkov photoelectrons, one could see that it could be a limiting factor for the dual readout calorimetry. Fluctuations in this small number could affect calorimeter performances. We will see in the next chapter that the RD52 collaboration found higher values of Cherenkov photoelectrons in a sampling dual readout calorimeter, that is nowadays the priority of the collaboration.

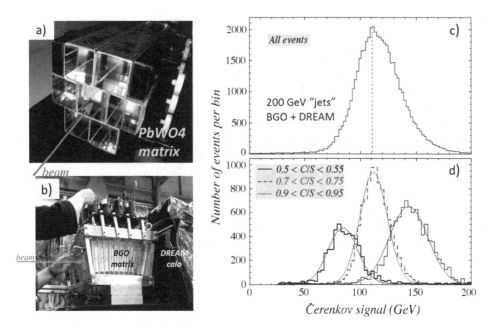

Fig. 16. Picture of the PbWO$_4$ matrix (a) and of the hybrid BGO matrix and fiber DREAM calorimeter (b). Cherenkov signal distribution for 200 GeV jet events detected in the BGO plus fiber calorimeter system (c), distributions for subsets of events selected on the basis of the ratio of the total Cherenkov and scintillation signals (d).

the C/S value of the selected event sample. This is precisely what was observed for the fiber calorimeter in stand-alone mode, and what allowed to eliminate the effects of fluctuations in the electromagnetic shower component in that calorimeter.

4.7. Future prospects for dual-readout crystal calorimeters

In conclusion to our R&D on crystals for dual readout calorimetry, we can say that our interest in studying high-Z scintillating crystals for the purpose of dual-readout calorimetry derived from the potential reduction of the contribution of stochastic fluctuations to the energy resolution of such calorimeters. Our goal in further developing this experimental technique is to reduce the contribution of stochastic fluctuations to the point where these are comparable to the irreducible effects of fluctuations in visible energy. Crystals were believed to offer good opportunities in this respect. However, as the results of this study show, things are not so easy.

For the application of crystals to the dual readout method, the detection of Cherenkov light is the crucial issue and particular care was taken in the precision with which the calorimeter performance can be measured using this signal component. Extracting sufficiently pure Cherenkov signals from these scintillating crystals with the use of optical filters implies a rather severe restriction to short wavelengths.

As a consequence, a large fraction of the potentially available Cherenkov photons needs to be sacrificed, but also, the light that does contribute to the Cherenkov signals is strongly attenuated, because of the absorption characteristics of the crystals. As a result, the remaining light yield is such that fluctuations in the detected numbers of photoelectrons become a significant contribution to the em energy resolution. This is an important difference with experiments in which the unfiltered light of such crystals is used for the electromagnetic calorimeter signals.

The conclusion reached after a long and in-depth study of crystal performances for dual readout calorimetry is that no such significant improvements in term of Cherenkov light yield seem to be offered by crystals in combination with filters in dual-readout calorimeters. The RD52 collaboration decided therefore to focus on the fiber option.

5. Summary and Future Plans

We have proved that the Dual-REAdout Method calorimeter could achieve the high-quality energy measurement for both electromagnetic particles and hadrons. Even, GEANT4 simulation anticipates that the dual-readout calorimeter can have $30\%/\sqrt{E}$ for the stochastic term and very small constant term. This performance will satisfy the important requirement of the future lepton collider, which is the separation of hadronically decaying W/Z bosons. The RD52 collaboration is make effort to build a large size of calorimeter with Cu and will try to prove that it would achieve the excellent energy measurement of hadrons and jets in the future.

The more excellent detector performance, the better physics results we have. We expect that the Dual-REAdout Method calorimeter will be able to achieve the high-precision energy measurement for all fundamental particles and open a new era of experimental particle physics, just as the high purity Germanium crystal detector achieved the excellent energy resolution in the nuclear spectroscopy and showed very detail energy levels of nuclei.

Acknowledgments

We thank CERN for making good particle beams available to our experiments in the H8 beam. This study was carried out with financial support from the United States Department of Energy, under contract DE-FG02-12ER41783, by Italy's Istituto Nazionale di Fisica Nucleare and Ministero dell'Istruzione, dell'Università e della Ricerca, and by the Basic Science Research Program of the National Research Foundation of Korea (NRF), funded by the Ministry of Science, ICT & Future Planning under contract 2015R1C1A1A02036477. In addition, we thank Korea University for the support received by their researchers who contributed to this project.

References

1. D. Acosta, *et al.*, *Nucl. Instrum. Methods A* **308**, 481 (1991).
2. N. Akchurin *et al.*, *Nucl. Instrum. Methods A* **536**, 29 (2005).

3. S. Agostinelli *et al.*, *Nucl. Instrum. Methods A* **506**, 250 (2003).
4. http://geant4.cern.ch/support/proc_mod_catalog/physics_lists/useCases.shtml.
5. N. Akchurin *et al.Nucl. Instrum. Methods A* **735**, 120 (2014).
6. RD1 Collab., *Nucl. Instrum. Methods A* **337**, 326 (1994).
7. http://fiberopticpof.com/pdfs/Product_Specs/SK-40_Product_Information.pdf.
8. N. Akchurin *et al.*, *Nucl. Instrum. Methods A* **595**, 359 (2008).
9. N. Akchurin *et al.*, *Nucl. Instrum. Methods A* **640**, 91 (2011).
10. N. Akchurin *et al.*, *Nucl. Instrum. Methods A* **638**, 47 (2011).
11. N. Akchurin *et al.*, *Nucl. Instrum. Methods A* **604**, 512 (2009); *Nucl. Instrum. Methods A* **621**, 212 (2010).
12. N. Akchurin *et al.*, *Nucl. Instrum. Methods A* **582**, 474 (2007), doi:10.1016/j.nima.2007.08.174.
13. N. Akchurin *et al.*, *Nucl. Instrum. Methods A* **610**, 488 (2009).
14. N. Akchurin *et al.*, *Nucl. Instrum. Methods A* **686**, 125 (2012).
15. S. Fracchia, Prototypes studies of fiber calorimetry with dual readout, Master Thesis, Pavia University (2012).

A Merged Quadrupole-Calorimeter for CEPC

Richard Talman

Laboratory for Elementary-Particle Physics,
Cornell University, Ithaca, NY 14853, USA
richard.talman@cornell.edu

John Hauptman

Iowa State University, Ames, IA 50011, USA
hauptman@iastate.edu

The luminosity $\mathcal{L}$ of colliding beams in a storage ring such as CEPC depends strongly on l^*, the half-length of the free space centered on the intersection point (IP). l^* is also the length from the IP to the front edges of the two near-in quadrupoles that are focusing the counter-circulating beams to the IP spot. The detector length cannot, therefore, exceed $2l^*$. Since $\mathcal{L}$ increases strongly with decreasing l^*, there is incentive for reducing l^*; but this requires the detector to be shorter than desirable. This paper proposes a method for integrating these adjacent quadrupoles into the particle detector to retain (admittedly degraded) active particle detection of those forward going particles that would otherwise be obscured by the quadrupole. A gently conical quadrupole shape is more natural for merging the quadrupole into the particle detector than is the analytically exact cylindrical shape. This is true whether or not the calorimeter is integrated. It will be the task of accelerator physicists to determine the extent to which deviation from the pure quadrupole field compromises (or improves) accelerator performance. Superficially, both the presence of strongest gradient close to the IP and largest aperture farther from the IP seem to be advantageous. A tentative design for this merged, quadrupole-calorimeter is given.

Keyword: Collision point detector optics.

1. Sketches of the Proposed Apparatus

The basic combined quadrupole/calorimeter components are indicated in the following series of figures. The complications and uncertainties are too great for us to attempt to draw a composite figure with all elements superimposed.

The trade-off of detector length versus luminosity is by now quite well understood — longer (presumably better) detector can be obtained at the cost of lower luminosity. The trade-off can be parametrized by l^* which, on the one hand, is the distance from the intersection point (IP) to the front edge of the innermost

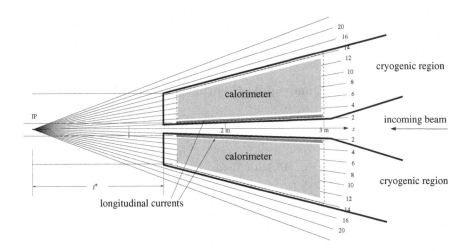

Fig. 1. (Color online) Figure showing the cryostatically isolated regions needed for the super-conducting $\cos 2\theta$-distributed longitudinal currents (shown in red). As drawn the inner and outer angles of the apparatus run from about $\pm 2°$ to about $\pm 14°$ but this is just for discussion purposes and does not correspond to any actual design.

quadrupole and, on the other hand, is the half-length of the small radius vertex de-tector. If the outer diameter of the quadrupole (with all its iron flux return and cryo-genic shell) is D_q, then a conventional quadrupole obscures a solid angle of roughly $(D_q/l^*)^2$. If this solid angle is a "black hole" for particles produced at the IP, there is a corresponding reduction from 4π of the detector's solid angle coverage. The point of the current proposal is to recover most of this blind solid angle, admittedly with somewhat inferior particle detection capability. In particular, it is proposed to instrument much of the volume taken with the calorimeters illustrated in Fig. 1.

2. Intersection Region Optics with Conical Innermost Quadrupole

2.1. Purely cylindrical quadrupole magnetic field

Our suggested quadrupole design employs cosine-2θ current density. High per-formance, large aperture quadrupoles using this design have been described by Brindza et al.[3]

Parameters for the ideal cylindrical magnetic field calculation are defined in Fig. 3. The goal is to produce a pure quadrupole field of given strength and orientation in region I, with negligible magnetic field in region III. Without loss of generality, we can assume the field is "erect" (meaning purely horizontal fields on the y-axis). "Longitudinal" currents flow parallel to the z-axis, into or out of the plane of the paper. Region I is the region of purely-transverse quadrupole field. Region II can be vacuum or any nonmagnetic material with vacuum magnetic per-meability μ_0. Region III is the region outside the quadrupole where the magnetic field can, if necessary, be designed to vanish.

The coordinates are related by

$$x = r\cos\theta \quad \text{and} \quad y = r\sin\theta, \tag{1}$$

and the differentials by

$$dx = -r\sin\theta\,d\theta \quad \text{and} \quad dy = r\cos\theta\,d\theta. \tag{2}$$

Two-dimensional magnetic fields inside a cylinder of radius a due to longitudinal currents on a can be expressed as superpositions of terms of the form

$$B_y + iB_x = A_n r^n e^{in\theta} = A_n(x + iy)^n(\cos n\theta + i\sin n\theta), \tag{3}$$

where n is a nonnegative integer. This field representation is applicable in the special case of fields that are purely transverse, (x, y), because the currents are purely longitudinal z. In this circumstance the only Maxwell conditions to be met are $\nabla \cdot B = 0$ and $\nabla \times B = 0$; both of which are satisfied by these fields. Strictly speaking these formulas apply only to purely parallel, longitudinal currents. But the subsequent discussion will apply them to the slightly conical currents proposed for the apparatus.

The various values of n correspond to pure multipole fields: 0 for dipole, 1 for quadrupole, 2 for sextupole, 3 for octupole, 4 for decapole, 5 for dodecapole, etc. For fixed n, the magnitude $|B(r)| = \sqrt{B_x^2 + B_y^2}$ is constant on circles of radius r centered at the origin. For $n = 1$, the fields are

$$B_x = Ar\frac{x}{r} = Ax \quad \text{and} \quad B_y = Ar\frac{y}{r} = Ay. \tag{4}$$

The main quadrupole currents flow in the cylinder (actually cone) of radius a. The circulating beams are affected only by the magnetic field in region I. The calorimeter is influenced (harmlessly) by fringe fields in region II. We tentatively assume that fields outside radius b are acceptably small to be neglected. If not, they could be cancelled by longitudinal currents in that cylinder, or by a ferromagnetic layer.

Surface current

$$\sigma_z^a = I_a \cos 2\theta, \tag{5}$$

on the inner surfaces provides the desired quadrupole field in region I. If required, there can be a similar current distribution on cylinder b. The full field pattern can be obtained by applying Ampèrian loop integrals along the long sides of the differential rectangle shown to match boundary conditions.

To simplify discussion of the construction of the quadrupole magnet and its mechanical compatibility with the calorimeter, we assume the exceedingly coarse approximation to $\cos 2\theta$ that can be obtained using just the simple pancake superconducting coils shown in Fig. 2. By symmetry the fields due to these coils contain only odd values of n in the field expansion of Eq. (3). Dipole, sextupole, and decapole terms are excluded by symmetry. Octupole terms can be excluded by judicious determination of the coil dimensions in Fig. 1. With the conical geometry, even this cancellation could be made perfect, independent of longitudinal position,

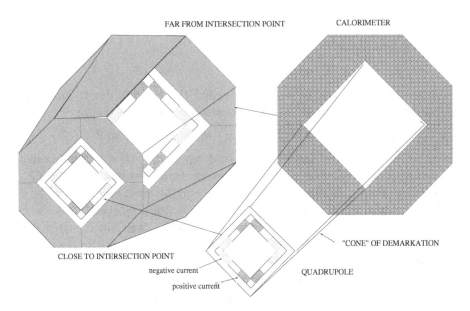

Fig. 2. (Color online) Cartoon view of the quadrupole running longitudinally from 1.5 m to 3 m in Fig. 1, and separated by a (rectangular) "cone of demarkation" from the enclosing calorimeter. Along this cone all physical quantities such as temperature, pressure, and magnetic field are necessarily identical for both the quadrupole and the calorimeter. But, in principle, there needs to be no other intercommunication between these regions. Positive quadrupole currents are red, negative, green.

only by making the coil dimensions expand proportionally with the cone diameter. This is unlikely to be practical. However, the octupole effect can, to the extent necessary, be canceled on the average, for the quadrupole as a whole.

This will leave a dodecapole field as the dominant deviation from an ideal quadrupole field. Canceling this field and/or other multipoles would require substantially more complicated current distributions. For now we trust that the field quality will be satisfactory with just the currents shown in Fig. 2.

2.2. *IR optics with conical innermost quadrupole*

Qualitative considerations suggest that the conical quadrupole geometry will actually be more favorable for accelerator operation than the conventional, discrete, pure cylindrical design. Of course this needs to be confirmed by actual accelerator modeling calculations. What is being suggested is a thick quadrupole in which the quadrupole gradient is greatest at the end where it is most needed.

For a fully relativistic electron of momentum p the inverse focal length $q = 1/f$ of a *thin* quadrupole of length l_q is given in MKS units by[1]

$$q = \frac{c\partial B_y/\partial x}{pc/e}\, l_q \,, \tag{6}$$

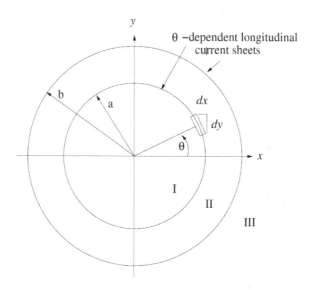

Fig. 3.   Figure defining parameters for the field calculation in a pure $2\cos\theta$, iron-free, quadrupole magnet.

where the denominator factor pc/e has MKS units of volt when pc is expressed in electron volts. For "thin" here to be applicable requires $l_q \ll f$. This inequality is clearly *not satisfied* for a conventional inner quadrupole, whose length l_q is longer than l^*, yet whose focal length needs to be only somewhat greater than l^* to focus a more or less parallel beam to a point at the IP. This means that thick lens geometric optics treatment is required or, more straightforward, direct solution of the differential orbit equation.

As it happens, though, for estimating its strength, it is not a terrible approximation to treat our conical quadrupole as a thin lens located at its center, with strength given by Eq. (6), using transverse dimensions valid at that point. This is because the product of greater lever arm but weaker gradient causes the focusing power per unit length to be more or less constant along the quadrupole's length. Accepting this, the effective position of our conical quadrupole of length l_q is at $s \approx l^* + l_q/2$, and its focal length is $f \approx l^* + l_q/2$. It is also a respectable approximation, when estimating the required quadrupole aperture, to still treat the conical quadrupole as thin, with transverse aperture evaluated at its longitudinal center.

We can concentrate on vertical orbits, because their tight focusing at the IP is more essential for maximizing the luminosity. The vertical r.m.s. spot size σ_y^* at the IP is given by

$$\sigma_y^* = \sqrt{\epsilon_y \beta_y^*}\,, \tag{7}$$

where ϵ_y is the all-important vertical emittance and β_y^* is the almost-equally important beta function value at the IP. In the region from $s = 0$ to $s = l^*$ the beta function dependence is given by

$$\beta_y(s) = \frac{s^2}{\beta_y^*} \quad \left(\approx \frac{(l^* + l_q/2)^2}{\beta_y^*} \text{ at quad center} \right). \tag{8}$$

Applying Eq. (7) at longitudinal position s the vertical beam size increases with s as

$$\sigma_y(s) = \sqrt{\frac{\epsilon_y}{\beta_y^*}} \, s \quad \left(\approx \sqrt{\frac{\epsilon_y}{\beta_y^*}} \, (l^* + l_q/2) \text{ at quad center} \right). \tag{9}$$

Perversely, but inevitably, as the luminosity is increased by decreasing β_y^*, the beam height downstream increases inversely as the square root.

Single beam vertical emittance as small as 10^{-11} m are achievable, but, at high luminosity, the beam–beam interaction can blow up the vertical beam emittance to a value as great as 10^{-10} m at $\beta_y^* = 0.003$ m at the Higgs energy of 120 GeV. To insure against particle loss in a single beam electron storage ring it is normal to require vertical apertures to exceed $10\sigma_y(s)$. This is based on the random walk associated with quantum fluctuations. (As such this limit may be overly conservative for calculating the aperture required for lossless containment of the beam blown up vertically by nonstochastic beam–beam forces.) Applying this condition would require the half-height of the quadrupole at its midpoint to exceed

$$y^{\max}(s) = 10\sqrt{\frac{\epsilon_y}{\beta_y^*}} \, (l^* + l_q/2)$$

$$\left(\approx 10\sqrt{\frac{10^{-10}}{0.003}} \, (2.5 + 0.75) = 6 \text{ mm at quad center} \right). \tag{10}$$

These parameters are taken from the PreCDR.[2] The required quadrupole field gradient can be calculated using Eq. (6). For a given maximum superconducting current density the minimum dimensions of the current carrying areas can then be estimated.

3. Integration of Quadrupole and Calorimeter

The proposed quadrupole current carrying elements are in four simple pancake coils running the 1.5 m length of the apparatus. The calorimeter has to be designed to restrain the magnetic forces tending to expand these elements. Conversely, though not strictly necessary in principle, the quadrupole can provide outward force to restrain the calorimeter elements from inward displacement. Other than these mutual dependencies, the quadrupole and calorimeter can be treated as if mechanically independent.

Fringe fields in region II, will cause charged particles to follow curved paths. This is not expected to significantly degrade the (coarse) particle direction measurement, nor the energy measurement in the calorimeter portion of our merged quadrupole/calorimeter. It may be necessary to correct for the stray fields in region III. They will be known to good accuracy by calculation and by DC measurement.

3.1. *Construction suggestions*

Conversations with Brindza[3] have yielded some suggestions concerning the design and fabrication of the instrumented conical quadrupole. He expects that individually wound cosine 2θ pancake coils, mechanically constrained within soft iron collars, would exhibit no training up to high magnetic fields. As well as eliminating stray magnetic fields, by minimizing the external field reluctance, this increases the achievable field gradient. This favors the use of iron over other absorbers for the calorimeter.

Coil construction should incorporate spacer material to more nearly approximate the desired cosine 2θ current distribution. The spacers could also be designed to provide the (very gently) fanning out of the current-carrying elements to keep the azimuthal current distribution matched to the coil radius at all longitudinal positions.

Even for a conical quadrupole, laminated construction is favored. Every lamination can be machined individually and economically using digitally controlled laser cutting, including very accurately placed circular holes for the active particle detection rods.

4. Detector Geometry and Performance

A natural geometry for the particle-sensing elements of a calorimeter built into the conical quadrupole is axial and parallel to the quadrupole superconducting strands. Optical fibers are an easy choice, and a well understood geometry from SPACAL and the RD52/DREAM calorimeters.[4]

For this conical quadrupole, the fibers are parallel to the conductors at radius b and, proceeding inwards radially, the upstream ends of the fibers would begin at radius values increasing with increasing z. All fibers emerge at the back of the quad enclosure, where they are coupled to photo-converters such as SiPMs, and are thus protected from the direct beam.

The fiber diameter and spacing are easy design parameters, but 1 mm diameter fiber on a 1.5 mm square grid is the RD52 design which achieves $\sigma/E \approx 10\%$–$13\%/\sqrt{E}$ electromagnetic energy resolution and, in GEANT FTFP_BERT_HP simulations, promises $\sigma/E \approx 30\%/\sqrt{E}$ hadronic energy resolution.[6] Timing information on the depth development of showers is also an asset[5] in this geometry. The geometrical complications of packing fibers into a quad volume will necessarily degrade both of these resolutions (the hadronic more than the electromagnetic since a hadronic shower will spread out farther spatially), so larger diameter fibers on a cruder spatial scale will also be acceptable. This is an easy optimization.

There is a choice between making an electromagnetic-only calorimeter or a full hadronic calorimeter. For a fiber calorimeter of the RD52 variety, this means clear (or radiation hard quartz) fibers for an electromagnetic-only calorimeter, and scintillating plus clear fibers for a hadronic calorimeter. The physics issue is the importance of measuring particles in jets, and thereby keeping nearly the whole 4π solid angle for event reconstruction, versus measuring just the electromagnetic particles in, predominantly, Bhabha events for luminosity measurement.

We think that maintaining good, but not excellent, hadronic particle coverage is very important, and therefore we would opt for a full hadronic calorimeter, even if not as deep in nuclear interactions lengths, λ_{Int}, as optimal.

The number of nuclear interactions lengths of the hadronic calorimeter in 1.5 m assumes a fiber "filling fraction" of 40%, the same as the RD52 calorimeters.[4]

Absorber material	X_0(cm)	λ_{Int} (cm)	# λ_{Int} in 1.5 m
Fe	1.76	16.8	5.4
Cu	1.44	15.3	5.9
W	0.35	9.9	9.1
Pb	0.56	17.6	5.1

Cu is best for hadronic calorimetry, W is difficult to form on the fine scales needed for a fiber calorimeter. Fe is acceptable and almost as good as Cu, and Pb is generally inferior for a hadronic calorimeter.

4.1. *Electromagnetic only*

With only clear fibers[a] in the absorber volume, the energy resolution will be (approximately) the RD52 clear fiber resolution divided by $\sqrt{2}$, or about

$$\sigma/E \approx 10\text{–}18\%/\sqrt{E} \to 1\text{–}2\% \text{ at } 100 \text{ GeV}.$$

For these fibers embedded in a copper matrix, a 30 X_0 deep electromagnetic calorimeter would be about 0.85 m deep, which fits well inside a 1.5 m long quadrupole.

4.2. *Electromagnetic and hadronic*

A hadronic-capable calorimeter would have both kinds of fibers (scintillation and clear) and would look like the fiber arrangement in Fig. 4 (one of the RD52 modules).

This RD52 geometry has a well-understood performance.[4–6] The electromagnetic performance will be nearly identical to the electromagnetic-only design since the mean and resolutions of the scintillating and clear fiber signals are similar.

The depth of the calorimeter will be the depth of the quad, about 1.5 m, and therefore the number of interaction lengths in this RD52 design will be only 5.9 λ_{int} for a Cu absorber and 5.4 λ_{int} for a Fe absorber, and consequently there will be frequent leakage of particles out at the back of the calorimeter. This is not good. However, the *debris* from this leakage has nowhere to go but into the main

---

[a]Mitsubishi SK-40 fibers, $NA = 0.50$, optical attenuation length $\lambda_{att} \approx 15$ m.

Fig. 4.   End of the copper module fibers illuminated from the far end.

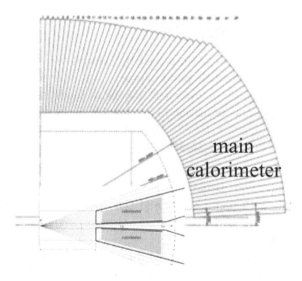

Fig. 5.   Detail of the main 4π calorimeter as a back-up for leakage from the quad-calorimeter.

calorimeter that closes on the beam pipe, as shown in Fig. 5. It may be possible to reconstruct energetic jets from the combined qual-calorimeter and main calorimeter.

Suppressing this leakage within a depth of 1.5 m would require, for instance, a tungsten (W) or uranium (U) absorber volume and possibly also a reduction in the fiber fractions. Although better for leakage, these would have smaller deleterious effects in energy resolution due to the e/mip variation below 5 GeV in high-Z absorbers, and also due to smaller sampling fractions.

5. Advantages and Disadvantages

(Ignoring solid angle imposed by irreducible cryogenic insulation layers — which may be so large as to defeat the scheme) the solid angle subtended by a quad/collimator of diameter D_{qc} is approximately $(D_{qc}/l^*)^2$. Without instrumentation this would cause a fractional loss $(D_{qc}/l^*)^2/(4\pi)$ in detector solid angle coverage.

Potential advantages of superimposing calorimeter onto the innermost quadrupole are:

- For fixed l^* this fractional loss of solid angle coverage is eliminated, admittedly with less than perfect calorimeter performance.
- At some (further) cost in reduced coverage imposed by cryogenic insulation, it may be possible to increase luminosity by reducing l^*.
- If the conical quadrupole design proves to be practical, there will be an increase in luminosity because the strongest quadrupole gradient is closest to the IP (where it is most needed) and the radial quadrupole becomes larger farther from the IP (where it is most needed). (This same advantage could be obtained even without integrating the quadrupole into the detector.)

There are also disadvantages:

- Quadrupole complexity will be increased by calorimeter considerations, which may cause impaired quadrupole performance.
- The detector performance improvement may be of too little value to justify the complication and possible reduction in quadrupole performance.

5.1. *Issues for further investigation*

Constructive comments from the reviewer emphasize further work and design on several issues. In particular, Brindza *et al.*[3] design is for a large aperture and low gradient quadrupole with high quality, and this note advocates an extrapolation to small aperture and high field gradient and therefore requires further study and work to assess feasibility.

The magnetic force on the quadrupole within the solenoidal tracking field will limit the mechanical precision and alignment and therefore the luminosity measurement from Bhabha scattering. The solid angular coverage in Fig. 5 is a small fraction of 4π but large enough that one-half of all $t\bar{t}$ events will populate this solid angle with at least one of its six jets.[8] This is one justification for a combined em and hadronic capability.

References

1. N. Malitsky and R. Talman, Text for UAL Accelerator Simulation Course (2005), http://uspas.fnal.gov/materials/materials-table.shtml.
2. The CEPC-SPPC Study Group, CEPC-SPPC Preliminary Conceptual Design Report, Volume II-Accelerator, IHEP-CEPC-DR-2015-01 and IHEP-AC-2015-01, 2015.

3. P. B. Brindza, S. R. Lassiter and M. J. Fowler, *IEEE Trans. Appl. Supercond.* **18**, 415 (2008), and private communication with Brindza on construction methods.

4. S. Franchino, Crystal and fiber dual-readout calorimeters: Building and understanding them, http://iasprogram.ust.hk/hep/2016/conf.html.

5. M. Cascella, Time structure in dual-readout calorimeters, http://iasprogram.ust.hk/hep/2016/conf.html.

6. S. Lee, Energy resolution and particle identification of the dual-readout calorimeter, http://iasprogram.ust.hk/hep/2016/conf.html.

7. N. Akchurin *et al.*, Combined forward calorimetry option for phase II CMS endcap upgrade, *J. Phys.: Conf. Ser.* **587**, 012015 (2015).

8. C. Damerell, private communication in ILC community.

Preliminary Conceptual Design About the CEPC Calorimeters

Haijun Yang

(on behalf of the CEPC Working Group)

Institute of Nuclear and Particle Physics, Department of Physics and Astronomy,
Shanghai Key Laboratory for Particle Physics and Cosmology,
Shanghai Jiao Tong University,
800 Dongchuan Road, Shanghai, 200240, P. R. China
haijun.yang@sjtu.edu.cn

The Circular Electron Positron Collider (CEPC) as a Higgs factory was proposed in September 2013. The preliminary conceptual design report was completed in 2015.[1] The CEPC detector design was using International Linear Collider Detector — ILD[2] as an initial baseline. The CEPC calorimeters, including the high granularity electromagnetic calorimeter (ECAL) and the hadron calorimeter (HCAL), are designed for precise energy measurements of electrons, photons, taus and hadronic jets. The basic resolution requirements for the ECAL and HCAL are about $16\%\sqrt{E}$ (GeV) and $50\%\sqrt{E}$ (GeV), respectively. To fully exploit the physics potential of the Higgs, W, Z and related Standard Model processes, the jet energy resolution is required to reach 3%–4%, or $30\%/\sqrt{E}$ (GeV) at energies below about 100 GeV. To achieve the required performance, a Particle Flow Algorithm (PFA) — oriented calorimetry system is being considered as the baseline design. The CEPC ECAL detector options include silicon–tungsten or scintillator–tungsten structures with analog readout, while the HCAL detector options have scintillator or gaseous detector as the active sensor and iron as the absorber. Some latest R&D studies about ECAL and HCAL within the CEPC working group is also presented.

Keywords: Circular electron positron collider; Higgs factory; electromagnetic calorimeter; hadron calorimeter; particle flow algorithm.

1. Introduction to Calorimeters

Many R&D researches are carried out within the CALICE collaboration in the past decade.[3] The majority of these studies aim to develop extremely fine granularity and compact imaging calorimeters with several technology options. Imaging calorimeter is a rapidly developing novel particle detector which has excellent spatial resolution. It is capable to provide enormous position information of incident and showering particles, which makes it possible to reconstruct every single particle cluster.

This is vital for Particle Flow Algorithm (PFA[4]) and help to significantly improve the energy resolution of hadrons. The basic idea of PFA is to distinguish charged ($\sim 65\%$) and neutral particles ($\sim 35\%$) inside the calorimeters. Charged particles measured in the inner tracker with high momentum resolution are matched to their energy depositions in the calorimeters. Energy depositions without matched inner tracks are considered to originate from neutral particles inside jets, among these neutral particles, about 25% of energy from photons are measured in the ECAL with good energy resolution, while the residual energy of merely 10% from neutral hadrons are measured by the calorimeters with poor energy resolution. Hence, the jet energy is determined by the charged track momenta of charged particles from inner tracker and energy depositions of neutral particles in the calorimeters. It has been demonstrated that significant improvement of the jet energy resolution is achievable based on MC simulations and test beam measurements. However, more efforts are needed to optimize the calorimeter design, to improve the PFA, and to develop the technologies for high granularity imaging calorimeters.

2. General Layout

The calorimeter system includes two sub-detectors, an electromagnetic calorimeter (ECAL) which is optimized for the measurement of photons and electrons, and a hadronic calorimeter (HCAL) which is employed to measure the energy deposit of the hadronic showers caused by the hadronic particles when they are absorbed in the HCAL detector. The two sub-detectors will be installed within the solenoid to minimize the inactive material in front of the calorimeters and to reliably associate tracks to energy deposits. The calorimeter system is divided into three parts, one cylindrical barrel and two end-caps.

The ECAL consists of layers of active sensors (such as silicon pads or pixels, or scintillator detector) interleaved with absorber tungsten plates. The HCAL is expected to have steel absorber plates with gaseous detectors such as Resistive Plate Chambers (RPC) or Gaseous Electron Multiplier (GEM or THGEM). Both ECAL and HCAL are sampling detectors with very fine granularity and segmentations of electronic readout which is driven by excellent separations requirement between charged and neutral particles for the particle flow algorithms.

From Fig. 1, there are more detector options with enormous worldwide R&D efforts ongoing within the CALICE collaboration. However, for this study, we have to be selective and focus on a few options. The CEPC detectors R&D is widely open for international collaboration with different detector options and new ideas.

3. ECAL Based on Scintillator–Tungsten (ScW ECAL)

A scintillator–tungsten sandwich sampling calorimeter (ScW ECAL) is proposed to build a fine-segmented calorimeter in a stable, robust and cost effective way. The structure of the ScW ECAL is shown in Fig. 2. The ScW ECAL consists of

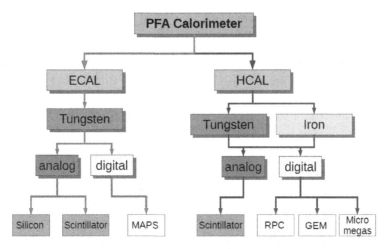

Fig. 1. PFA: Overview of imaging calorimeters with different absorber materials, readout technologies and active sensors which are under development for future lepton colliders.

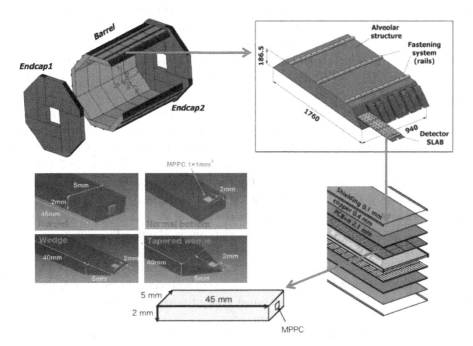

Fig. 2. Structure of the ScW ECAL.

a cylindrical barrel system and two large end caps with 25 super-layers. A super-layer is made of a tungsten plate (3 mm thick), scintillator strips (2 mm thick), and a readout/service layer (2 mm thick). The thickness of a super-layer is 7 mm. The total ScW ECAL thickness is 175 mm, and 21.4 X_0 in radiation length. The

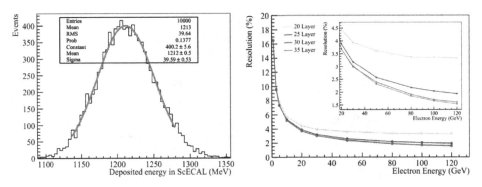

Fig. 3.   Left: plot shows energy spectrum of 25 GeV electron; Right: plot shows energy resolution.

active layers are plastic scintillators consisting of 5 × 45 mm² scintillator strips. The scintillator strips in adjacent layers are perpendicular to each other to achieve a 5×5 mm² effective transverse size. Each strip is covered by a reflector film to increase collection efficiency and improve uniformity of the scintillation light. Photons from each scintillator strip are readout by very compact photon sensor, SiPM, attached at the center of the strip. We are carrying out performance tests of SiPM, optimization of scintillator strip structure and shape to increase photon acceptance 2. Production of the scintillator strips can be performed with low costs. Considering to the strip structure, number of readout channels can be significantly reduced.

The performance of the ScW ECAL has been studied with a GEANT4 package. A standalone simulation program has been built to fulfill this work. In the Monte Carlo program, there are only plastic scintillator and tungsten plate, no readout layer included. The thickness of plastic scintillator and tungsten plate is 2 mm and 3 mm, respectively. The plastic scintillator is covered with 0.1 mm reflector film. The lateral size of the scintillator is 60×60 cm. The energy spectrum of 25 GeV electron is shown in Fig. 3 (left). The distribution is fit with a Gaussian function, and the energy resolution of the 25 GeV electron is about 3.3%, which coincides with CALICE collaboration beam test results.[3] The energy resolution vary with layer number is shown in Fig. 3 (right). To achieve adequate energy resolution, the number of layers should be around 25–30. The optimization of readout cell size (e.g. 5 × 5 mm or 10 × 10 mm) needs detailed study based on full detector simulation and reconstruction using PFA software.

The ScW ECAL consists of about 8 million channels of scintillator strip units. The stability of the light output has to be monitored. A light distribution system is under study to monitor possible gain drifts of SiPM by monitoring photo-electron peaks. The system consists of a pulse generator, a chip LED, and a notched fiber. The pulse generator circuit and the chip LED are arranged on a thin ($\sim 200\ \mu$m) FPC board. The chip LED is directly connected to the notched fiber to distribute lights to ~ 80 strips through its notches. An effective active cooling system is essential to the ScW ECAL, due to the high density of the channels. A heat exchanger will be coupled to a copper drain at the part of the ECAL layers. The FPGAs mounted at the end of the modules are a major source of heat and are directly connected to the cooling pipes. Thin copper plates will ensure evacuation of residual heat from the inner parts of the detector layers.

4. The Hadron Calorimeter (HCAL)

HCAL are sampling calorimeters with steel as absorber and scintillator tiles or gaseous devices with embedded electronics for the active part. The active detector element has very finely segmented readout pads, with 1×1 cm size, for the entire HCAL volume. Each readout pad is read out individually, so the readout channel density is approximately $4 \times 10^5/m^3$. For the entire HCAL, with $\sim$100 m^3 total volume, the total number of channels will be 4×10^7 which is one of the biggest challenges for the HCAL system. On the other hand, simulation suggests that, for a calorimeter with cell sizes as small as 1×1 cm^2, a simple hit counting is already a good energy measurement for hadrons. As a result, the readout of each channel can be greatly simplified and just record "hit" or "no hit" according to a single threshold (equivalent to a "1-bit" ADC), instead of measure energy deposition, as in traditional HCAL.

For the CEPC project, we propose gaseous detectors for the HCAL active layers: The Resistive Plate Chamber (RPC) or Thick Gas Electron Multiplier (THGEM). This is motivated by the excellent efficiency and very good homogeneity the gaseous detectors could provide. Another important advantage of gaseous detectors is the possibility to have very fine lateral segmentation. Indeed, in contrast to scintillator tiles, the lateral segmentation of gaseous devices is determined by the electronics readout used to read them. Active layer thickness is also of great importance, highly efficient gaseous detectors can be built with a thickness of less than 6 mm. Other important assets of the gaseous detectors are their cost-effectiveness and the simplicity with which they can be produced in large amount.

A drawing of the HCAL structure is shown in Fig. 4, the barrel part is made of five independent and self-supporting wheels along the beam axis. The segmentation of each wheel in eight identical modules is directly linked with the segmentation of the ECAL barrel. A module is made of 40 stainless steel absorber plates with independent readout cassettes inserted between the plates. The absorber plates consist of a total of 20 mm stainless steel: 10 mm absorber from the welded structure and 10 mm from the mechanical support of the detector layer. Each wheel is independently supported by two rails on the inner wall of the cryostat of the magnet coil. The cables as well the cooling pipes will be routed outside the HCAL in the space left between the outer side of the barrel HCAL and the inner side of the cryostat. Figure 5 shows the jet energy resolution obtained from GEANT4 simulation as a function of HCAL thickness. For jet energies of 100 GeV or less, leakage is not a major contributor to the jet energy resolution provided the HCAL is approximately 4.7 λ_I thick (40 layers).

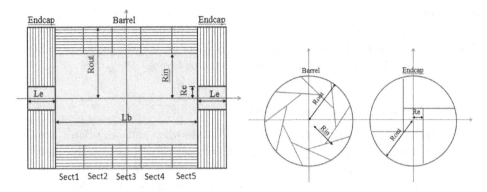

Fig. 4.   Longitudinal profile (left) and transverse section (right) of the HCAL.

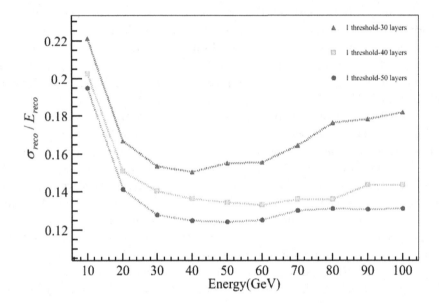

Fig. 5.   Jet energy resolutions for the detector model with different numbers of HCAL layers.

The structure of glass RPC (GRPC) as an active layer of the HCAL proposed for CEPC is shown in Fig. 6. Several test beam studies related to HCAL prototypes using GRPC have been carried out within CALICE collaboration.[5–8] It is made out of two glass plates of 0.7 mm and 1.1 mm thickness. The thinner is used to form the anode while the thicker forms the cathode. Ceramic balls (or cylindrical spacers) of 1.2 mm diameter are used as spacers between the glass plates. The gas volume is closed by a 1.2 mm thickness and 3 mm wide glass-fiber frame glued on both glass plates. The glue used for both the frame and the spacers was chosen for its chemical passivity and long term performance. The resistive coating on the glass plates which is used to apply the high voltage and thus to create the electric field in the gas

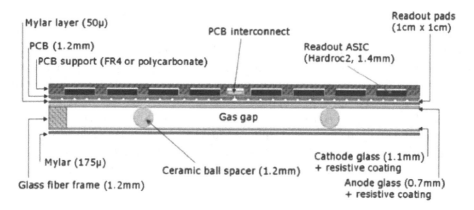

Fig. 6. A schematic cross-section of the active layer of DHCAL.

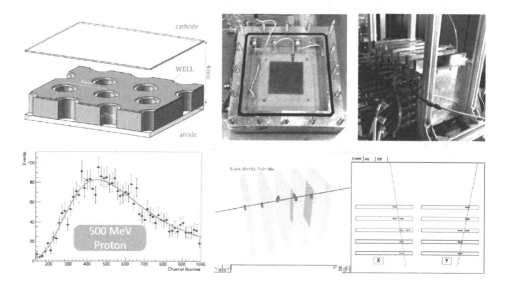

Fig. 7. A five THGEM modules setup for beam test at IHEP in October, 2015.

volume was found to play an important role in the pad multiplicity associated to a minimization ionization particle (MIP). The GRPC and its associated electronics are housed in a special cassette which protects the chamber and ensures that the readout board is in intimate contact with the anode glass. The cassette is a thin box consisting of 2.5 mm thick stainless steel plates separated by 6 mm wide stainless steel spacers. Its plates are also a part of the absorber. The electronics board is assembled thanks to a polycarbonate spacer which is used to fill the gaps between the readout chips and to improve the overall rigidity of the detector.

Another option for HCAL active layer is Gas Electron Multipliers (GEM or Thick GEM) which can be built in large quantities at relative low cost. THGEM detectors can provide flexible configurations which allow small anode pads for high granularity. It is robust and fast with only a few nano-second rise time, and has a short recovery time which allows a higher rate capability than other detectors. It operates at a relatively low voltage across the amplification layer with high gain and is stable. The ionization signal from charged tracks passing through the drift section of the active layer is amplified using a WELL THGEM layer structure. The amplified charge is collected at the anode layer with pads at zero volts. The structure of gaseous detector with thinner thickness is currently under design and optimization to keep the sensitive layer as compact as possible. A beam test for five THGEM modules using 500 MeV proton beam was successfully carried out at IHEP, Beijing in October 2015, as shown in Fig. 7.

Acknowledgment

The author would like to thank the CEPC study working group for numerous efforts related to the preliminary conceptual design of CEPC calorimeters.

References

1. The CEPC Pre-CDR, http://cepc.ihep.ac.cn/preCDR/volume.html.
2. International Large Detector Letter of Intent, the ILD concept group, February (2010), http://ilcild.org/.
3. The CALICE Collab., https://twiki.cern.ch/twiki/bin/view/CALICE/WebHome.
4. M. A. Thomson, Nucl. Instrum. Methods A **611**, 25 (2009).
5. M. Bedjidian et al., J. Instrum. **6**, P02001 (2011).
6. I. Laktineh et al., CALICE Analysis Note CAN-037 (2012).
7. B. Bilki et al., J. Instrum. **4**, P06003 (2009).
8. B. Bilki et al., J. Instrum. **4**, P10008 (2009).

Printed in the United States
By Bookmasters